Statistical Methods
for Social Scientists

This is a volume of

Quantitative Studies in Social Relations

Consulting Editor: Peter H. Rossi, University of Massachusetts, Amherst, Massachusetts

A complete list of titles in this series appears at the end of this volume.

Statistical Methods
for Social Scientists

ERIC A. HANUSHEK
Department of Economics
Yale University
New Haven, Connecticut

JOHN E. JACKSON
Department of Government
Harvard University
Cambridge, Massachusetts

ACADEMIC PRESS   New York   San Francisco   London
A Subsidiary of Harcourt Brace Jovanovich, Publishers

ACADEMIC PRESS, INC.
111 Fifth Avenue, New York, New York 10003

United Kingdom Edition published by
ACADEMIC PRESS, INC. (LONDON) LTD.
24/28 Oval Road, London NW1

Library of Congress Cataloging in Publication Data

Hanushek, Eric Alan,   Date
          Statistical methods for social scientists.

          (Quantitative studies in social relations series)
          Bibliography: p.
          1.   Estimation theory.   2.   Least squares.   3.   Social
sciences—Statistical methods.   I.   Jackson, John Edgar,
joint author.   II.   Title.
QA276.8.H35   519.5'4   76-9158
ISBN 0–12–324350–5

Contents

Chapter 4 **Ordinary Least Squares in Practice**

Chapter 5 **Multivariate Estimation in Matrix Form**

Chapter 6 **Generalized Least Squares**

Preface

The origins of this book go back several years. At that time there were few introductory or intermediate books on estimation methods available, and none that met our needs for the advanced undergraduate and graduate courses we were teaching. The courses were designed to develop practitioners able to apply empirical methods and to recognize the pitfalls in misapplying the same methods. In teaching these courses, we were looking for a text which combined a theoretical development with a discussion of actual applications. Such a blending is of course what most texts claim they have achieved. However, in surveying available books, we found that they tended toward one of the extremes: they were either too theoretical for our students or they did not include enough theoretical foundation to ensure proper applications in a wide variety of circumstances. This situation led us to develop extensive lecture notes which could be distributed to our students.

Since we began that project, a significant number of new texts have been published. Many have been aimed at the level of advanced undergraduate or introductory graduate students. However, in reviewing these new additions, we still felt there was a gap between the promises in terms of the blending of theory and application and the execution. Therefore we returned to our lecture notes to develop this book.

In retrospect, the lecture notes comprised a smaller proportion of the overall project than we had originally estimated. Also, both our notions of what topics should be covered and the material available in the field changed significantly during the time between our original teaching and the publication of this book. The end product represents material that has been extensively pretested on a wide variety of students. This material has been presented to undergraduate and graduate students in economics, political science, and sociology; this presentation has been done both by us and by others. The result, we believe, is a book which recognizes the difficult areas and attempts to provide some intuitive discussion of what to students appears to be quite mystical developments.

Through the long development and pretesting of this material we came to appreciate the need for a variety of different courses. These courses are differentiated in terms of the specific subject matter considered and the length and the level of difficulty of the course. In this regard we might relate our experiences with teaching the material.

The book itself presumes that the students have had some introduction to statistical methods. This introduction would include at least a discussion of probability distributions, random variables and their expectations, and hypothesis testing. A review of basic statistical methods is included in Appendix I, but this is intended as a review and not something by which this material could be introduced. A one-semester course following immediately after introductory statistical work would include Chapters 1–4 and possibly an introduction to one of the remaining topics in the book. Chapters 1–4 take the student through regression analysis with two independent variables. They develop the basic least squares estimators, their statistical properties, and an extensive discussion of practical single equation problems. These problems, contained in Chapter 4, include model specification, multicollinearity, functional forms, and dummy variables. The topics which can be followed at the end of such a one-semester course would include an introduction to matrix notation and least squares analysis (Chapter 5), problems with single equation models such as autocorrelation or heteroskedasticity (Chapter 6), problems with discrete dependent variables (Chapter 7), or multiequation models (Chapter 8).

If students have not been introduced to matrix notation previously, it may be introduced through Appendix II. This appendix, in contrast to the statistical appendix, is developed to introduce students to matrix notation and the elements of matrix algebra that are used within the book. With the exception of Chapter 8, all the chapters after Chapter 5 make use of matrix algebra. However, Chapters 6–10 are organized so that individual problems are discussed without matrix algebra; thus students can be introduced to these topics without extensive matrix algebra treatment.

A two-semester course following after introductory statistics would include Chapters 1–5 and probably two of the applications chapters in depth. The other applications of specific problems could be introduced through the first parts of each of the remaining chapters.

More advanced students—those who have had a more extensive introduction to statistical methods, say, through bivariate regression models, those with more mathematical sophistication, or those with some exposure to matrix methods—can review Chapters 1–3 quickly. For them, beginning with Chapter 5, the general matrix presentation of ordinary least squares, is probably appropriate. They would then return to Chapter 4 before proceeding to more specific applications and problems.

In Chapters 6–10, the presentational style involves a thorough discussion of the estimation problem and the consequences for estimation by ordinary least squares. This is followed by simple diagnostics and, generally, by the most common corrective measure. In terms of corrective actions, or alternative estimators, no effort is made to be exhaustive. Instead, the focus is on developing one alternative to the extent that a student could reasonably be presumed to have the tools for applying it and on referencing more advanced treatments. As such, the interest is more in application than in ultimate sophistication.

The book also weights topics somewhat differently than the majority of texts currently available. There is, we believe, a more even-handed treatment of problems associated with cross-sectional analysis and microunits as opposed to time series problems and large systems of equations which frequently involve aggregate data. This is evident in Chapters 7 (Discrete Dependent Variables) and 10 (Errors in Variables and Unobserved Variables).

The presentation of materials throughout includes proofs of only central parts of the estimation. These proofs occasionally include a minimal amount of calculus. However, if the proofs are not considered in detail, calculus is not required or used extensively.

The aspects of this text which we believe are novel, at least in degree, include: an effort to motivate different sections with practical examples and an empirical orientation; an effort to intersperse several easily motivated examples throughout the book and to maintain some continuity in these examples; and the extensive use of Monte Carlo simulations to demonstrate particular aspects of the problems and estimators being considered. In terms of material being presented, the unique aspects include the first chapter which attempts to address the use of empirical methods in the social sciences, the seventh chapter which considers models with discrete dependent variables, and the tenth chapter which introduces recent work in errors in variables and unobserved variables. Clearly these last two topics in particular are quite advanced—more advanced than material generally found in introductory texts. The notion behind these chapters is to introduce advanced topics in a way that provides both an appreciation of the problems and a roadmap to the material that is currently available on the subject. These last two topics are also currently experiencing rapid development and are not adequately described in most other texts.

Problems are included at the end of most chapters. These problems generally include one or two review questions, one or two questions that call for actual application of the material and usually some estimation, and one or two equations that are aimed at extending the material developed in the chapter.

Acknowledgments

Any undertaking of this sort incurs numerous debts, both intellectual and personal. We want to acknowledge several of the larger debts we incurred. Frank Campbell, George Farkas, Tom Gustafson, John Kain, Kim Peck, and Doug Price read all or parts of the manuscript. Their comments, advice, and encouragement were invaluable, proving once again the value of colleagues. Numerous classes have by now played the role of human subjects as we tried to develop the various examples, simulations, and arguments. The comments and reactions were particularly important in the evaluation of this text.

We are grateful to the literary executor of the late Sir Ronald A. Fisher, F.R.S., to Dr. Frank Yates, F.R.S., and to Longman Group Ltd., London, for permission to reprint Tables III-V from their book *Statistical Tables for Biological, Agricultural, and Medical Research* (6th edition, 1974).

Joyce Watson's typing at an early stage was much appreciated. Jean Singer's skills as well as her unfailing patience, humor, and understanding made the authors' tasks substantially easier. Finally, we wish to thank our wives for their tolerance of numerous working "weekends" and "vacations" during the past several years.

1 | Empirical Analyses in the Social Sciences

1.1 Introduction

Social science is the analysis of human behavior. How a rise in status changes people's racial attitudes, the increase (or decrease) in total revenues expected from a five percent price increase, or the anticipated differences in governmental expenditures between states with competitive and noncompetitive political parties are all concerns of social scientists. This list can be extended almost indefinitely with very diverse examples. The common thread is the concern with developing and using systematic, causal explanations of human behavior.

Explanations are conventionally referred to as models, which are merely statements about the most important determinants of the behavior being studied. Models, by design, are simplifications and abstractions of reality. They describe the most important systematic aspects of behavior seen in a wide variety of circumstances rather than completely account for specific events.

There are two different approaches to model building, deductive and inductive. The deductive method begins with a few simple axiomatic statements and from these derives a large number of predictions about expected behavior. For example, simple economic theory makes the basic assumptions that people prefer more of all goods to less and that the value attached to additional units of any one good declines as the amount consumed increases. Combining these assertions with a budget constraint and maximizing individual satisfaction or utility yields the well-known downward sloping demand curve. This demand curve and its expected negative relationship between amount demanded and price are used to make extensive predictions about observed consumer behavior.

However, most social science models are inductive, and not so rigorously deductive. Modeling is more often based on previous empirical observations and intuitive hunches about the behavior expected in various circumstances. Notions about the relationship between status and attitudes or the effect of party structure on governmental behavior are not deduced from a set of simple axiomatic statements but are instead derived from other experiences. Hypotheses and predictive statements derived in this manner perform the same function and are as necessary a part of applied social science as formally and rigorously derived theories.

Models derived in either manner are essential to empirical work because they identify key variables and make predictions about behavior given certain specified conditions. Without them empirical work is simply a collection of variables and statistical tests incapable of meaningful interpretation.

We continually refer to the existence of a model throughout this book, and assume it exists prior to the application of the statistical procedures to be developed here. The focus of this book is on the development of certain statistical methods that have proved useful in analyzing observed behavior. However, the techniques themselves are virtually meaningless in the absence of theory.

The purpose of this first chapter is to establish the links between models and subsequent empirical research, to discuss the purposes and objectives for statistical analysis, and to provide the setting that makes certain types of statistical tools appropriate. The remainder of the book develops a class of analytical tools and discusses the assumptions required for their use and their limitations if these assumptions are not satisfied. The emphasis throughout is on the application of the tools and not on their development.

1.2   Social Science Theory and Statistical Models

Systematic, causal relationships of behavior form the cornerstone of any social science. Models are explicit statements of these causal relationships.

Empirical studies have a very important role in the development and refinement of model building efforts. The first, and fundamental, objective of all empirical work is to find situations that may refute the model by showing that behavior is inconsistent with its predictions. Observed behavior cannot "prove" theories or hypotheses, but can only refute or disprove them. Consistency between data and the predictions of a given theory does not rule out the possibility that the same observations are also consistent with alternative explanations. Competing hypotheses commonly give similar predictions across a wide range of circumstances. A model's power and acceptance are derived from the variety of circumstances that do not disprove it. Thus, as we accumulate evidence consistent with the predictions of a model, we are more justified in acting *as if* that model were true. At the same time, if we can show that behavior is inconsistent with other theories, it further increases the apparent validity of the explanations whose predictions are consistent with observed behavior.

When the predictions of a model are empirically refuted, considerable time and energy is then spent considering the model and possibly other explanations that can account for the deviant observations. The process then begins anew with attempts to find behavior that can possibly refute the revised theories and hypotheses. This process results in the refinement and improvement of that model or the development of new ones. In this process, quantitative empirical work is advantageous because it requires and permits greater precision in stating and testing expected relationships.

The second important function of quantitative empirical studies is to estimate the magnitudes of the expected relationships. If a model is going to be useful for dealing with empirical problems, we must predict the values of certain variables under assumed circumstances. To make predictions and evaluate alternative courses of action, we must know the magnitude of the relationships and the causal effects implied by the model.

The statistical procedures discussed in this book are designed to analyze hypothesized causal relationships. The techniques are valuable because they both test implications of different models and estimate the size of the predicted causal effects.

Let us reemphasize that meaningful empirical work must be based upon explicit hypotheses and statements about predicted behavior. These hypotheses are necessary to guide the many decisions required in any research. The causal relationships derived from a model determine the specific variables needed, the behavior one must sample to test and estimate the model, and the interpretations one can give the results and their implications. Without the analytical structure provided by a theoretical base, we have little assurance that any statistical results are more than a collection of random numbers. Prior theory and hypotheses permit one to interpret and attach meaning to observed behavior.

The discussion up to this point has been quite general. It traces the relationship between theoretical and empirical analysis and the value of both of these in the development of a discipline. The subject matter of this book is considerably narrower however. It is concerned with developing an important and powerful class of statistical techniques that can be used for the quantitative analysis of behavior.

We first wish to restrict ourselves to *quantitative models*—ones where the behavioral relationships can be expressed in terms of mathematical relationships and evaluated for a numerical solution. However, as we shall see, the restriction to quantitative models is not too limiting; many qualitative events can be represented in precise, quantitative terms.

In quantitative models, behavior is represented and measured by numerical variables. The variables in a model are divided into two groups: *endogenous* (or dependent) variables and *exogenous* (or independent) variables. The endogenous variables represent the event(s) that are explained by the model. The exogenous variables have values determined outside of the model and are taken as given, i.e., the model does not explain the levels of these variables.

The model is a series of hypotheses about how an endogenous variable (Y) is related to (is a function of) one or more exogenous variables (X). This may begin with a general form such as $Y = F(X)$, where $F(\cdot)$ merely indicates a functional relationship. However, to do actual empirical work, $F(X)$ must be made operational by specifying a functional form relating the variables. The most common form is the linear model,

$$Y = \beta_1 + \beta_2 X, \tag{1.1}$$

where β_1 and β_2 are unknown *parameters* of the model.[1]

The task we generally face is to collect data for values of Y and X from which estimates of the parameters of the model (β_1 and β_2) can be computed. We can also test certain hypotheses about the behavior measured by Y with these estimates. For example, we can test whether or not X appears to influence Y. [In model (1.1) this is equivalent to testing whether β_2 appears to be zero or not. If $\beta_2 = 0$, X has no influence on Y.]

Each step of the process has its own questions and problems. The specification of the model, the definition of variables, the collection of relevant data, and the selection of statistical techniques each represents a set of often difficult decisions. The remainder of this book is designed to give guidance on these matters.

Some Examples

The transition from a theoretical framework to explicit statistical models is perhaps best developed through some examples. The first is the case of a microeconomic demand relationship between price of a good and the quantity purchased. The simple demand model of microeconomic theory begins with a theoretical development based upon consumers maximizing their satisfactions. The hypothesis of a negative relationship between the price of a commodity and the amount of that commodity people will demand follows from this theory. One simple specification of the price–quantity relationship is a linear model such as

$$Q = \beta_1 + \beta_2 P, \tag{1.2}$$

where Q is the quantity, P its price, and β_2 measures the changes in demanded quantity we expect to result from a unit change in price, assuming no other changes. A further hypothesis is that β_2 is negative. The advantage of this simple microeconomic model is that the theory and data collection activities are well defined and provide a good illustration of how empirical work should proceed.

---

[1]We shall return to the reasons for frequently using linear models, and present some alternatives in Chapter 4. The term linear model simply implies that the effect of a variable (X) is constant and does not depend upon any other variable. In this model, a unit change in X yields a change of β_2 in Y; this does not depend on either the magnitude of X or any other variables. In calculus terms, the partial derivative of Y with respect to X is a constant.

Unfortunately the example is too good. Other areas of the social sciences have much less fully developed theories, and even the data requirements are not as well defined. Yet it is these fields that both need research and provide some of the most interesting opportunities. Throughout this textbook we use examples that are more representative of social science research to illustrate the development of our statistical techniques. One example is an analysis of presidential voting; the other is an analysis of a series of interrelated issues involving learning behavior in schools, the relationships between schooling and income, and the occupational and educational aspirations of young adolescents.

The example of presidential voting is a case where alternative explanations for behavior can be included within a single model. By estimating the influence of the variables associated with each explanation, we are in a position to evaluate alternative hypotheses. One theory views voting primarily as a psychological phenomenon and focuses on the idea that people observe and interpret political information through a perceptual filter. The most important element of this filter is party identification. Voters' party identifications are hypothesized to determine their perceptions and opinions of candidates and thus their votes. An alternative theory holds that citizens in democratic societies vote to choose among public policy actions. In this model, votes are determined by the candidates' positions on certain public issues relevant to each voter, and voters choose the candidate whose positions most closely parallel their own.

We can combine these two explanations into a single model

$$V = \beta_1 + \beta_2 E + \beta_3 P, \tag{1.3}$$

where V measures the person's vote, E is an evaluation of the relative distance of each candidate from the voter's own issue positions, and P is the person's party identification. In subsequent discussions we shall develop measures for V, E, and P and try to estimate the coefficients β_1, β_2, and β_3 from a sample of U.S. voters taken during the 1964 presidential election campaign between Lyndon Johnson and Barry Goldwater.

The first of the education examples estimates the effect of various teacher characteristics on the achievement of third graders. A number of teacher characteristics are commonly thought to affect teacher ability and thus student learning. For example, teacher experience and graduate education are factors that determine pay and, thus, would be expected also to determine skills. Do they? And, if so, what is the magnitude of the effect? Previous tests of the effects of these factors on achievement indicate that they do not have a significant effect on achievement, and thus we turn to other models. We begin with the simple model that student's achievement A is considered to be a function of their teacher's verbal ability T and the number of years since last formal education Y. The precise model is

$$A = \beta_1 + \beta_2 T + \beta_3 Y. \tag{1.4}$$

We want to use our statistical analysis to estimate the values for β_2 and β_3 to provide a measure of the importance of teacher characteristics in the educational process. These estimates should then guide the development of educational policy with respect to improving teachers and, one hopes, achievement of future students.

Third, we examine the value of additional years of schooling. Few people question that earnings tend to rise with schooling level, but this knowledge by itself does not imply that individuals should go on for more schooling. More schooling is usually costly—there are both direct payments (tuition) and indirect payments (foregone earnings). Thus, the actual magnitude of the increased earnings is important. Estimating the change in earnings of an additional year of schooling is not completely straightforward however. Merely sampling incomes for individuals with different levels of schooling and comparing mean incomes is not sufficient. One must also consider the effects on income of other factors that might vary within a sample of individuals, e.g., experience, age, ability, race, and sex. These other factors might also be correlated with schooling levels and thus confound simple analysis. Alternative ways of including the effects of these factors will be considered through the course of this book.

These four examples begin to illustrate the conversion of explanations of social behavior into the precise formation of a statistical model. Each of these simple equations contains all the elements of the predictions derived from the theoretical propositions about consumer behavior, voting decisions, student achievement, and individual earnings.

Model development and the statement of hypotheses and predictions precede data analysis. A good rule of thumb to follow in evaluating the theory behind any empirical analysis is whether you believe the model, the hypotheses it generates, and their application to the observed behavior *prior to seeing the empirical results*. The hypotheses must be able to stand on their own either by the logic of the theory or through previous empirical findings before the data are analyzed. Only then can we decide whether the behavior and data being analyzed constitute a rigorous and useful test of the model. If the model is doubtful prior to seeing the data, it is difficult to see what purpose there is to the analysis. Since we cannot "prove" explanations with observed data, simply showing that behavior is consistent with a weak model is not useful; and showing that the observations refute a weak model is even more useless.

As the data are analyzed, we may find inadequacies in the hypotheses and perform additional empirical work to refine and reformulate the structure. However, these refinements now constitute hypotheses for future investigation.

1.3 Fitting Models to Data

The previous discussion has aimed at developing ways of looking at problems and a suitable methodology to allow data about actual behavior to be introduced into analyses. The first order of business is consideration of what data should be used. Most theoretical and empirical models assume common causal relationships. The statistical techniques discussed in this book not only share this assumption, they require it.

Many analyses will be concerned with the behavior of individuals, but this does not mean individualistic behavior. Instead, we shall be interested in the common behavior of a group of different individuals. Following the same behavioral relationships does not imply that all units exhibit identical behavior, instead it means that *under similar circumstances* we expect the average behavior to be the same. In terms of the previously presented models, this means that the parameters of the model (the β's) are the same across individuals.

The set of all units that follow the same causal relationship are referred to as the *population*. This population may consist of individuals (as in a model of voting), years (as in models of the United States economy), or elementary schools (as in a description of education performance). The precise population is defined by the behavior being studied. For example, males and females may together constitute a homogeneous population for studying elections. They may not constitute a homogeneous population for modeling the schooling and income relationship because of job market differences. The model describing the behavior of the members of the population is called the population model. This should indicate that it refers to and describes all members of the population.

In the case of the presidential voting model, all individuals included in the analysis have the same *expected* voting behavior, given the same evaluation and party identification, i.e., that β_1 and β_2 describe the voting behavior of all people in our sample. Similarly, for the education example, all children with the same teacher inputs should have equal *expected* achievements.

Assuming we have collected measurements of the voting intentions of a sample of voters or the verbal achievement levels for a group of elementary students, the object of our statistical analysis is to understand better these population models. In the simplest case, we might be concerned with whether the observed behavior is consistent with the simple expression given in Eq. (1.1). This is the hypothesis testing objective. Beyond this, however, we want to estimate the strength of the relationship. The first issue is called "statistical significance," or whether the data appear consistent with the model. The second issue might be labeled "empirical significance." In statistical terms, both objectives require us to make as accurate estimates as possible about the magnitude of the coefficients β_1 and β_2.

The requirement that all observations come from the same population is apparent in the above examples. If some voters choose candidates solely on the basis of party identification and others strictly on their issue positions, or if one group of children respond differently to their teachers than other children, each group of voters or children needs to be described by a different model. By combining observations from the different populations into one sample, we cannot get good estimates for the coefficients for either group. If we then try to use the model to predict voting behavior in another election or to assign teachers, we would make erroneous decisions because we would be using an inappropriate model for either population. In an extreme case, where members of two populations behave in completely opposite fashions, combining members of each population into one sample could completely disguise the behavioral relationship being estimated and lead to the erroneous conclusion that the hypothesized relationship does not exist. Consequently, in all of our statistical work we shall assume that all data are collected from the same population. (At a later point in Chapter 5 we shall see that with appropriate modifications our statistical techniques can be modified to handle observations from different populations, but one must still decide beforehand whether two or more populations are present in the data.)

A further question we need to consider is the comparability of the particular sample and its population to other situations where the estimated model will be applied. It should be remembered that social scientists want to learn about behavior generally, not about the behavior in a particular sample. It is highly unlikely that we shall re-create the 1964 presidential election. Similarly, we are not solely concerned with the education of third graders in one California school district. Knowledge of that one election or school district is useful only if it teaches us about future elections and other school districts. Consequently, we want to consider how similar the circumstances and population generating our data are to other situations. We also want statistics that tell us about population characteristics and behavior and not about one particular sample.

It is not enough to specify an explicit model of behavior and to obtain an appropriate sample. Any statistical analysis requires explicitly defined measures of the conceptual variables in the model. In some instances it is fairly easy to obtain such measures. The microeconomic model is very explicit about what variables are needed—prices and quantity—and it is usually not too difficult to determine the values for these variables for the observations in a sample. It may require extensive work to find the data, but conceptually the problem is clear. Many other models are not so simple. Problems arise both in defining the appropriate variables and in obtaining measures that fit the definitions. Although beyond the scope of this book, this is an important area of social science research. One of the real challenges in much empirical work is finding new and better ways to define and measure the theoretical concepts in social science research.[2]

The education and voting examples illustrate the various measurement and sample problems encountered and how they are resolved in the course of an

---

[2]Simon Kuznets, for example, received the Nobel prize in economics largely for his efforts to structure macroeconomic data into the national income accounts so the information could be used to estimate models of national economies.

empirical study. In the education model, the learning outcomes of the student A and the ability of the teacher T are hard to define and measure. The approach followed here is to use the tests of verbal skills developed by psychologists. These can provide standardized measures of the ability and achievement levels of both children and adults.

The appropriateness of any variable definitions clearly depends upon the purpose of the analysis. If one is interested in how to affect the cognitive development of a student, test measures seem to be a plausible way to proceed. On the other hand, if one is interested in how schools affect the socialization of children, this is probably not the way to proceed without information that socialization is highly related to cognitive test scores.

If we could find a sample that presented complete test data for a group of students and their teachers, we could possibly estimate the hypothesized relationship. However, the data requirements for such an analysis are quite large. The abstract model in Eq. (1.4) does not contain any explicit consideration of time factors, but surely they should exist. The relationship between teacher characteristics and student achievement is not instantaneous but evolves over time. And the school system ensures that teachers (for any individual student) change over time. Thus, we must not measure current teachers but a series of past teachers. These are difficult data to obtain, but it may be feasible with some modification in the specification of the model.

Consider student achievement at the end of the third grade. If we can administer the test to children at two points in time, say in the first grade and then in the third grade, we can reduce the data requirements considerably. Using the third grade score as the dependent variable and including the first grade score as an additional explanatory variable transforms the model into one predicting third grade achievement levels on the basis of initial achievement levels and the characteristics of the child's teachers in the intervening two years:

$$A_3 = \beta_1 + \beta_2 T_2 + \beta_3 T_3 + \beta_4 Y_2 + \beta_5 Y_3 + \beta_6 A_1, \qquad (1.5)$$

where the subscripts refer to the different grade levels. For example, T_2 is the verbal facility of the second grade teacher, obtained by using teachers' scores on a standardized test administered to teachers in the school system where the observations were collected . In this form, the entire history of a student's teachers is not needed; only that of the previous two years is needed.[3]

Finally, there is reason to suspect that the effect of teachers on achievement will vary for different socioeconomic and ethnic groups; i.e., the β's may not be the same for different populations. This could arise from education in the home, the motivations of various groups, or a variety of

---

[3]See Hanushek (1972) for a complete discussion of the model and data. The value of altering the formulation of the model would be more evident if later grades in school were being studied. However, even by the third grade, this procedure eliminates the need for observing first grade, kindergarten, and preschool aspects of the student's experiences.

other factors that vary systematically by socioeconomic status. An implication of this is that care must be taken in data collection to provide information about the student's background so that homogeneous populations can be defined for the estimation of the model.

The voting model presents even more difficult measurement problems. The dependent variable in this case is quite straightforward, people indicate they are either going to vote for Johnson or for Goldwater, or not vote. The particular study considers only voters, and the vote variable is a simple dichotomy with zero indicating a Goldwater vote and one a Johnson vote.[4] (Even this simple measure presents statistical problems, as discussed in Chapter 7.) The definition and measurement of both of the exogenous variables (party identification and issue preference) are complicated. For the present we shall illustrate the sort of issues that often arise by discussing the party identification variable. The measurement of issue preference will be discussed later when some actual estimates are presented.

Information on party identification is obtained in a survey. People were asked to classify themselves as independents, or weak or strong members of either the Republican or Democratic parties. This information is first used to stratify (or divide) the sample. Because there are strong reasons to think that people who are independents and people who identify with a specific party constitute different populations with respect to voting decisions, the model is estimated only for those identifying with one of the two major parties.

This is a clear example of an analytical decision. Instead of attempting to model differences between independents and people with party identification, independents are excluded. This simplifies the model, but at the same time limits the scope of the analysis. Statements about the voting behavior of independents can no longer be made.

Four classes of party identifications, or affiliations, remain after excluding independents. We can envision a scale of party membership going from most Republican (least Democratic) to most Democratic, with all respondents located somewhere along this scale. Our information, however, does not re-create this scale, but simply groups people into four ordinal categories, based on their own perceptions of party affiliation. Each grouping can be considered as an aggregation of a section of the true scale, plus whatever error is introduced by respondents' own perceptions. We can either view these classifications as an ordinal scale and construct four separate dichotomous party identification variables (one for each group), or we can arbitrarily assign values to each group and hope that these values correspond to the relative location of each group along the true party identification scale.[5]

---

[4]Note that voting for a particular person is usually thought of as a qualitative variable, but it is possible to define a quantitative variable that captures the concept and which can be used in the quantitative study.

[5]Ordinal variables are ones where the values signify order but not magnitude of differences; cardinal variables are ones where the magnitudes are also interpretable. With cardinal variables, there is a scale of measurement such as dollars, tons, or grade level of achievement.

Neither method takes into account the fact that we have aggregated large numbers of presumably different people with different "true" party identifications into only four groups. To select one of the two alternatives, we must consider the statistical economy of dealing with only one rather than four variables (this consideration is discussed in Chapter 4) and the additional error introduced if the values assigned to each differ substantially from the "true" values for the members of that group. (We devote considerable time in Chapter 10 to the problems created by introducing errors of measurement into explanatory variables.)

These examples are meant to illustrate some variable creation and definition problems that social scientists face and the decisions researchers must make in the course of a study. These problems will vary with the behavior being studied, the type of data and sample available, and the measuring tools. It is one thing to say that we want the most accurate measures available, and another to determine the accuracy of actual variables and to know the consequences of any erroneous measures. The questions throughout this text will be the statistical implications of the decisions a researcher must make, and particularly the implications if the "wrong" choice is made.

Most of the development of statistical tools for social scientists deals with the problems of weak data and how best to make inferences about social behavior from available data.

1.4   The Development of Stochastic Models

Once we have collected a sample of observations and constructed the necessary variables, if we could observe the true behavior of the population, we would quickly notice that the observations do not fit the model exactly. We often find that a portion of the observed behavior is unexplained by the hypothesized model and, for that reason, we shall modify the simple models described earlier. The previous model [Eq. (1.1)] is called *deterministic* because it implies that knowledge of X is sufficient to predict Y exactly. The alternative, which is much more useful for empirical analysis in the social sciences, is a *stochastic* model where a random, or stochastic, variable is explicitly added. Thus, instead of modeling the observed behavior Y solely as a function of $X, Y = F(X)$, we shall use an alternative specification that includes a random variable U such that $Y = F^*(X, U)$. This stochastic element is always unobserved; only Y and X are observed. The implication of adding this unobserved component is that social scientists must concentrate on analytical tools that are appropriate for probabilistic models. Further even if we know the value of X and the functional form $F^*(\ )$, we cannot predict the value of Y with certainty because of the presence of U.

Beginning with the next chapter, we shall see that the addition of this stochastic error term complicates the analytical task. We are going to have to make some assumptions about the nature of these errors in order to proceed.

Prior to that, it is important to understand the various factors contributing to the errors so that we can realistically consider the assumptions.

The deviations from the exact relationship, the U's in the observed variables, result from a variety of factors. In most social science work, a primary factor is that models are not sufficiently precise to indicate every possible influence on the behavior in question. Social science explanations are, at best, partial theories that indicate only the few important potential influences. Even if one could establish the list of all influences, this list would be sufficiently long to prohibit most analysis. Generally, researchers are trying to estimate accurately the importance of the influences that are central to policy making or to theoretical developements. To do this, one hopes only to obtain a data set where the influences excluded from the formal statistical analysis are sufficiently small and random so as to warrant exclusion.

A second contributor is imprecise data collection methods.[6] The problem of measurement error is persistent regardless of how simple the task. (For example, take ten measures of the width of your desk with a yardstick and see if all ten are equal to within a sixteenth of an inch.) Measurement techniques in the social sciences are still very imprecise, in most cases considerably worse than those in the physical sciences. Consider the efforts expended in trying to measure attitudes on political matters, or examine the aggregate economic data collection in many countries. If one were to administer the same attitude survey to the same respondents or to try to collect the same economic data a second time, it is extremely unlikely that the two data sets would be identical (just as the measures of the width of the desk do not all agree). Many of the differences can, it is hoped, be attributed to random errors in the data collection, errors that would cancel if many measurements were taken.

Finally, one may not be specifying the exact functional relationship. For estimation, we are not restricted to linear relationships between the variables, but the functional form specified by the researcher may still not be the proper relationship for the population in question. Errors are introduced in those observations where the true relationship deviates from the one specified.

All of the preceding explanations for the presence of implicit errors in the observed variables are likely to be present in one's data. We can illustrate this with reference to the voting example. Our model includes the effects of several salient issues and the person's party affiliation. Excluded are the effects of neighbors' opinions; the personal characteristics of the candidates, such as age, race, or religion, which might make them more or less attractive to some voters; and the many other policy issues that may concern individual voters but which were not deemed of sufficient general importance to be included in the survey. All of these potential influences, plus many others you can think of, operate among the voters one might sample and lead to deviations from the hypothesized relationship between voting and party affiliation and issues.

---

[6]For a classic discussion of this, see Morgenstern (1963).

There is also considerable measurement error in the voting variable (errors in the explanatory variables will be covered later.) Besides the obvious possibility that some respondents simply gave the "wrong" answer to the vote intention question, voting may be a probabilistic event. If so, the question about people's intended vote results in the wrong variable since the appropriate variable is the probability the person will vote for one candidate rather than the other. The binary vote intention variable can be considered as a combination of this probability variable plus a measurement error equal to the difference between the true probability and either zero or one, depending upon the response given to the vote intention question. Thus two people who indicate they intend to vote for the same candidate may actually have different probabilities of voting for that candidate, and thus contain different measurement errors in the observed vote variable. Again these errors lead to deviations from any exact relationship between voting and issue and party.

Finally, the functional relationship between voting and party identification and evaluations may not be linear, but rather some curvilinear function. This will be particularly so if one had measures of the actual voting probabilities since these probabilities are bounded by zero and one (a condition not implied by linear functions). If one chooses to estimate a linear function, or the wrong nonlinear one, the observed variable will deviate from the hypothesized relationship. All of these possible effects are included in and accounted for by the error term in the model relating voting in the 1964 presidential election to party affiliations and issue evaluations.

1.5   The Analysis of Nonexperimental Data and the Selection of a Statistical Procedure

All research, regardless of discipline, must deal with the stochastic problems described above. What makes empirical work in the social sciences different and often more difficult than in the physical sciences is the lack of an experimental methodology by which to control these factors. Chemists, physicists, and even some psychologists work with controlled environments. This allows the researcher to exclude the effects of variables that are not of interest but might confound the results simply by designing experiments where these factors are held constant. Thus the chemist performs certain experiments in an environment of constant pressure and temperature since variations in these variables may influence the results and make it difficult to observe the desired effects. In the experimental sciences, considerable work has gone into experimental design and the development of methods that permit external influences to be controlled.

In some research, certain influences cannot be explicitly controlled, but the experiment can be designed so the presence of uncontrolled factors is randomly distributed among the observational units in the study. Medical researchers and psychologists cannot control the life history of white mice prior to an experiment. What they can do, however, is replicate the experiment on a large number of mice and assign subjects to treatment and nontreatment groups on a random basis. Again this type of control is available through the experimental nature of the studies.

Experimental researchers can also determine the amount of treatment applied to various subjects. Thus they can ensure adequate variation in the factors whose influence they are studying. We shall see that obtaining a large range of values for the variable whose influence we are trying to estimate is an important factor in getting a good estimate of its effect.

Experimental researchers can also take advantage of the ability to repeat an experiment many times. Thus even with stochastic components in the results, there is a good chance of overcoming their presence by averaging the results of these many replications. In this way, stochastic errors in one direction will cancel those in other directions, leaving, one hopes, a good estimate of the true effects.

The difficulties facing the social scientists should be obvious at this point. Economies are not run for the benefit of economists; elections are run to choose leaders and public policies, not to satisfy the intellectual whims of political scientists; and teachers are assigned to classrooms on the basis of set assignment rules and not by education researchers. We refer to the type of situations that generate the data available to social scientists as "nonexperimental" because the researcher has no control over the behavior being observed. Economists, sociologists, and political scientists must observe behavior in its natural setting and deal with the results of these "natural" experiments.

The lack of an experimental basis for research precludes many techniques that other scientists use to refine their results. As social scientists, we cannot dictate the treatments our subjects receive, control the environment in which the behavior takes place, replicate the event many times, or even randomly assign cases to treatment and nontreatment groups. Consequently, we must use statistical procedures to try to achieve what others can do experimentally. In this procedure, we must carefully examine the circumstances in which our nonexperiment took place, how the data were collected, and the attributes of our statistical procedures.

We first need to be sure there is adequate variation in the explanatory variables. When we speak of estimating the relationship between two variables, we are really talking about how one variable is affected by changes in another variable. If we do not observe any changes in an exogenous variable, we have no information on which to base our estimate. We also want to be sure we have identified and obtained information about all the factors that systematically influence the observed behavior. Finally, we want statistical procedures that are appropriate to the data and model and that provide the best inferences about the behavior of the population being studied.

In the best of situations, where the researcher at least controls the collection of the data and the selection of a sample, some of the problems can be dealt with at the data collection stage. The correct sample can ensure the needed variation in variables important to the theory being tested and possibly reduce the variations in other influences.

In most research, the investigator does not even control the collection and organization of the data. In such situations, selection of a sample, the collection of information, the choice of variables, and the dissemination of the data are done by public agencies, private organizations, or other researchers. These agencies and individuals may have had quite different objectives in mind as they performed their task than does the current researcher. A good example is the vast amount of data collected by federal agencies in the process of monitoring their various programs. These data are presumably collected to aid the agencies' evaluation processes, not to facilitate research. Yet, such sources can prove very valuable to social scientists for studying certain types of behavior. The researcher must be very cautious, however, in using data collected for other purposes to make sure that all the problems mentioned above, plus others we shall mention in the course of this text, do not destroy our ability to make inferences about observed behavior.

In situations where sample selection cannot be used to control all other influences or where the researcher must rely on data collected by someone else for reasons unrelated to the immediate research project, it is necessary to apply the required controls through the use of statistical procedures which examine and estimate the influences of several factors simultaneously and deal with the other problems one usually encounters. We implicitly introduced the notion of statistical controls when we discussed the example of determining the monetary rewards for additional schooling. The models that we consider will include the age or experience of the individual. We do this not because we are interested in the effect of experience for these purposes but because we have samples where the effects of differential experience might confuse our estimates of schooling effects.

Experimental research has one additional important advantage over the nonexperiments of social scientists. All the models discussed and certainly all the applications of behavioral models imply a cause and effect relationship from explanatory to dependent variables. However, it has long been recognized that covariations among observed variables cannot establish or prove causation. Let us consider a simple example. We observe that achievement in school and absences from school are negatively correlated: the more frequently a student is absent, the lower is the student's observed achievement. Does the low attendance cause low achievement (in the sense that raising an individual's attendance would lead to higher achievement)? Or does a student's low achievement lead to discouragement and thus nonattendance? Simply observed, the correlation between attendance and achievement cannot sort this out.

The picture is actually more complicated than indicated above. Perhaps there is some third factor such as parental education and interest affecting both achievement and attendance. It might even be that there is no linkage between achievement and attendance, that the observed correlation is simply a spurious correlation.

Causation must be justified on theoretical grounds and cannot be "proved" by either experimental or nonexperimental research. However, experimental methods provide a much greater opportunity for testing alternative causal models. For example, in an experimental setting, it is possible to design different experiments that explicitly vary only one variable and measure what changes in the second variable, if any, follow. Such experiments cannot prove that one variable "causes" another, but is may disprove certain hypothetical causal statements, and thus certain models. In nonexperimental research it is much harder to find cases that permit such tests. In Chapters 9 and 10 we discuss certain limited cases where the implications of different causal statements can be tested.

1.6   Simple Methods

Nearly everyone is familiar with some quantitative methods long before taking a formal methodology course. Newspapers and popular magazines frequently present cross tabulations of data in order to describe associations between different factors. The correlation coefficient is not quite so widely used but still is familiar to most social science students. As a way of introducing the direction and focus of this book, we shall examine these elementary methods and use them as a starting point in the development of more sophisticated techniques.

Contingency Tables/Cross Tabulations

Conceptually, one of the easiest methods for analyzing the relationships among variables is to construct a contingency table or cross tabulation of the variables. These tables display the joint distribution across all categories of the variables being analyzed.

To begin, consider the previously discussed voting analysis. There are three variables: intended vote, party identification, and the issue evaluation for each party. Voting and party identification fall into clear categories (vote for Johnson or Goldwater; strong Republican, weak Republican, weak Democrat, and strong Democrat). The issue evaluation variable is an aggregation of an individual's view about which party the person thinks is best able to handle a series of issues. There are a large number of possible values for this variable (representing different combinations of evaluations on seven issues). Here we simply divide them into three categories indicating whether on average the individual thinks that the Republicans are closer to their positions on the issues, the Democrats are closer, or there is no difference between the parties. Since the implicit dependent variable (vote) is dichotomous, we can combine all variables into a single table by showing the proportional vote in each of the cells defined by the explanatory variables.[7] Such a cross tabulation for the voting example is shown in Table 1.1.

TABLE 1.1

Cross Tabulation of 1964 Voting by Issue Evaluation and Party Identification

	Issue evaluation						
	Republican[a]		Indifferent		Democrat		Total
Party identification							
Strong Republican	0.102[b]	(109)[c]	0.115	(35)	0.214	(33)	0.126 (177)
Weak Republican	0.258	(75)	0.544	(50)	0.667	(52)	0.461 (177)
Weak Democrat	0.606	(70)	0.890	(76)	0.975	(189)	0.879 (335)
Strong Democrat	0.727	(31)	0.893	(56)	0.990	(344)	0.958 (431)
Total	0.334	(285)	0.687	(217)	0.917	(618)	0.724 (1120)

[a] Defined as those voters for whom the Republican party is closer to their own issue position.
[b] Proportion voting for Johnson.
[c] Number of people with specified party identification and evaluation.

The table entries are the proportions of the voters with the given characteristics voting for Johnson. The numbers in parentheses are the number of voters in the sample with the given party and evaluation characteristics.[8] For example, there are 109 strong Republicans assessed as perceiving the Republicans as being closer to their positions on the issues. Of these 109 people, 10.2% voted for (or said they would vote for) Johnson. We can see from the table that Johnson's strength increases as people become more Democratic and as their evaluations become more favorable toward the Democrats.

More complicated models—i.e., more categories or more variables—are accommodated by extending the size of the tables or by producing a series of tables that correspond to specific values of the additional variables. A variety of possible statistical tests can be used to judge the degree of association between all or some of the variables (Weisberg, 1974).

Perhaps the primary advantage of contingency table analysis is its simplicity and understandability. It does not take lengthy statistical training to read and digest a contingency table.[9] Another aspect that leads to the popularity of

[7] When the implicit dependent variable is continuous, we can also collapse the joint frequency by presenting a mean value of the dependent variable for each cell defined by the explanatory variables.

[8] These numbers are adjusted by the weighting variable suggested by the Survey Research Center for its 1964 sample.

[9] This describes the actual cell entries and not the statistical methods and tests which relate to contingency tables. These latter methods are quite sophisticated and complicated.

contingency tables is that no assumptions need be made about the functional form of the relationship under consideration; it is not necessary, for example, to assume a linear relationship.

However, these attributes are not without their costs. First, analyzing either a few variables with many categories or many variables leads to a large number of cells. The number of cells is simply the product of the number of categories for each variable; thus, cross tabulating four variables with three categories each yields 3^4 or 81 cells. With even moderate size models, such tables are difficult both to display and to interpret. Even with large samples, many cells will contain only a few individual observations. Small cell sizes substantially reduce the reliability of the estimated behavior for given characteristic and make comparison among cells difficult and risky. These problems provide an inherent limitation to the complexity of the model that can be considered.

Second, while contingency tables and their related statistical tests may be used to analyze the degree of association between variables, they do not adequately provide information about the magnitude of any effects or the strength of any relationship. This is a particularly serious problem since one of the two central tasks of empirical analysis is the determination of the magnitude of any relationship.

Third, the use of contingency tables tends to obscure the analysis of causal models. Contingency tables do not require any explicit consideration of causal structure and, in fact, make it somewhat inconvenient to treat multidimensional (several variable) models as causal models.

Finally, contingency tables can handle only categorical variables. Yet many variables in social science research are continuous in nature. For example, income, racial composition of a community, price, tax rates, and fertility rates are all continuous variables. Other variables such as occupational classification are categorical in nature but contain a large number of categories. In both of these cases, it is possible to group the variable values together into a few categories. However, this throws out any information about within-category differences. The errors introduced in this manner may seriously affect our ability to interpret the magnitude of the cell entries.

There is a considerable tradition of contingency table analysis in the social sciences. In fact, as discussed in Chapter 7, some powerful statistical methods for treating the analysis of qualitative dependent variables have been developed in the contingency table framework.[10] Nevertheless, the limitations inherent in this type of analysis are serious. And, they are becoming more constraining as social science models become more complex and as larger and

---

[10]Qualitative dependent variables are variables that in general represent two or more classifications where the classifications do not have a standard measure associated with them. Presidential voting is one example. Others would include decision variables such as move/not move, buy/not buy, or married/not married.

more comprehensive data sources become available. Consequently, the empirical methods developed here follow along different lines in the tradition of correlation and regression analysis.

Simple Correlation and Regression

The use of simple correlation coefficients in the social sciences is almost as common as the use of contingency tables. The basic conceptual approach underlying correlation coefficients is the development of a standardized measure of association between variables.

For two variables (X and Y), consider plotting their values as in Fig. 1.1 and dividing the plots into four quadrants (denoted I,...,IV) according to the mean values of the variables ($\bar{X}, \bar{Y}$). To develop the correlation coefficient, we start by calculating the product of the deviations of each variable from its mean, i.e., $(X_t - \bar{X})(Y_t - \bar{Y})$ for each observation t. For observations in the first and third quadrants, this product will be positive; for observations in the second and fourth quadrant, this product will be negative. If we sum these products over all observations, a positive sum indicates that the observations tend to fall in the first and third quadrants, while a negative sum indicates that they are more heavily weighted toward the second and fourth quadrants. Thus, the sign corresponds to the direction of any relationship: a positive sign indicates that larger values of Y tend to be associated with larger values of X; a negative sign indicates that smaller values of Y are associated with larger values of X.

The signs tell the direction of association but say nothing about the strength of the relationship. In order to obtain that information, we make two

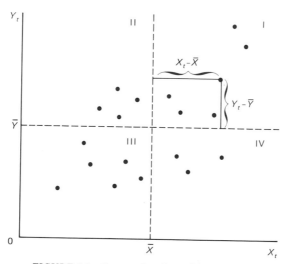

FIGURE 1.1   Scatter plot of variables X and Y.

adjustments: (1) divide the sum by the number of observations to eliminate the influence of sample size[11]; (2) divide this "average product" (or covariance) by the standard deviation of each variable to allow for the degree of scatter and units of measurement. This yields

$$r_{yx} = \frac{(1/T) \sum_{t=1}^{T} (X_t - \bar{X})(Y_t - \bar{Y})}{S_x S_y} = \frac{\text{cov}(X, Y)}{S_x S_y}, \qquad (1.6)$$

where T is the number of observations, S_x the standard deviation of $X = \sqrt{\Sigma(X_t - \bar{X})^2/T}$, S_y the standard deviation of $Y = \sqrt{\Sigma(Y_t - \bar{Y})^2/T}$, and $\text{cov}(X, Y)$ the covariance of X and $Y = (1/T)\Sigma(X_t - \bar{X})(Y_t - \bar{Y})$.

In point of fact, the simple correlation coefficient corresponds completely to a linear model relating X to Y along with a term representing the deviations of actual observed points from a straight line. The deviations of the correlation coefficient from a linear model are considered in Chapters 2 and 3. For now, we shall discuss only the interpretation of the correlation coefficient in the context of a linear model and the uses of empirical analysis.

The correlation coefficient has a range from -1 to $+1$. If $r = 1$, all observations fall on a straight line going through quadrants I and III. If $r = -1$, all points fall on a straight line going through quadrants II and IV. If $r = 0$, the points in quadrants I and III and the points in quadrants II and IV are such that no linear relationship between the variables is descernible. These alternative situations are shown in Fig. 1.2. The correlation coefficient thus gives an indication of direction of relationship and degree of association.

The interpretation of the magnitude of the correlation coefficient when it falls between the extreme values should be considered explicitly. The correlation coefficient implicitly assumes an underlying linear relationship such as the one introduced earlier:

$$Y_t = \beta_1 + \beta_2 X_t + U_t \qquad (1.7)$$

where β_1 and β_2 are parameters of a linear relationship and U_t is a random error term. If we think in terms of Eq. (1.7) as the true population model that generates the observed values of Y, we see that the size of the correlation coefficient between X and Y will vary for a number of reasons—some are related to the nature of the sample of observations and some are related to the underlying population model.[12] In the context of Eq. (1.7), *ceteris paribus*, r will be larger (in absolute value): (1) the smaller is the variance of U_t; (2) the larger is the variance in X_t; and (3) the larger is the absolute value of β_2. These alternatives follow because the correlation coefficient essentially compares the variance of the systematic component $\beta_1 + \beta_2 X_t$ of the behavior

---

[11]This new quantity should be recognized to be simply the covariance of the two variables.

[12]We shall have considerably more to say about the nature of the population model and the error term later.

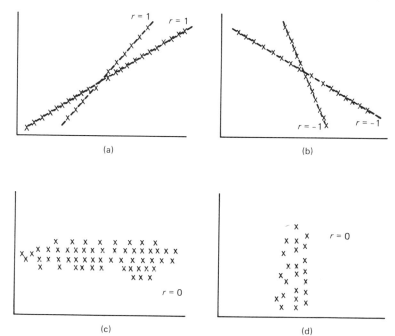

FIGURE 1.2   Extreme values of simple correlation coefficients.

represented by Y_t to the variance in the stochastic component U_t. For a given variance in U, the variance in the systematic portion of the model is larger when either β_2 is larger or the variance in X_t is larger; this implies that r is larger. Note that the magnitude of the intercept β_1 has no effect on r.

Consideration of the determinants of the size of the correlation coefficient suggests some interpretive problems. In our empirical efforts, we are interested in knowing the strength of any relationship; this is the size of β_2 because β_2 tells how differences in the size of X translate into differences in the size of Y. Yet this is only one of the factors that affect the size of the simple correlation coefficient. The other factors relate directly to the specific sample that is collected. For different samples, the variation in the exogenous variable or the stochastic term could vary, implying a different correlation coefficient even if the behavioral parameters (β_1 and β_2) are the same across samples.

An alternative to the simple correlation coefficient is to estimate directly the parameters of the linear model β_1 and β_2. This is called bivariate or simple regression. Since bivariate regression analyzes a very similar linear model as simple correlation analysis, it is not surprising that they are closely related. The derivation of the method for estimating the individual coefficients is presented in the next chapter. Here we present just the estimating formula and a few comments on the use of this technique. The parameters β_1 and β_2 of Eq. (1.7) are estimated as

$$b_2 = \frac{\text{cov}(X, Y)}{\text{var}\, X}, \tag{1.8}$$

$$b_1 = \bar{Y} - b_2 \bar{X}, \tag{1.9}$$

where $\text{cov}(X, Y)$ is the covariance between X and Y, $\text{var}\, X$ the variance of $X = \sum_{t=1}^{T}(X_t - \bar{X})^2 / T$, $\bar{Y}$ the mean value for Y, and $\bar{X}$ the mean value for X. [Note that $b_2 = r_{YX}(S_Y / S_X)$, $S_X = \sqrt{\text{var}(X)}$, and $S_Y = \sqrt{\text{var}(Y)}$.] The estimated coefficients b_1 and b_2 correspond directly to the parameters of the linear model β_1 and β_2. Further, as we shall see in Chapters 2 and 3, while the accuracy of b_1 and b_2 may be affected by characteristics of the sample, the parameters that we are estimating are not sample specific, and estimated values of the parameters can be readily compared across samples.

Estimation of these parameters also allows predictions about the magnitude of Y for a given value of X or the change expected in Y from a given change in the value of X. Such predictions cannot be derived from simple correlation coefficients. Simple correlation coefficients measure the degree of association between variables but not the magnitude or strength of any such relationship.[13]

Even though bivariate regression coefficients are approaching what we want, they are not without problems. In particular, in order to interpret b_2 as a "good" estimate (to be defined later) of β_2, certain conditions about the relationship between X and U must be satisfied. X and U must be independent. If this is not the case, we shall tend to be misled about the true value of the model parameter.

The reasoning behind this condition is perhaps best seen by considering why X and U might not be independent. Suppose that the behavior of Y is strongly influenced by an additional variable Z. If Z and X are correlated in our sample so that they tend to move together, estimation of the bivariate model will attribute some of the influence of Z to X. In other words, if the true model is

$$Y_t = \beta_1 + \beta_2 X_t + \beta_3 Z_t + U_t^*,$$

estimation of the bivariate model that excludes Z_t will yield a coefficient that is an amalgamation of both β_2 and β_3. This estimated coefficient could then be quite misleading.

The point of the previous discussion of the nonexperimental nature of the social sciences is simply that there are frequently instances where it is necessary to consider multiple explanatory variables. Because we cannot eliminate the influence of Z_t through experimental design or sample selection,

---

[13]One way to see this is to observe the units of r and b_2. r is dimensionless, while b_2 has dimensions of units of Y/unit of X.

we are often faced with the necessity of considering multivariate models. We may be driven to multivariate models even though we are interested only in the relationship between Y and X and not in the relationship between Y and Z. Thus, our first task is the expansion of the simple bivariate regression model to allow for more exogenous variables.

REVIEW QUESTIONS

1. Describe three examples of behavioral relationships that can be analyzed by statistical methods. What are the variables that would be indicated by theoretical considerations? What are the parameters of the models.

2. What is the implied equation for analyzing the relationship between earnings and schooling described in the text? How should the variables be measured?

3. Assume that you are studying expenditures by state governments and that you have data for all 50 states. Does this constitute the whole population? Why not?

4. For one of the models described in Question 1, indicate what factors might lead to a stochastic, or random, term.

5. Assign the following values to the categories of variables in the voting example:

Vote: Johnson = 1; Goldwater = 0
Issue evaluation: Republican = 0; Indifferent = .5; Democrat = 1.0
Party identification: strong Republican = 0; weak Republican = .25; weak Democrat = .75; strong Democrat = 1.0

Based upon the data in Table 1.1:
(a) compute the simple correlations between vote and party identification and between vote and issue evaluation;
(b) compute the bivariate regression coefficients for vote regressed upon party identification and for vote regressed upon issue evaluation (i.e., with vote as the dependent variable).

2 | Estimation with Simple Linear Models

2.1 Introduction

Social scientists are often, by the very nature of their problems and their data, forced to consider multivariate models. The bivariate models of Chapter 1 do not adequately describe most observed social phenomena. The simplest extension of the bivariate model is a trivariate model—one with two independent variables. Many of the issues of multivariate analysis can be illustrated in this case without the necessity of using more complicated matrix notation and matrix algebra. After developing most of the properties of regression analysis for this trivariate case, we shall generalize to larger and more complex models and at that time (Chapter 5 on) introduce matrix concepts.

The techniques that are developed in this chapter and in subsequent chapters have quite general applications. For this reason, we shall use a general notation to develop and discuss the statistical techniques. Y_t is the value of the dependent, or endogenous, variable that we are examining for observation t, where t is simply the index of observations and can take on values from 1 to T (T is the total number of observations in any given sample). The observations come from different years, from different political jurisdictions, or from different individuals. The independent, or exogenous, variables are denoted by $X_{t1}, X_{t2}, \ldots, X_{tK}$. t is the same observational index, and the second index $(1, 2, \ldots, K)$ refers to a specific (and different) variable. In each case, all variables for the same observational unit have the same index t. Thus in the example of presidential voting, X_{t2} might be the party affiliation of the tth person, and X_{t3} might be the same tth person's evaluation of the parties' positions on different issues, and Y_t is that individual's intended vote. (Frequently, when we are not interested in identifying specific observations, we shall drop the observational subscript and refer to variable X_2 or X_3.)

One simplification we shall employ in this chapter and through most of the rest of the book is the emphasis upon linear models. Why do we choose a linear model? After all, theory seldom suggests such a functional form, and at times theory might even suggest that linear relationships are inappropriate. The emphasis upon linear relationships can be explained in a number of ways. As we shall see in Chapter 4, many nonlinear relationships can be transformed into linear relationships for the purpose of statistical analysis. Linear models are also easily understood and are mathematically and statistically tractable. Finally, over limited ranges, many complicated functional forms can be approximated by linear relationships. Thus, in the absence of strong reasons for not using linear relationships, assumption of a linear model is usually the starting point for analysis. However, as is obvious from the above discussion, linear models may not always be the best choice. Therefore, some attention will be given to testing the appropriateness of linear models and to developing nonlinear alternatives for more common cases where linear models are clearly inappropriate.

2.2 The Basic Model

Throughout this chapter, we consider the situation where the true behavioral relationship (or population model) is represented by[1]

$$Y_t = \beta_1 + \beta_2 X_{t2} + \beta_3 X_{t3} + U_t. \tag{2.1}$$

This equation says the behavior measured by the variable Y_t is composed of two elements: there is a systematic component arising from a linear behavioral relationship with X_{t2} and X_{t3} ($\beta_1 + \beta_2 X_{t2} + \beta_3 X_{t3}$), and a random component (U_t). X_{t1} is presumed to have the value 1 for all observations and is used to generate the constant term, which is $\beta_1 X_{t1} = \beta_1$.

Our interest is in estimating the parameters of the systematic portion of the population model. The only observable quantities are the values of the exogenous or explanatory variables X_{t2} and X_{t3} and the final outcome for the dependent or endogenous variable Y_t. It is not possible to observe the random component (U_t) or the behavioral parameters ($\beta_1, \beta_2, \beta_3$). From our observations of Y_t and the X_t's, we wish to estimate the parameters β_1, β_2, and β_3. We shall denote parameters in the population model by Greek letters (e.g., $\beta_1, \beta_2, \beta_3$). Coefficient estimates obtained from sample data are denoted by corresponding lower case italic letters (e.g., b_1 is the sample estimate of β_1).

The systematic component contains all the hypotheses about the variables influencing the behavior we are explaining. The parameters of the system (the β_k) are constant for the whole population; they do not change for different

---

[1]We shall use the three variable model for expository purposes and for most illustrations in this and the next chapter. However this model is readily extendable to additional variables, in which case the equation is $Y_t = \beta_1 + \beta_2 X_{t2} + \beta_3 X_{t3} + \beta_4 X_{t4} + \cdots + \beta_K X_{tK} + U_t$. Chapter 5 concentrates on the extensions in this more general multivariate formation.

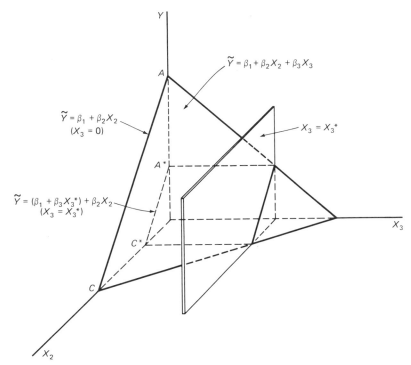

FIGURE 2.1   Projection of trivariate relationship on Y_1X_2 plane for $X_3 = X_3^*$.

cases or observations. Moreover, they represent the independent effect of each exogenous or explanatory variable on the endogenous or dependent variable when all the other variables in the equation are held constant. Thus, β_2 is the amount of change in Y that corresponds to a unit change in X_2 with "other things equal": i.e., with X_3 held constant; similarly for β_3.

The implications of the model and the interpretation of the coefficients can be illustrated by Fig. 2.1. This displays the plane in Y, X_2, and X_3 space traced out by the systematic relationship between X_2, X_3, and Y:

$$\tilde{Y}_t = \beta_1 + \beta_2 X_{t2} + \beta_3 X_{t3}, \tag{2.2}$$

where $\tilde{Y}$ indicates the systematic part of Y ($\tilde{Y}_t = Y_t - U_t$). (Implicitly β_2 and β_3 are negative in Fig. 2.1, and we show $\tilde{Y}$ only for positive values of the variables.)

Consider fixing the value of X_3 at some constant, say X_3^*. We can then observe how different values of X_2 affect $\tilde{Y}$. This is equivalent to adding a plane in Fig. 2.1 perpendicular to the X_3 axis at the value of X_3^*. A similar result holds for $X_3 = 0$. Figure 2.2 plots the bivariate relationships between X_2 and $\tilde{Y}$. The slopes[2] of these bivariate relationships equal β_2, regardless of the

---

[2]An alternative way of seeing this is to put the question in terms of derivatives. We wish to know what is the change in $\tilde{Y}$ that arises from a change in X_2, all other variables held constant. This is the partial derivative $\partial \tilde{Y} / \partial X_2$ and in the linear case equals β_2.

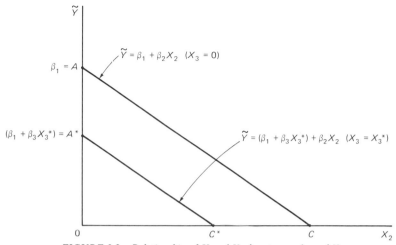

FIGURE 2.2   *Relationship of Y and X_2 for given values of X_3.*

value for X_3. (This fact is a consequence of the linear functional form. It is not necessarily true if the model under consideration is nonlinear.) The $\tilde{Y}$ intercept is the constant in the model, β_1, plus the constant effect of $X_3, \beta_3 X_3$. Thus, $A = \beta_1 + \beta_3 \cdot 0 = \beta_1$ and $A^* = \beta_1 + \beta_3 X_3^*$.

The random part U_t represents all other factors besides the explicit systematic part contained in the measured variable Y_t: random behavior, excluded influences, measurement errors, and errors in specifying the systematic relationship. The presence of this error term U_t dictates the use of statistical methodology, and the composition of this error term and its effects on our attempts to estimate the systematic relationships are the central issues of this book.

The estimation would be simple if we did not have a stochastic term in the model. If the U_t were identically zero, it would be possible to determine β_1, β_2, and β_3 exactly from observing Y_t for three different values of X_2 and X_3. If U_t is always zero, the coefficients can be found by solving the simultaneous equations

$$\beta_1 + \beta_2 X_{12} + \beta_3 X_{13} = Y_1,$$

$$\beta_1 + \beta_2 X_{22} + \beta_3 X_{23} = Y_2, \tag{2.3}$$

$$\beta_1 + \beta_2 X_{32} + \beta_3 X_{33} = Y_3.$$

The actual observations provide numerical values for the X's and Y's, and we can then solve for the unknowns β_1, β_2, and β_3.

This process is illustrated graphically for the two variable case in Fig. 2.3. In this case, we have the points (X_1, Y_1) and (X_2, Y_2) and are fitting the line $\beta_1 + \beta_2 X$.

However, when the error term is added, the observation of two points is no longer sufficient to ensure accurate estimates of β_1 and β_2. One may find a

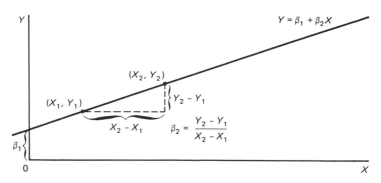

FIGURE 2.3   *Estimation of deterministic bivariate model.*

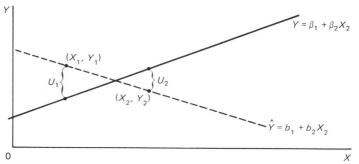

FIGURE 2.4   *Estimation of stochastic bivariate model with two observations.*

situation like Fig. 2.4, where $U_1 > 0$ and $U_2 < 0$. The errors U_1 and U_2 could even reverse the sign of our estimate of β_2. Similar results hold when we add error terms to our three variable model.

The problem created by the stochastic term is lessened by collecting many observations on the X_t's and Y_t. Through many observations, it is possible to improve the estimates of β_1, β_2, and β_3. In the following pages, we shall consider in detail how to make best use of a large number of observations of X and Y in estimating β_1, β_2, β_3. We shall concentrate on a technique called least squares regression. The reasons for this emphasis, along with its assumptions, procedures, and results, will be developed as we proceed.

2.3   Least Squares Estimators

Our problem is to find estimators[3] for the unknown parameters β_1, β_2, and β_3. These estimators are simply procedures for making guesses about the unknown parameters on the basis of the known sample values of Y, X_2, and

---

[3]An estimator is a formula for calculating a guess for the magnitude of an unknown population parameter on the basis of sample data. An estimate is the value of the estimator calculated from any one data set. Throughout the text we denote both estimator and a specific estimate by lower case italic letters, e.g., *b*, with the interpretation determined by the context.

X_3. For any estimates of the parameters, denoted b_1, b_2, and b_3, we can estimate the value for Y_t by $\hat{Y}_t = b_1 + b_2 X_{t2} + b_3 X_{t3}$. The difference between the actual (observed) and predicted values for each observation is

$$e_t = Y_t - \hat{Y}_t = Y_t - b_1 - b_2 X_{t2} - b_3 X_{t3}. \tag{2.4}$$

e_t is called the residual for the tth observation and is the vertical distance between the estimated plane $b_1 + b_2 X_{t2} + b_3 X_{t3}$ and the actual observation Y_t. Figure 2.5 illustrates this for the two variable case.

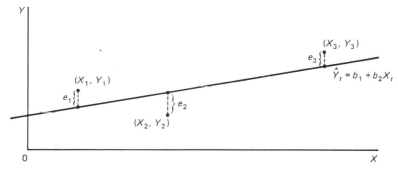

FIGURE 2.5   *Estimated bivariate regression model and residuals.*

One natural way to choose among alternative estimators of β_1, β_2, and β_3 would be to compare the values of e_t each produces when applied to our sample. This is an obvious approach, since the e_t indicate how closely an estimated plane comes to describing the data points: the larger the absolute values of the e_t, the worse the estimated plane does at representing the data. In fact, we shall develop the estimator for each β based upon these residuals.[4]

The estimation procedure is developed for a given sample of T observations of Y_t, X_{t2}, and X_{t3}. The criterion for the estimators we propose is to find

[4]There are, of course, alternatives to using the vertical distance for developing the estimators of the parameters. For example, we could consider the perpendicular distance to the estimated plane; this would be more consistent with common notions of distance. However, this estimator has two serious drawbacks. First, it is considerably more difficult to compute. Secondly, and more important, such an estimator would not be invariant to the units of measurement for the variables. When vertical distances are used, changing the units of measurement of one of the exogenous variables in Eq. (2.3), say multiplying X_2 by c, will (as shown below) leave the estimators for β_1 and β_3 unchanged and will yield the previous estimator of β_2 divided by c. In other words, there is no change except for the change in units. Estimators derived from considering the perpendicular distances do not have such a property—the estimator for all parameters will change with a change in units of one of the exogenous variables. This is undesirable since we generally do not a priori have any way of choosing between units of measurement. Another alternative to considering the vertical distances between observed points and the estimated plane is to concentrate on the horizontal distances with respect to a given exogenous variable. We concentrate upon the vertical distances because that is consistent with the form of the basic model [Equation (2.1)] and with the assumptions which we will make about the stochastic term U_t.

b_1, b_2, and b_3 minimizing the sum of squared errors (residuals), i.e., "least squares." In other words, we want to minimize

$$\sum_{t=1}^{T} e_t^2 = \sum_{t=1}^{T} \left(Y_t - \hat{Y}_t \right)^2 = \sum_{t=1}^{T} \left(Y_t - b_1 - b_2 X_{t2} - b_3 X_{t3} \right)^2 \qquad (2.5)$$

with respect to b_1, b_2, and b_3.

At first blush, a criterion involving the sum of errors rather than the sum of *squared* errors seems more natural. However, minimizing the sum of errors gives no assurance of accurate estimates since very large positive errors can be balanced by very large negative errors. In Fig. 2.6, for example, we show a set of four sample observations and a line that minimizes the sum of the errors for the bivariate case. In this case $e_4 = -e_1$ and $e_3 = -e_2$, so $\Sigma e_t = 0$. Our criterion is minimized but we have badly estimated[5] β_1 and β_2.

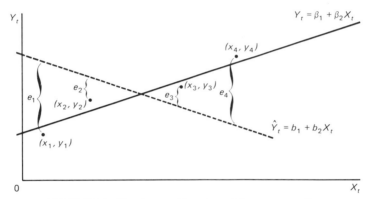

FIGURE 2.6   *Bivariate models estimated by minimizing* Σe_t.

An alternative, of course, is to choose b_1, b_2, and b_3 to minimize the sum of absolute errors, i.e., $\Sigma_{t=1}^{T} |e_t|$. This procedure presents problems because it is difficult to calculate the values of b_1, b_2, and b_3 minimizing this expression.

Minimizing the sum of absolute errors clearly attaches different importance to calculated errors than that attached by a least squares criterion. Large errors in the absolute error case are treated the same as many small errors, while the least squares criterion weights large errors much more heavily. In certain situations, giving less weight to large deviations may be more desirable. Nevertheless, leaving aside the computational difficulties, the statistical properties of estimators that minimize absolute errors are not well understood, while those of least squares estimators are both understood and often desirable.

---

[5]The estimated coefficients derived from simply minimizing the sum of the errors also are not unique. An infinite number of estimated relationships, corresponding to rotations of the line in Fig. 2.6, yield the same sum of residuals.

With the least squares criterion a minimum is easily found. More importantly, under certain conditions it is possible to demonstrate that these estimates have several desirable statistical properties. Consider the minimization problem. Once a sample of observations on the variables in the model (Y_t and the X_t's) has been drawn, the data are fixed and cannot be adjusted in order to minimize $\sum_{t=1}^{T} e_t^2$. The only "variables" for solving the minimization problem are the possible coefficient estimates b_1, b_2, and b_3.

Deriving the estimators for β_1, β_2, and β_3 is straightforward using elementary calculus. The problem is to pick the values for b_1, b_2, and b_3 giving the minimum sum of squared errors $(SSE) = \sum_{t=1}^{T} e_t^2$. To determine the minimum, we set the first derivative of SSE with respect to each estimated coefficient equal to zero.[6] The derivatives of SSE are found from the application of two properties: the derivative of the sum is the sum of the derivatives,

$$\frac{\partial SSE}{\partial b} = \frac{\partial \left(\sum_{t=1}^{T} e_t^2 \right)}{\partial b} = \frac{\partial \left(e_1^2 \right)}{\partial b} + \frac{\partial \left(e_2^2 \right)}{\partial b} + \cdots + \frac{\partial \left(e_T^2 \right)}{\partial b} = \sum_{t=1}^{T} \frac{\partial \left(e_t^2 \right)}{\partial b}; \quad (2.6)$$

and the chain rule,

$$\frac{\partial \left(e_t^2 \right)}{\partial b} = \frac{\partial \left(e_t^2 \right)}{\partial e_t} \frac{\partial e_t}{\partial b} = 2e_t \frac{\partial e_t}{\partial b}. \quad (2.7)$$

By Eq. (2.5),

$$\frac{\partial e_t}{\partial b_1} = \frac{\partial \left(Y_t - b_1 - b_2 X_{t2} - b_3 X_{t3} \right)}{\partial b_1} = -1, \quad (2.8)$$

$$\frac{\partial e_t}{\partial b_2} = \frac{\partial \left(Y_t - b_1 - b_2 X_{t2} - b_3 X_{t3} \right)}{\partial b_2} = -X_{t2}, \quad (2.9)$$

$$\frac{\partial e_t}{\partial b_3} = \frac{\partial \left(Y_t - b_1 - b_2 X_{t2} - b_3 X_{t3} \right)}{\partial b_3} = -X_{t3}. \quad (2.10)$$

Thus combining Eqs. (2.7)–(2.10) and setting the derivatives equal to zero for minimization, we have

$$\frac{\partial SSE}{\partial b_1} = 2 \sum_{t=1}^{T} e_t \frac{\partial e_t}{\partial b_1} = 2 \sum_{t=1}^{T} \left(Y_t - b_1 - b_2 X_{t2} - b_3 X_{t3} \right)(-1)$$

$$= -2 \sum_t Y_t + 2 \sum_t b_1 + 2 b_2 \sum_t X_{t2} + 2 b_3 \sum_t X_{t3}$$

[6]The principles of the calculus minimization are discussed in Appendix 2.2. The basic operations involving summations are discussed in Appendix 2.1. It is possible to derive these estimators without resorting to calculus. Such a derivation can be found in Wonnacott and Wonnacott. The solution, no matter what solution method is used, will be identical to Eqs. (2.14)–(2.16).

or

$$-\sum_t Y_t + Tb_1 + b_2 \sum_t X_{t2} + b_3 \sum_t X_{t3} = 0; \qquad (2.11)$$

$$\frac{\partial SSE}{\partial b_2} = 2\sum_{t=1}^{T} e_t \frac{\partial e_t}{\partial b_2} = 2\sum_{t=1}^{T}(Y_t - b_1 - b_2 X_{t2} - b_3 X_{t3})(-X_{t2})$$

or

$$-\sum_t Y_t X_{t2} + b_1 \sum_t X_{t2} + b_2 \sum_t X_{t2}^2 + b_3 \sum_t X_{t3} X_{t2} = 0; \qquad (2.12)$$

$$\frac{\partial SSE}{\partial b_3} = 2\sum_{t=1}^{T} e_t \frac{\partial e_t}{\partial b_3} = 2\sum_{t=1}^{T}(Y_t - b_1 - b_2 X_{t2} - b_3 X_{t3})(-X_{t3})$$

or

$$-\sum_t Y_t X_{t3} + b_1 \sum_t X_{t3} + b_2 \sum_t X_{t2} X_{t3} + b_3 \sum_t X_{t3}^2 = 0. \qquad (2.13)$$

Solving Eqs. (2.11)–(2.13) yields the least squares estimators for β_1, β_2, and β_3. For any given data set, the solution to these equations will provide specific estimates. There are three linear equations in the three unknowns b_1, b_2, and b_3. Because the estimation equations are linear in b, the estimator is referred to as a linear estimator. Although we could compute b_1, b_2, and b_3 from Eqs. (2.11)–(2.13), the properties of these estimators will be more apparent if we proceed with additional transformations.

We divide Eqs. (2.11)–(2.13) by T to give

$$b_1 + a_1 b_2 + a_2 b_3 = c_1, \qquad (2.11a)$$

$$a_1 b_1 + a_3 b_2 + a_4 b_3 = c_2, \qquad (2.12a)$$

$$a_2 b_1 + a_4 b_2 + a_5 b_3 = c_3, \qquad (2.13a)$$

where

$$a_1 = \frac{1}{T}\sum X_{t2} = \bar{X}_2, \qquad a_2 = \frac{1}{T}\sum X_{t3} = \bar{X}_3, \qquad a_3 = \frac{1}{T}\sum X_{t2}^2,$$

$$a_4 = \frac{1}{T}\sum X_{t2}X_{t3}, \qquad a_5 = \frac{1}{T}\sum X_{t3}^2,$$

$$c_1 = \frac{1}{T}\sum Y_t = \bar{Y}, \qquad c_2 = \frac{1}{T}\sum Y_t X_{t2}, \qquad c_3 = \frac{1}{T}\sum Y_t X_{t3}.$$

This shows the three simultaneous equations in more conventional format.

The solution of these equations for b_1, b_2, and b_3 provides the set of coefficient estimates that minimizes the sum of squared errors, $\sum_{t=1}^{T} e_t^2$. By means of calculus, we effectively searched the infinite number of possible estimates of the parameters and found that set satisfying the estimation criterion.

By the proper substitutions, Eq. (2.11) is

$$\bar{Y} = b_1 + b_2 \bar{X}_2 + b_3 \bar{X}_3 \quad \text{or} \quad b_1 = \bar{Y} - b_2 \bar{X}_2 - b_3 \bar{X}_3; \quad (2.14)$$

and (2.12) and (2.13) are

$$\bar{X}_2 b_1 + \left(\frac{1}{T} \sum X_{t2}^2 \right) b_2 + \left(\frac{1}{T} \sum X_{t2} X_{t3} \right) b_3 = \frac{1}{T} \sum Y_t X_{t2}, \quad (2.12b)$$

$$\bar{X}_3 b_1 + \left(\frac{1}{T} \sum X_{t2} X_{t3} \right) b_2 + \left(\frac{1}{T} \sum X_{t3}^2 \right) b_3 = \frac{1}{T} \sum Y_t X_{t3}. \quad (2.13b)$$

Substituting for b_1 from (2.14) gives

$$b_2 \left(\frac{1}{T} \sum X_{t2}^2 - \bar{X}_2^2 \right) + b_3 \left(\frac{1}{T} \sum X_{t2} X_{t3} - \bar{X}_2 \bar{X}_3 \right) = \left(\frac{1}{T} \sum Y_t X_{t2} - \bar{Y} \bar{X}_2 \right),$$

$$b_2 \left(\frac{1}{T} \sum X_{t2} X_{t3} - \bar{X}_2 \bar{X}_3 \right) + b_3 \left(\frac{1}{T} \sum X_{t3}^2 - \bar{X}_3^2 \right) = \left(\frac{1}{T} \sum Y_t X_{t3} - \bar{Y} \bar{X}_3 \right).$$

By definition of the variance of the variable X $[V_X = (1/T)\sum X_t^2 - \bar{X}^2]$ and the covariance of two variables X and Y $[C_{XY} = (1/T)\sum X_t Y_t - \bar{X}\bar{Y}]$, these two equations are[7]

$$b_2 V_{X_2} + b_3 C_{X_2 X_3} = C_{YX_2}, \quad b_2 C_{X_2 X_3} + b_3 V_{X_3} = C_{YX_3}.$$

For notational convenience, from this point on we shall suppress the X subscript. Thus, V_2 is the variance of X_2, C_{23} is the covariance of X_2 and X_3, C_{Y2} is the covariance of Y and X_2, etc. These equations can be solved to give our estimates for b_2 and b_3:

$$b_2 = (V_3 C_{Y2} - C_{23} C_{Y3}) / (V_2 V_3 - C_{23}^2), \quad (2.15)$$

$$b_3 = (V_2 C_{Y3} - C_{23} C_{Y2}) / (V_2 V_3 - C_{23}^2). \quad (2.16)$$

Our estimated slope coefficients b_2 and b_3 that minimize the sum of squared residuals are functions of the variances and covariances of the observed variables. The estimated intercept b_1 is the value that places the estimated relationship through the point of means $\bar{Y} = b_1 + b_2 \bar{X}_2 + b_3 \bar{X}_3$.

Equations (2.15) and (2.16) are somewhat complicated. A transformation of these equations brings out some important aspects of the least squares estimator. Let us concentrate on (2.15); this is done without loss of generality since the estimator of β_3 is subject to identical considerations and the results from (2.15) are symmetrical with the results that could be obtained from (2.16). Equation (2.15) can be rewritten in terms of the simple correlation coefficients between two variables. The transformations are accomplished by

---

[7] See Appendix I if you do not see that

$$\frac{1}{T} \sum (X_t - \bar{X})^2 = \frac{1}{T} \sum X_t^2 - \bar{X}^2 \quad \text{and} \quad \frac{1}{T} \left[\sum (X_t - \bar{X})(Y_t - \bar{Y}) \right] = \frac{1}{T} \sum X_t Y_t - \bar{X}\bar{Y}.$$

multiplying and dividing Eq. (2.15) by terms involving the variances and standard deviations of the variables in the model.

Since the simple correlation of X_2 and Y is $r_{Y2} = C_{Y2}/S_2 S_Y$ where $S_2 = \sqrt{V_2}$ and $S_Y = \sqrt{V_Y}$ (the standard deviation), multiplying and dividing Eq. (2.15) by $S_Y/V_2 V_3$ yields

$$b_2 = \frac{r_{Y2} - r_{23} r_{Y3}}{(1 - r_{23}^2)} \frac{S_Y}{S_2}. \tag{2.15'}$$

This states the estimator of β_2 in terms of the simple correlation coefficients. In the bivariate case, the estimator for β_2 was simply $r_{Y2}(S_Y/S_2)$. In the multivariate case, this estimator is modified to reflect both the sample correlation between the exogenous variables r_{23} and the relationship between the other exogenous variable and the endogenous variable r_{Y3}. A fundamental point introduced by Eq. (2.15) is that the least squares estimator of a parameter in the trivariate case equals the bivariate estimator when $r_{23} = 0$. In other words, when the sample correlation between the exogenous variables is zero, there is no need to use multivariate models or multivariate estimators. This feature is central to experimental design in the physical sciences: by either holding constant other variables or randomizing the experiment so that other exogenous variables are uncorrelated with the variable of interest (X_2), it is possible to look at just the bivariate model. Unfortunately, the nonexperimental world of the social scientist generally precludes designing an analysis so that r_{23} equals zero. Because people, counties, states, and countries do not group themselves randomly and are subjected to many uncontrolled influences, causal factors not of interest will be correlated with those of interest; and consequently even when we are interested only in the value of β_2, we must also include other exogenous variables such as X_3.

In general, the bivariate regression coefficients do not equal the multivariate estimates of the same population term. Since, as demonstrated in Chapter 3, there is a direct correspondence between the multivariate regression coefficient and the true model parameter, Eq. (2.15') implies that the simple bivariate estimators tend to give inaccurate indications of the true parameters. Therefore, searching the simple correlation table to understand the effects of various variables may be grossly misleading, depending upon the structure of the entire model and the degree of correlation among the exogenous variables. The problem of what variables to include in the multivariate model parallels this discussion and is so central to the entire question of estimation techniques and research design that it will occupy a major portion of Chapter 4.

The development of the estimator for b_3 is entirely symmetrical to that for b_2. In Eq. (2.15'), one needs only to interchange X_3 for X_2 and vice versa.

Estimation of the model parameters by Eq. (2.15') is not always possible. A necessary condition for estimation of a coefficient is the presence of some variance in the associated variable. For example, if our observed values for

X_{t2} are all the same so that $V_2 = 0$, then C_{23} and C_{Y2} also equal zero, and we have no information with which to estimate β_2.

The second necessary condition, and a relatively important one, is that $D = V_2 V_3 - C_{23}^2$ must not equal zero. If D does equal zero, then the required divisions cannot be executed, and we have no estimates. The simplest way to see how D might equal 0 is to divide D by $V_2 V_3$ as in Eq. (2.15'). As long as $V_2 \neq 0$ and $V_3 \neq 0$, $D = 0$ if and only if $1 - r_{23}^2 = 0$, or if $|r_{23}| = 1$. $|r_{23}| = 1$ only if X_2 and X_3 are perfectly correlated, that is, only if X_2 and X_3 are linearly related to each other as in $X_2 = d_1 + d_2 X_3$. For example, if X_2 is income in dollars and X_3 is income in thousands of dollars ($X_2 = 1000 X_3$) or if X_2 is a variable that is one for all males and X_3 is a similar variable for all females ($X_2 = 1 - X_3$), D would equal zero, and we could not estimate such a model. Thus, not only must there be some variance in X_2 and X_3, but they must also exhibit some independent variation ($|r_{23}| \neq 1$) in order for us to estimate our model.[8] This condition corresponds to the heuristic notion that b_2 and b_3 are computed on the basis of the independent variation that X_2 and X_3 exhibit.

Equations (2.14)–(2.16) give us *one* way of estimating the coefficients in our behavioral relationship. There are, of course, many other possible estimators or ways of using sample data to estimate the β's. For example, we could calculate the deviations of each variable about its mean, giving a set of T observations $X_{t2} - \bar{X}_2$, $X_{t3} - \bar{X}_3$, and $Y_t - \bar{Y}$. From these we could estimate β_2 by taking the average of the ratios $(Y_t - \bar{Y})/(X_{t2} - \bar{X}_2)$, giving the estimate

$$b_2' = \frac{1}{T} \sum \frac{Y_t - \bar{Y}}{X_{t2} - \bar{X}_2} \qquad \text{and} \qquad b_3' = \frac{1}{T} \sum \frac{Y_t - \bar{Y}}{X_{t3} - \bar{X}_3}.$$

b_1' can be calculated in a fashion similar to Eq. (2.14), $b_1' = \bar{Y} - b_2' \bar{X}_2 - b_3' \bar{X}_3$, which puts the plane through the point of means.

We have concentrated on the estimators of (2.14)–(2.16) rather than other perhaps simpler methods because of the statistical properties of the *OLS* estimators. These properties are based on several assumptions about the nature of the U_t. These assumptions, the resulting statistical properties, and their implications and use are discussed in detail in the next chapter. However, we shall give two simple examples of estimating trivariate equations before beginning this statistical discussion.

2.4   Two Examples

The following equation states a simple model relating earnings to schooling levels:

$$\text{Earnings} = \beta_1 + \beta_2 \text{Schooling} + U. \qquad (2.17)$$

---

[8]For the time being, it is not important to consider why two independent variables might be correlated. As long as there is not a perfect linear relationship between them (a point to be discussed soon), the exact mechanism that leads to the correlation does not affect our arguments. It could be historical accident, coincidental associations, or a causal structure that links the exogenous variables.

In this model, the additional earnings associated with one more year of schooling is simply the estimated coefficient β_2. However, this result might be quite misleading because we also expect other things to affect earnings. In particular, additional experience in the labor market would be expected to increase earnings also, and we are unlikely to find a sample where schooling varies but experience does not. Thus, a model such as

$$\text{Earnings} = \beta_1 + \beta_2 \text{Schooling} + \beta_3 \text{Experience} + U \qquad (2.18)$$

seems more appropriate.

Let us observe both how the estimation can be accomplished and the effect of estimating the more complete model. Using the 1970 U.S. Census of Population data, we have gathered a sample of black males living in Boston and working full time in 1969. For these individuals, we know their earnings and years of schooling and can estimate their experience on the basis of their age and schooling.[9]

For these individuals, we have calculated the means and standard deviations for the variables (Table 2.1), and the simple correlation matrix (Table 2.2). On the basis of this information, we can estimate Eqs. (2.17) and (2.18). First, consider just the bivariate relationship between schooling and earnings. Using Eq. (1.7), we see that b_2^* (the bivariate regression coefficient from Y regressed on only X_2) is found from

$$b_2^* = \frac{0.24 \times \$4277}{2.9 \text{ yr}} = 354 \quad \text{dollars/yr.}$$

The intercept (b_1^*) would be $b_1^* = \$7711 - (354 \text{ \$/yr})(11.4 \text{ yr}) = \$3676$. Thus, estimation of just the bivariate model indicates that each additional year of schooling is worth \$354 in annual earnings.

However, what is the picture when we allow for the effects of experience on earnings? This is simply illustrated by using Eq. (2.15') to estimate b_2 as

$$b_2 = \frac{0.24 - (-0.40)(0.07)}{(1 - 0.16)} \times \frac{4277}{2.9} = 470 \quad \text{dollars/yr.}$$

Estimation of the rest of the trivariate models yields $b_1 = \$1045$ and $b_3 = 67$ dollars/yr. The lesson is clear: the estimated effects of the multivariate model are quite different from those in the simple bivariate model. The revised estimates indicate that the earnings value of an additional year of schooling—holding constant the amount of experience the individual has—is \$470. After correcting for the effect of experience, our estimate of the value of schooling rises by 33%. A similar calculation for experience shows that the estimate for the value of an additional year of experience rises by over 150% (from 24 to 67 dollars/yr), when one uses a trivariate rather than bivariate model.

---

[9]Experience is estimated as Age − Schooling − 6. This assumes that all time not spent in school after a certain age was spent in the labor force.

TABLE 2.1

1969 Earnings, Schooling, and Experience of Full-Time Black Workers in Boston, T = 105

Variable	Mean	Standard deviation
Earnings (dollars)	7711	4277
Schooling (years)	11.4	2.9
Experience (years)	19.5	12.6

TABLE 2.2

Simple Correlations

	Earnings	Schooling	Experience
Earnings	1.00	0.24	0.07
Schooling		1.00	−0.40
Experience			1.00

The estimates of Eq. (2.18) provide information about the effects of schooling and experience on earnings for black males in Boston. What is more, the estimation procedure has separated the effects of these two explanatory variables. Since we could not run an experiment that, say, looked at the earnings obtained by individuals with different amounts of schooling but the same amount of experience, we had to resort to statistical methods to sort out the effects of each factor.[10] The model indicates that an additional year of schooling (holding experience constant) is worth \$470, while an additional year of experience (holding schooling constant) is worth \$67.

The estimated equation also says that somebody with no schooling and no experience would be expected to earn \$1045 (the intercept level). This latter result should probably not be taken too seriously, however, since nobody in the sample comes close to having no schooling and no experience. This implies that such an individual would be far away from any of the observed data. While the true model may be linear over the range of observed schooling and experience, it may be quite nonlinear outside of that range. Thus, interpreting such predictions too literally is inappropriate.

The second application of this estimation model is the presidential voting example discussed in the introductory chapter. In that chapter, we presented a table showing the voting intentions of individuals in the 1964 presidential election contest between Barry Goldwater and Lyndon Johnson. We now want to try to answer the question, What effect did party affiliation and issue preference have on voting decisions in that election? This means that we need to state a simple model relating voting intention to party affiliation and issue

---

[10]In using statistical methods rather than experimental methods to control for the effects of experience, we must assume a particular functional form. In this case, we assumed a linear additive relationship between earnings and schooling and experience.

preference. This model is shown as

$$V = \beta_1 + \beta_2 P + \beta_3 E + U, \tag{2.19}$$

where V is the intended vote with Johnson $= 1$ and Goldwater $= 0$, P is the party identification of each respondent, and E is their evaluation of the perceived party positions averaged for the several issues in the survey.

Converting party affiliation to a single variable requires that we assign arbitrary values to each party preference. This study sets values of the variable P at 0.0, 0.25, 0.75, and 1.0 for strong Republicans, weak Republicans, weak Democrats, and strong Democrats, respectively. By arbitrarily assigning the above values, we are saying that weak Republicans are more like strong Republicans and weak Democrats are more like strong Democrats than they are like each other. This is clearly an arbitrary scale (at this point, there perhaps is as much justification for choosing the values 0.0, 0.33, 0.67 and 1.0 as any other set), but such decisions must be made many times during an analysis. Subsequent chapters will discuss alternatives to this procedure and ways of improving such decisions, but they cannot be avoided.

The evaluation variable is derived from voters' perceptions of which party is closest to their preferred position on seven different public policy issues. Each evaluation is coded as a 1.0 if the Democrats are closer and 0.0 if the Republicans are. The overall evaluation variable E is created by averaging the evaluations on all issues where the person has a preference and sees a difference in the parties, thus omitting all issues where no party preference is expressed. The difference between people perceiving the Democrats as being closer on all issues and those perceiving the Republicans as closer is one. People judged to be indifferent on all seven issues are assigned the value 0.5.

The interpretation of Eq. (2.19) is fairly straightforward. The expected vote is β_1 for a strong Republican who favors the Republicans on all issues and $\beta_1 + \beta_2 + \beta_3$ for a strong Democrat favoring the Democrats on all issues. For people with the same issue preferences, $0.5\beta_2$ is the expected voting difference between a weak Republican and a weak Democrat. For those having opposing evaluations, the expected change in voting intention is β_3. Comparisons of this sort between β_2 and β_3 provide us with an estimate of the relative importance of party affiliation and issue evaluations on voting decisions. Our purpose in estimating the β's is so we can proceed with such substantive interpretations.

The estimated equation calculated from the continuous variables rather than from Table 1.1 is

$$V = 0.087 + 0.636 P + 0.360 E. \tag{2.20}$$

The expected voting intentions range from 0.087 for a strong Republican preferring the Republicans on all issues to 1.083 for a strong Democrat with the opposite preferences. In a model such as this, the predicted values can be interpreted as the probability that an individual with given characteristics will vote Democratic. (A prediction greater than 1.0 clearly does not make sense for a probability; at this point, we will simply interpret the 1.083 as a probability of 1.0.) The individual coefficients then can be interpreted as marginal probabilities, i.e., how the probability of a Democratic vote changes with each variable. The coefficient of 0.636 on the party variable indicates that expected voting intentions differ by 0.16 between weak and strong members of the same party and by 0.32 between weak members of the opposing parties, and by 0.636 between strong members of opposite parties. These compare in magnitude to an expected difference in probability of voting for Johnson of 0.36 between people preferring one party or the other on all relevant issues. (These calculations make use of the coefficient magnitudes and the values of the variables that correspond to different party identification and issue evaluation. For example, the difference in the value of P for a strong and weak Democrat is 0.25; this times the coefficient 0.636 yields a predicted difference in voting of 0.16.)

We shall have considerably more to say about these two models. Their specifications, the variable construction, and the problems associated with the estimation in Eqs. (2.18) and (2.20) will be discussed throughout the book. For the time being it is important to understand why we formulated the models in this fashion, how we related the models to observable data and estimated the coefficients, and finally what interpretations we gave these estimates. The preceding discussion has been a critical first step in our exposition of estimation procedures and difficulties. If the logic and motivation behind the discussions and examples in this chapter are not well understood, comprehending succeeding chapters will be extremely difficult.

2.5 Conclusion

This chapter has the limited objective of introducing the linear behavioral model and the linear estimation procedure commonly referred to as ordinary least squares (OLS). It is important to understand the distinction between these two "linear" models. The first refers to the functional form of the hypothesized relationship between the dependent and explanatory variables. Section 2.2 and Figs. 2.1 and 2.2 presented the implications of this behavioral model. However, in Chapter 4 we shall see that we are not limited to this strict linear relationship in the hypothesized behavioral model.

The term linear estimation model refers to the set of equations solved in the process of estimating the coefficients in the hypothesized model. Equations (2.11)–(2.13) constitute a set of three *linear* simultaneous equations with three unknowns. Hence the name linear estimation model. This use of the term linear model makes no direct reference to the form of the behavioral model being estimated, but simply to the manner in which the estimates are calculated once the data are collected.

The particular linear estimation model developed here is called least squares for the simple reason that we pick values for the estimated coefficients minimizing the sum of squared estimation errors, or residuals. The expression for the sum of squared errors is written as Eq. (2.5), and contains in algebraic form the coefficients we want to estimate. We then develop a method for choosing the values that minimize this expression for any given set of variables and data. So far this is strictly an arithmetic exercise. Given any set of variables, we can compute a set of least squares regression coefficients, provided that no explanatory variable is perfectly correlated with any others. However, this is only one of many ways to "estimate" these coefficients. In the next chapter we shall discuss why the least squares method is preferred and what assumptions about the model and data are required to make inferences about the population relationships from the estimated values.

APPENDIX 2.1

Properties of Summations

The operators applied to summations are simply generalizations of basic algebra. The simple properties that we shall make use of are, for constants a, b, and c,

$$(a+b)+c = a+(b+c), \tag{A}$$

$$a(b+c) = ab + ac. \tag{B}$$

Consider variables X and Y that are indexed by a subscript i, any constant C, and integers l and m. The basic summation definition is

$$\sum_{i=1}^{m} X_i = X_1 + X_2 + X_3 + \cdots + X_m, \qquad \text{where} \quad m \geq 1.$$

Property 1 $\displaystyle\sum_{i=1}^{m} CX_i = C \sum_{i=1}^{m} X_i$

since

$$\sum_{i=1}^{m} CX_i = (CX_1 + CX_2 + \cdots + CX_m) = C(X_1 + X_2 + \cdots + X_m).$$

Property 2 $\displaystyle\sum_{i=1}^{m} C = mC$

since

$$\sum_{i=1}^{m} C = (C + C + \cdots + C).$$

Property 3 $\displaystyle\sum_{i=1}^{m} (X_i + Y_i) = \sum_{i=1}^{m} X_i + \sum_{i=1}^{m} Y_i$

since

$$\sum (X_i + Y_i) = (X_1 + Y_1) + (X_2 + Y_2) + \cdots + (X_m + Y_m)$$
$$= (X_1 + X_2 + \cdots + X_m) + (Y_1 + Y_2 + \cdots + Y_m).$$

It is often convenient to use the notation of double summations where each summation pertains to a separate index. The definition of the operation is

$$\sum_{i=1}^{m} \sum_{j=1}^{n} X_{ij} = \sum_{i=1}^{m} (X_{i1} + X_{i2} + \cdots + X_{in})$$

$$= (X_{11} + X_{12} + \cdots + X_{1n}) + (X_{21} + X_{22} + \cdots + X_{2n}) + \cdots + (X_{m1} + X_{m2} + \cdots + X_{mn}).$$

Property 4 $\displaystyle\sum_{i=1}^{m} \sum_{j=1}^{n} (X_{ij} + Y_{ij}) = \sum_{i=1}^{m} \sum_{j=1}^{n} X_{ij} + \sum_{i=1}^{m} \sum_{j=1}^{n} Y_{ij}.$

Property 5 $\displaystyle\sum_{i=1}^{m} \sum_{j=1}^{n} X_i Y_j = \sum_{i=1}^{m} X_i \sum_{j=1}^{n} Y_i$

since

$$\sum_{i=1}^{m} \sum_{j=1}^{n} X_i Y_j = \sum_{i=1}^{m} (X_i Y_1 + X_i Y_2 + \cdots + X_i Y_n) = \sum_{i=1}^{m} X_i (Y_1 + Y_2 + \cdots + Y_n).$$

Property 6 $\displaystyle\left(\sum_{i=1}^{m} X_i \right)^2 = \sum_{i=1}^{m} X_i^2 + 2 \sum_{i=1}^{m-1} \sum_{j=i+1}^{m} X_i X_j$

since

$$\left(\sum_{i=1}^{m} X_i \right)^2 = (X_1 + X_2 + \cdots + X_m)^2$$

$$= X_1^2 + X_2^2 + \cdots + X_m^2 + 2(X_1 X_2 + X_1 X_3 + \cdots + X_1 X_m)$$

$$+ 2(X_2 X_3 + X_2 X_4 + \cdots + X_2 X_m) + \cdots + 2X_{m-1} X_m.$$

Summations can also be generalized to arbitrarily begin and end the range of summation. For example,

$$\sum_{i=l}^{l+m} X_i = X_l + X_{l+1} + X_{l+2} + \cdots + X_{l+m}.$$

APPENDIX 2.2

Calculus and the Minimization of Functions

Elementary and partial differential calculus provide a way of analytically locating the maximal or minimal values of a large class of mathematical functions. We write such a function as $Y = f(X)$, where $f(\cdot)$ denotes the operations determining the value of Y for given values of X. The question is, What value of X gives the largest and/or smallest values for Y?

The key to locating these maxima and minima is the slope of the function for different values of X, where the slope is defined as the ratio of the change in Y to the change in X, for very, very small changes in X. For the linear function $Y = BX$, such as we estimate in this chapter, the slope of Y with respect to X is simply B. This means that for any change in X, no matter how small, Y changes by B times this amount. We denote these changes in X and Y as ΔX and ΔY, so that $\Delta Y = B \Delta X$, or the slope equals $\Delta Y / \Delta X = B$.

For nonlinear functions, such as shown in Fig. 2.A1, the slope of the function is defined as the slope of the linear function (a line in this two-dimensional case) that is tangent to the function at the point being considered. We can see that this slope differs at each value of X. In this illustration, the slope is negative for $X < X_1$ and positive for $X > X_1$. This means that where the slope is negative, Y decreases as X increases; and where the slope is positive, Y increases as X increases.

The key to finding a maximum or minimum for a function is to find the point (or points) where the slope equals zero, meaning that Y neither increases nor decreases with very, very small changes in X. This is illustrated by X_1 in Fig. 2.A1. The figure is drawn so that the slope is zero at the point X_1. This means that for $X < X_1$, Y decreases as X increases (as X moves toward X_1); and that for $X > X_1$, Y increases as X increases (as X moves away from X_1). Thus X_1 is the minimum. The behavior at a maximum would be just the opposite, with a positive slope for $X < X_1$ and a negative

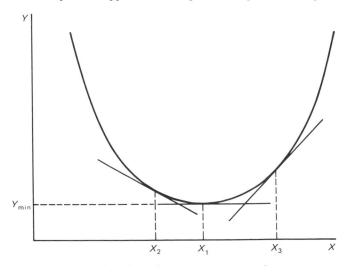

FIGURE 2.A1   Slope of quadratic function at minimum and nonminimum points.

slope for $X > X_1$. The purpose then is to find an expression for the slope of the function, and then determine the value of X that equates this expression to zero.

The process by which we obtain a general expression for the slope of a function is referred to as the differential calculus. This expression, denoted as dY/dX, is called a derivative. For our purposes we concentrate on quadratic functions since that is the form of the sum of squared residuals function we are trying to minimize. The general form for a quadratic function in two dimensions is $Y = a + bX + cX^2$. The rule for determining the derivative of any quadratic function is

$$\frac{dY}{dX} = \frac{d(a + bX + cX^2)}{dX} = b + 2cX.$$

Thus the slope of the quadratic function at any point X^* is $b + 2cX^*$.

To locate the value of X that minimizes (or maximizes) this quadratic function, we need only find the value of X for which the derivative equals zero:

$$\frac{dY}{dX} = b + 2cX = 0 \quad \text{or} \quad X = -\frac{b}{2c}$$

Thus we can find the minimum (or maximum) of this quadratic function quite easily. (More complicated differential calculus is required to determine whether this point is a maximum or a minimum for any given function, but that need not concern us here since it will be a minimum for the sum of squared residuals function.)

We can expand this treatment to functions of several unknowns, such as we have in the sum of squared residuals expression minimized in this chapter. We merely take the derivative of the function with respect to each unknown, treating the other variables as constants. Each of these derivatives is set to zero, and the minimum (or maximum) value of the function is found at the values that set all derivatives to zero for each variable.

The general quadratic function in two variables X and Z is

$$Y = a + bX + cX^2 + dXZ + fZ + gZ^2.$$

The derivatives of Y with respect to X and Z, referred to as partial derivatives and denoted as $\partial Y/\partial X$ and $\partial Y/\partial Z$ are obtained from

$$Y = (a + fZ + gZ^2) + (b + dZ)X + cX^2 = a' + b'X + cX^2,$$

$$Y = (a + bX + cX^2) + (f + dX)Z + gZ^2 = a'' + b''Z + gZ^2,$$

so that

$$\partial Y/\partial X = b' + 2cX = b + dZ + 2cX \quad \text{and} \quad \partial Y/\partial Z = b'' + 2gZ = f + dX + 2gZ.$$

The minimum (or maximum) value of Y is found by solving two linear simultaneous equations in two unknowns:

$$b + dZ + 2cX = 0,$$

$$f + 2gZ + dX = 0.$$

This process can be extended to quadratic functions with any number of unknowns. Each time we add an unknown, we add another equation to the system, which is the partial derivative of Y with respect to this new variable, providing as many linear equations as unknowns.

REVIEW QUESTIONS

1. The accompanying table gives data on birthrates, income, and racial composition of geographic regions in the United States.

Region	Live births per 1000 population (1968)	Per capita income ($) (1969)	Percent black (1970)
New England	22.5	3998	3.3
Middle Atlantic	21.5	4193	10.6
East North Central	24.2	3928	9.6
West North Central	24.0	3492	4.3
South Atlantic	24.2	3304	20.9
East South Central	24.4	2724	20.3
West South Central	25.4	3122	15.8
Mountain	27.3	3267	2.2
Pacific	23.6	4134	5.7

(a) Estimate the model: $BIRTHS = b_1 + b_2 INCOME + b_3 RACE$.

(b) What are the units for each of the estimated coefficients?

(c) If median income was measured in thousands of dollars instead of dollars, what would the new coefficient estimates of the model in part (a) be?

(d) Calculate the residuals associated with the different regions. Does the pattern of these residuals suggest any other hypotheses about the determinants of birthrates?

2. The U.S. Bureau of the Census annually publishes a compendium of data about various social and economic factors for the U.S. This is called the *Statistical Abstract of the United States*. Using a recent edition of this source book, collect the data on robbery rates, income levels, and number of policemen per thousand population by state. Use these data to estimate the effect of income and police protection on crime. What relationship (i.e., sign of effect) would you expect between crime and the two exogenous variables? Why do you expect this? (This exercise, involving 50 observations, is most efficiently done on a computer.)

3. The accompanying matrix describes the simple correlation between earnings, years of schooling, and years of experience for black males living in Cleveland during 1969. The means and standard deviations for each of the variables are shown below the matrix.

	Simple correlations			Mean	Standard deviation
	Earnings	Schooling	Experience		
Earnings	1.00	0.28	0.14	$10,761	$7163
Schooling		1.00	−0.38	12.29 yr	2.90 yr
Experience			1.00	22.43 yr	12.75 yr

(a) Calculate the bivariate regression coefficient for earnings regressed on schooling.

(b) Calculate the coefficient for schooling in a model that also includes experience as an exogenous variable and compare this estimate to the bivariate coefficient.

(c) How does the value of an additional year of schooling in Cleveland compare to the value of an additional year of schooling in Boston?

3 | Least Squares Estimators: Statistical Properties and Hypothesis Testing

3.1 Introduction

Least squares is far and away the most commonly used estimating procedure. One reason for this is the simplicity of the computations required, particularly with modern digital computers. These machines are very easily programmed to solve the linear equation system outlined in Eqs. (2.11)–(2.13). There is a second, and more important, reason for this popularity however. With certain assumptions about the characteristics of each error term, it is possible to derive statistical properties for the least squares estimators. From these properties, such as the mean and variance of the possible estimates, we hope to make inferences about the values of the coefficients in the population that generated our data. Furthermore, if these assumptions are met, the statistical properties of least squares estimators are generally preferred to those of other linear estimators.

This effort at statistical inference and the accompanying substantive discussions based on the inferred population model sets serious and productive research apart from simple numerical computing. Anyone, or at least anyone with access to a computer, can calculate a set of coefficients for almost any set of variables. However, substantial understanding of the process being modeled and of the available data is required to consider the appropriateness of the assumptions we are about to discuss, to evaluate the statistical results in light of these assumptions and the hypothesized model, and to make the correct behavioral inferences. This is the hardest part of any empirical research and its importance cannot be understated.

This chapter is central to the entire book. One of the threads going through the development is that there are times when some modification of the basic least squares technique is appropriate. The need for a particular modification depends crucially upon the structure of the model and the error term. The ability to recognize this need requires complete understanding of the basic

45

linear estimation model. Therefore, this chapter will discuss situations where ordinary least squares (OLS) estimators are appropriate. Subsequent chapters (4 and 6–10) delve into alternatives when certain features of the model make OLS estimators inappropriate.

3.2   Properties of Least Squares Estimators

The heart of this discussion of the properties of least squares estimators is the fact that the estimators themselves are random variables, distributed about some mean with a given variance. (The concept of random variables and their distributional properties are used throughout the remainder of the book. Appendix I reviews these concepts.) They are random variables because of the presence of the error term in the observed values for Y that are used to calculate the estimated coefficients. If we were to observe a different set of values for Y, based upon the same population model with the same values for the X's but with different error terms, we would get a different set of estimated coefficients. (We shall demonstrate this property in detail later in this chapter.) However, we usually have only one data set, so we must consider the one estimated value for each coefficient as a draw out of a distribution of possible values. Statistical methods combine a single estimated value with information about the distribution of all possible coefficients to make inferences about the true population values. All of the assumptions discussed in this chapter are directed toward estimating certain parameters of the coefficient distributions, parameters from which we make inferences about the population model.

The most important distributional parameters are the mean and variance of the estimators. In particular, under the assumptions about the distribution of the error terms to be discussed here, least squares estimators b_1, b_2, and b_3 are unbiased estimators of β_1, β_2, and β_3. [Remember that unbiased implies $E(b) = \beta$, where $E(\ )$ denotes expected value.]

The numerical estimates of b_1, b_2, and b_3 in any sample will generally *not* (the probability is zero) equal the true values exactly; the closeness to the actual values (in probability terms) will depend upon the distribution of the estimators and, in particular, upon the variance of the estimators. Again, under certain assumptions about the distributions of the U_i, it is possible to estimate these variances. Further, with these assumptions the least squares estimator can be shown to have minimum variance among all estimators that are linear functions of the observed Y's and X's and that are unbiased.[1]

---

[1] Having the least variance among unbiased estimators does not ensure that the mean squared distance of the OLS estimator to the true population parameter is a minimum. It might be possible to find a biased estimator with a smaller expected mean squared error. However, in general, the mathematics of finding such an estimator becomes intractable. Thus, we content ourselves with looking at just the set of unbiased estimators of the parameters. This is discussed in Appendix I.

Unbiased estimators with minimum variance are said to be the *best*, or most *efficient*, estimators. Thus, the least squares estimator is called *BLUE* (best linear unbiased estimator).

We shall consider each of these properties in turn.

It will be necessary to make various assumptions in the form of restrictions on the model and the U_t. These assumptions are very important; learning them now will be to your long-run advantage since they are referred to repeatedly. In fact, most of this book (Chapters 4 and 6–10) is organized in terms of failures of these specific restrictions and their consequences for our ability to make proper inferences. First, however, we shall develop the basic or ideal case where all the assumptions are satisfied.

This development is based upon the population model of Eq. (2.1),

$$Y_t = \beta_1 + \beta_2 X_{t2} + \beta_3 X_{t3} + U_t,$$

which relates the observed dependent variable to the hypothesized explanatory variables X_2 and X_3.

We have already discussed in Chapter 2 our first assumption about the model and the data. This is

A.1 $\boxed{|r_{23}| < 1.0,}$

the requirement that each of the explanatory variables exhibit some independent variation. (Implicitly, we also require that the variances of X_2 and X_3 cannot be zero.) The explanatory variables may be correlated—in fact we expect this in any data set—but they may not be linearly related to each other. This assumption is required to compute the OLS estimators.

Unbiasedness

One intuitively desirable property of an estimator is unbiasedness—or, that the expected value of the estimator equals the true population value: $E(b) = \beta$. If we could draw many samples (that is, observe many sets of T values of Y for our X's) and estimate the parameters for each sample, then the means of the estimator would equal the true population value in the unbiased case.

This concept of obtaining repeated samples of observations for Y for the same values of the explanatory variables is referred to as the assumption of

A.2 $\boxed{\text{Fixed } X.}$

This assumption, that we could obtain many different samples for the same values of X and estimate coefficients for each sample, is analogous to the procedure of a physical scientist who repeats a controlled experiment under laboratory conditions several or many times. Each of these samples is referred to as a replication of the experiment. Although in a nonexperimental situation such exact replications of an "experiment" are impossible, we want to act as if these replications are possible for the sake of discussing the distributions of possible coefficients. The one data sample being analyzed then is considered one of these potentially infinite number of replications.[2]

Since only one sample of observations is available, i.e., one replication, it is appealing to have estimators of β_1, β_2, and β_3 whose distributions are, among other things, centered about the true values, or are unbiased.

To find the expected values of b_1, b_2, and b_3, we must first put the estimators from Eq. (2.15) and (2.16) in terms of the true but unknown values β_1, β_2, β_3, and U_t. Again, relying on the symmetry of the estimators, we shall concentrate upon b_2. Rearranging (2.15) to separate Y yields

$$b_2 = \frac{V_3 C_{2Y} - C_{23} C_{3Y}}{D}$$

$$= \frac{(1/T) \sum_{t=1}^{T} \left\{ \left[V_3 (X_{t2} - \bar{X}_2) - C_{23} (X_{t3} - \bar{X}_3) \right] (Y_t - \bar{Y}) \right\}}{D}$$

where $D = V_2 V_3 - C_{23}^2$. From the true model for Y and from averaging the Y_t over the sample, we know that

$$Y_t - \bar{Y} = \beta_1 + \beta_2 X_{t2} + \beta_3 X_{t3} + U_t - \left(\beta_1 + \beta_2 \bar{X}_2 + \beta_3 \bar{X}_3 + \bar{U} \right)$$

$$= \beta_2 (X_{t2} - \bar{X}_2) + \beta_3 (X_{t3} - \bar{X}_3) + (U_t - \bar{U}),$$

where $\bar{U}$ is the mean of all error terms implicit in the sample. By substitution, we have

$$b_2 = \frac{1}{TD} \left\{ \sum_{t=1}^{T} \left[V_3 (X_{t2} - \bar{X}_2) - C_{23} (X_{t3} - \bar{X}_3) \right] \right.$$

$$\left. \times \left[\beta_2 (X_{t2} - \bar{X}_2) + \beta_3 (X_{t3} - \bar{X}_3) + (U_t - \bar{U}) \right] \right\}$$

$$= \frac{1}{D} \left\{ \frac{\beta_2}{T} V_3 \sum (X_{t2} - \bar{X}_2)^2 - \frac{\beta_2}{T} C_{23} \sum (X_{t3} - \bar{X}_3)(X_{t2} - \bar{X}_2) \right.$$

$$+ \frac{\beta_3}{T} V_3 \sum (X_{t2} - \bar{X}_2)(X_{t3} - \bar{X}_3) - \frac{\beta_3}{T} C_{23} \sum (X_{t3} - \bar{X}_3)^2$$

$$\left. + \frac{1}{T} \sum \left[V_3 (X_{t2} - \bar{X}_2) - C_{23} (X_{t3} - \bar{X}_3) \right] (U_t - \bar{U}) \right\}.$$

The first summation can be written as $\beta_2 V_3 (1/T) \sum (X_{t2} - \bar{X}_2)^2 = \beta_2 V_3 V_2$.

---

[2]The importance of this assumption is that, with assumption A.3, it assures independence of the X's and U's. Independence of X and U is sufficient to develop unbiasedness.

Similar treatment of the succeeding terms gives

$$b_2 = \frac{\beta_2 V_3 V_2 - \beta_2 C_{23}^2 + \beta_3 V_3 C_{23} - \beta_3 C_{23} V_3}{D} + \frac{V_3 C_{2U} - C_{23} C_{3U}}{D}$$

$$= \frac{\beta_2 (V_2 V_3 - C_{23}^2)}{D} + \frac{V_3 C_{2U} - C_{23} C_{3U}}{D} = \beta_2 + \frac{V_3 C_{2U} - C_{23} C_{3U}}{D}. \quad (3.1)$$

Similar treatment of Eq. (2.16) gives

$$b_3 = \beta_3 + \frac{V_2 C_{3U} - C_{23} C_{2U}}{D}. \quad (3.2)$$

These two expressions indicate that our estimated coefficients b_2 and b_3 equal the population coefficients β_2 and β_3, respectively, plus a term involving the sample covariances of U_t and X_{t2} and X_{t3}.

Equations (3.1) and (3.2) show that the estimated coefficient is itself a random variable because it is a function of the error term in the population model. (In the nonstochastic case, where U is identically zero, the last term in each of these equations is also zero, and the estimated coefficient equals the true value.) The distribution of the estimated coefficients is then related to the distributions of the individual error terms.

Taking the expected value of b_2 and b_3 (see Appendix I) to get the mean of the coefficient distributions, we have

$$E(b_2) = E(\beta_2) + E\left[\frac{V_3 C_{2U} - C_{23} C_{3U}}{D} \right],$$

$$E(b_3) = E(\beta_3) + E\left[\frac{V_2 C_{3U} - C_{23} C_{2U}}{D} \right].$$

β_2 and β_3 are constants, so $E(\beta_2) = \beta_2$ and $E(\beta_3) = \beta_3$. Since we treat the values for the X's as nonstochastic, or as fixed during all replicated experiments, V_2, V_3, C_{23}, and D can be treated as constants in taking expected values. Thus

$$E(b_2) = \beta_2 + \frac{V_3 E(C_{2U})}{D} - \frac{C_{23} E(C_{3U})}{D},$$

$$E(b_3) = \beta_3 + \frac{V_2 E(C_{3U})}{D} - \frac{C_{23} E(C_{2U})}{D}.$$

Our estimates will be unbiased if $E(C_{2U}) = 0 = E(C_{3U})$. The development concentrates upon these terms, where

$$E(C_{2U}) = E\left[\frac{1}{T} \sum_t (X_{t2} - \bar{X}_2)(U_t - \bar{U}) \right],$$

$$E(C_{3U}) = E\left[\frac{1}{T} \sum_t (X_{t3} - \bar{X}_3)(U_t - \bar{U}) \right].$$

Since the expected value of any sum is the sum of the expected values and since the X's are fixed,

$$E(C_{2U}) = \frac{1}{T}\sum(X_{t2} - \bar{X}_2)E(U_t - \bar{U}) = \frac{1}{T}\sum(X_{t2} - \bar{X}_2)(\bar{U}_t - \mu) = C_{2\bar{U}_t}$$

(3.3)

$$E(C_{3U}) = \frac{1}{T}\sum(X_{t3} - \bar{X}_3)E(U_t - \bar{U}) = \frac{1}{T}\sum(X_{t3} - \bar{X}_3)(\bar{U}_t - \mu) = C_{3\bar{U}_t},$$

(3.4)

where $\bar{U}_t$ is the mean of the error terms associated with observation t from all possible replications and μ refers to the mean of all error terms, taken over all observations and all replications, $\mu = E(\bar{U}) = E((1/T)\sum\bar{U}_t)$.

The necessary condition for the estimates to be unbiased is that Eqs. (3.3) and (3.4) equal zero, or that the expected values of each error term be uncorrelated with the values of each explanatory variable. If we can make this statement for the particular data set in question, then the expected values of the estimated coefficients equal the true population values, $E(b) = \beta$.

Unfortunately, the above condition relies on the sample distribution of the explanatory variables to demonstrate the unbiasedness of the estimator. With a different set of values for the explanatory variables it may well be that the X_t and $\bar{U}_t$ will not be uncorrelated. What we want to do is make the attribute of unbiasedness dependent solely upon the characteristics of U_t so we do not have to worry about the sample values of the explanatory variables. A sufficient condition is the assumption that each U_t has the same mean, $E(U_t) = \mu$, or that

A.3

$$\boxed{E(U_t - \bar{U}) = 0.}$$

From Eqs. (3.3) and (3.4), the estimates are unbiased regardless of the explanatory variables if this assumption is met.

We can summarize the above discussions quite easily. All of our observations on Y implicitly include an unmeasurable error term. The mean of the distribution of the estimated slope coefficients will equal the true or population value if the observations' mean error terms are either equal or uncorrelated with the values of X. This is probably the most important assumption we shall make; we shall return to it repeatedly.

We have not talked about the expected value of b_1 so far. It is a simple exercise to show that if $E(U_t) = E(\bar{U}) = 0$, then b_1 is an unbiased estimate of β_1. In simple terms, if U_t is distributed independently of the X's and has a zero mean, then the estimated intercept is unbiased:

$$b_1 = \bar{Y} - b_2\bar{X}_2 - b_3\bar{X}_3 = \beta_1 + \beta_2\bar{X}_2 + \beta_3\bar{X}_3 + \bar{U} - b_2\bar{X}_2 - b_3\bar{X}_3$$

$$= \beta_1 + (\beta_2 - b_2)\bar{X}_2 + (\beta_3 - b_3)\bar{X}_3 + \bar{U},$$

so that

$$E(b_1) = E(\beta_1) + E(\beta_2 - b_2)\overline{X}_2 + E(\beta_3 - b_3)\overline{X}_3 + E(\overline{U}).$$

This expression will equal β_1 if two conditions hold. If U_t is distributed independently of X_2 and X_3, then our estimates b_2 and b_3 are unbiased so that $E(\beta_2 - b_2) = E(\beta_3 - b_3) = 0$. Secondly, the expected value of $\overline{U}$ must equal zero, $E(\overline{U}) = 0$. If both conditions hold, $E(b_1) = \beta_1$. Thus, for all estimated coefficients to be unbiased, we modify assumption A.3 to be $E(U_t) = 0$. Note that this is a stronger assumption than is needed for unbiasedness of the slope coefficients.[3]

Arithmetic Properties

Unfortunately, we cannot test either of these assumptions, $E(\overline{U}) = 0$ and $\Sigma(X_{tk} - \overline{X}_k)(U_t - \overline{U}) = 0$, with the observed data. Intuitively, one would test these assumptions by examining the mean of the residuals from the estimated equations and the correlation between these residuals and X. Our intuition, however, is wrong in this case. The mean of the residuals is

$$\frac{1}{T}\sum_{t=1}^{T}(Y_t - \hat{Y}_t) = \frac{1}{T}\sum(Y_t - b_1 - b_2 X_{t2} - b_3 X_{3t})$$

However, this is simply $-1/T$ times Eq. (2.11) which *by construction* equals zero. Thus the mean of the estimated residuals must be zero, whether or not the mean of the true errors is zero.

Since $\bar{e} = 0$, the numerator of the correlation between the residuals and X_2 is

$$\frac{1}{T}\sum e_t(X_{t2} - \overline{X}_2) = \frac{1}{T}\sum(Y_t - \hat{Y}_t)(X_{t2} - \overline{X}_2)$$

$$= \frac{1}{T}\sum(Y_t - \hat{Y}_t)X_{t2} - \frac{1}{T}\sum(Y_t - \hat{Y}_t)\overline{X}_2$$

$$= \frac{1}{T}\sum(Y_t - b_1 - b_2 X_{t2} - b_3 X_{t3})(X_{t2}) - \frac{1}{T}\overline{X}_2\sum e_t.$$

The first summation in this expression is simply $-1/T$ times Eq. (2.12) which *by construction* is zero. The second summation is the average of the residuals, which we have already shown to be zero. A similar expression applies for X_3. Thus the numerators in the correlations between e_t and the X_t's are zero, implying that the correlation will be zero. Consequently, we cannot use the calculated residuals to test either of the assumptions made so far—that the expected value of the error terms is zero and that the error terms and X are uncorrelated.

---

[3]Unbiasedness of the intercept term is usually not too important. Most behavioral hypotheses concern the slope parameters and not the intercept. Therefore, the assumption of errors with mean zero—while untestable—is generally innocuous.

Variances of b_2 and b_3

It cannot be said too often that unbiasedness does not imply that the estimate obtained in every sample (or, for that matter, *any* sample) equals the true value. The social scientist generally has but one sample, and unbiasedness merely implies there is no reason to expect overestimates any more than to expect underestimates. Thus, when a numerical estimate is desired, an unbiased estimator has some intuitive appeal, but it is not an assurance of being particularly close to the true value in any given case. For this reason the dispersion of the estimates, or their variance, is important. Even though any estimate might not equal the true value exactly, we might take solace when there is a high probability that any estimate is "close" to the true parameter, i.e., has a small variance.

Assuming that b_2 is unbiased (or that assumption A.3 holds), the variance of our estimator b_2 can be found from a rearrangement of Eq. (3.1):

$$b_2 - \beta_2 = \frac{V_3 C_{2U} - C_{23} C_{3U}}{D}, \tag{3.1'}$$

$$\text{var}(b_2) = E\left[b_2 - E(b_2) \right]^2 = E(b_2 - \beta_2)^2 = E\left(\frac{V_3 C_{2U} - C_{23} C_{3U}}{D} \right)^2$$

$$= \frac{1}{D^2} E\left(V_3^2 C_{2U}^2 - 2 V_3 C_{23} C_{2U} C_{3U} + C_{23}^2 C_{3U}^2 \right)$$

$$= \frac{1}{D^2}\left[V_3^2 E(C_{2U}^2) - 2 V_3 C_{23} E(C_{2U} C_{3U}) + C_{23}^2 E(C_{3U}^2) \right]. \tag{3.5}$$

(Note, by assumption A.2, the X values are fixed.)

We need to investigate the three expected value terms in detail:

$$E(C_{2U}^2) = E\left\{ \frac{1}{T^2}\left[\sum (X_{t2} - \bar{X}_2)(U_t - \bar{U}) \right]^2 \right\}$$

$$= \frac{1}{T^2} E\left[(X_{12} - \bar{X}_2)(U_1 - \bar{U}) + (X_{22} - \bar{X}_2)(U_2 - \bar{U}) \right.$$

$$\left. + \cdots + (X_{T2} - \bar{X}_2)(U_T - \bar{U}) \right]^2$$

(from expanding the summation)

$$= \frac{1}{T^2} E\left[\sum_{t=1}^{T} (X_{t2} - \bar{X}_2)^2 (U_t - \bar{U})^2 \right.$$

$$\left. + 2 \sum_{t=1}^{T-1} \sum_{s=t+1}^{T} (X_{t2} - \bar{X}_2)(X_{s2} - \bar{X}_2)(U_t - \bar{U})(U_s - \bar{U}) \right]$$

(from regrouping terms)

(equation continues)

$$= \frac{1}{T^2}\left[\sum_{t=1}^{T} \left(X_{t2}-\bar{X}_2\right)^2 E\left(U_t-\bar{U}\right)^2 \right.$$

$$\left. +2\sum_{t=1}^{T}\sum_{s=t+1}^{T}\left(X_{t2}-\bar{X}_2\right)\left(X_{s2}-\bar{X}_2\right)E\left(U_t-\bar{U}\right)\left(U_s-\bar{U}\right) \right]$$

(from fixed X's)

$$= \frac{1}{T^2}\left[\sum\left(X_{t2}-\bar{X}_2\right)^2\sigma_t^2 +2\sum\sum\left(X_{t2}-\bar{X}_2\right)\left(X_{s2}-\bar{X}_2\right)\sigma_{ts} \right], \quad (3.6)$$

where $\sigma_t^2 = E(U_t-\bar{U})^2$ and $\sigma_{ts} = E(U_t-\bar{U})(U_s-\bar{U})$. Similar expressions for $E(C_{2U}C_{3U})$ and $E(C_{3U}^2)$ are

$$E\left(C_{3U}^2\right) = \frac{1}{T^2}\left[\sum\left(X_{t3}-\bar{X}_3\right)^2\sigma_t^2 +2\sum\sum\left(X_{t3}-\bar{X}_3\right)\left(X_{s3}-\bar{X}_3\right)\sigma_{ts} \right], \quad (3.7)$$

$$E\left(C_{2U}C_{3U}\right) = \frac{1}{T^2}\left[\sum\left(X_{t2}-\bar{X}_2\right)\left(X_{t3}-\bar{X}_3\right)\sigma_t^2 \right.$$

$s \neq t$
$$\left. +2\sum\sum\left(X_{t2}-\bar{X}_2\right)\left(X_{s3}-\bar{X}_3\right)\sigma_{ts} \right]. \quad (3.8)$$

These expressions are quite complicated, but they can be simplified with two more assumptions about the distribution of the errors. These assumptions, however, are not employed just to simplify the algebra. They are also important for developing the properties of the OLS estimator relative to other estimators. If all error terms have the same variance (we have already assumed they have the same mean), then

A.4a

$$\boxed{E\left(U_t-\bar{U}\right)^2 = \sigma_t^2 = \sigma^2 \qquad \text{for all } t.}$$

Further, if all the error terms are drawn independently of each other so that all the possible error terms associated with one observation are independent of, and thus uncorrelated with, the error terms at other observations, then

A.4b

$$\boxed{E\left(U_t-\bar{U}\right)\left(U_s-\bar{U}\right) = \sigma_{ts} = 0 \qquad \text{for } t \neq s.}$$

With these two assumptions, or restrictions on the error terms, the above simplify to

$$E\left(C_{2U}^2\right) = \frac{1}{T^2}\sum\left(X_{t2}-X_2\right)^2\sigma^2 = \frac{\sigma^2 V_2}{T}, \quad (3.6a)$$

$$E\left(C_{3U}^2\right) = \frac{\sigma^2 V_3}{T}, \quad (3.7a)$$

$$E\left(C_{2U}C_{3U}\right) = \frac{\sigma^2 C_{23}}{T}. \quad (3.8a)$$

When these are substituted into Eq. (3.5), we get[4]

$$\text{var}(b_2) = \frac{\sigma^2}{TD^2}\left[V_3^2 V_2 - 2V_3 C_{23}^2 + C_{23}^2 V_3\right] = \frac{\sigma^2 V_3}{TD^2}\left[V_3 V_2 - C_{23}^2\right]$$

$$= \frac{\sigma^2 V_3}{TD} = \frac{1}{T}\sigma^2\left[\frac{V_3}{V_2 V_3 - C_{23}^2}\right] = \frac{\sigma^2}{T}\left[\frac{1/V_2}{1 - r_{23}^2}\right]. \tag{3.5a}$$

The similar expression for $\text{var}(b_3)$ is

$$\text{var}(b_3) = \frac{1}{T}\sigma^2\left[\frac{V_2}{V_2 V_3 - C_{23}^2}\right] = \frac{\sigma^2}{T}\left[\frac{1/V_3}{1 - r_{23}^2}\right].$$

[A useful exercise for the student is to show for the bivariate case that $\text{var}(b_2) = \sigma^2 / \Sigma(X_t - \bar{X})^2 = \sigma^2 / T\text{var}(X)$.]

The expressions for the variance of our estimators illustrate several important properties. The variance of b_2 increases as the variance of the error terms (σ^2) and as the correlation between explanatory variables (r_{23}) increases in absolute value; it decreases as the variance of X_2 (V_2) and the number of observations (T) increases. We can improve the precision of our estimates (reduce their variance) by using data that contain relatively small error terms (low σ^2), have a large number of observations, have a relatively large variance for the explanatory variables, and/or have low sample correlations among the explanatory variables.

The presence of T, the number of observations, in the denominator of the expression for the variance of our estimators [Eq. (3.5a)] results in an important statistical property. We have commented on the fact that the variance of our estimates decreases as the number of observations in the sample increases. If we could envision the sample size increasing infinitely, i.e., T going to infinity, the variance of the estimators goes to zero,[5] $\lim_{T \to \infty}$ $\text{var}(b_2) = 0$. This property of the estimated coefficients is referred to as "consistency," and we say that the least squares estimates are consistent estimators.[6] This term simply means that as T increases, the variances decrease and the distributions of the estimated coefficients collapse about the true values. If we had an infinite sample size, the estimators would have zero variance, and the coefficients calculated for each replication would equal the true value, assuming the criteria for unbiasedness were also satisfied.

---

[4]Remember $D = V_3 V_2 - C_{23}^2$.

[5]This property, defined when the estimator has a finite mean and variance, also requires the condition that $\lim_{t \to \infty} b_2 = \beta_2$ and $\lim_{t \to \infty} V_2 = V_2^*$ where V_2^* is a finite limit. The previous assumptions of unbiasedness ensures this however. A description of consistency can be found in Appendix I.

[6]Consistency is an important property because it can be viewed as a minimal requirement for an estimator. In many cases (discussed in the latter chapters of this book), unbiased estimators either do not exist or have not been developed even though consistent estimators are available. In such cases attention turns to consistency by necessity. (Note that unbiased estimators can be consistent.) Consistency, while not too critical at this stage, will become more important later.

Best

One of the important characteristics of the least squares estimators is that they have the smallest variance among a large class of estimators. The class is the set of all linear, unbiased estimators.[7] This characteristic is referred to as "best." This attribute of our estimator is an important one for social scientists who typically have only one replication of an experiment. Given one estimate of each coefficient, corresponding to one draw of a random number from a given distribution, it is somewhat comforting to know it is obtained from a distribution more concentrated about the true value than is the distribution for any other linear unbiased estimator. In other words, the least squares method gives estimates closer, in a probabilistic sense, to the true values than other methods in the same class.

The proof that b_2 is best is only sketched here. A complete proof is shown in Appendix 5.1. The proposition to be demonstrated is that, among all *linear* and *unbiased* estimators of β_2 and β_3, the least squares estimators b_2 and b_3 have the minimum variance when assumptions A.1–A.4 hold. We first define an arbitrary linear estimator $b_2^{\#}$. Linear refers to the fact that the estimator is a linear function of the Y_t, $b_2^{\#} = \Sigma C_{t2}^{\#}(Y_t - \bar{Y})$, where $C_{t2}^{\#}$ is any set of weights. (The weights are

$$C_{t2} = \frac{1}{T} \frac{V_3\left(X_{t2} - \bar{X}_2\right) - C_{23}\left(X_{t3} - \bar{X}_3\right)}{D}$$

for the least squares estimator of β_2.) With complete generality, we can write $C_{t2}^{\#}$ as the least squares weight plus an arbitrary number g_{t2}, $C_{t2}^{\#} = C_{t2} + g_{t2}$. The restriction of unbiasedness implies that $E[\Sigma C_{t2}^{\#}(Y_t - \bar{Y})] = \beta_2$. However, for OLS we showed that $E[\Sigma C_{t2}(Y_t - \bar{Y})] = \beta_2$. This implies that $E[\Sigma g_{t2}(Y_t - \bar{Y})] = 0$ since $E[\Sigma C_{t2}^{\#}(Y_t - \bar{Y})] = E[\Sigma C_{t2}(Y_t - \bar{Y}) + \Sigma g_{t2}(Y_t - \bar{Y})]$. Using this restriction and assumption A.4, the variance of $b_2^{\#}$ is

$$\text{var}(b_2^{\#}) = \text{var}(b_2) + \sigma^2 \sum_{t=1}^{T} g_{t2}^2, \tag{3.9}$$

where $\text{var}(b_2)$ is the variance of the least squares estimator. Since $g_{t2}^2 \geqslant 0$, $\text{var}(b_2^{\#})$ cannot be less than the variance of the least squares estimator b_2. Further, it can equal $\text{var}(b_2)$ only if each perturbation (g_{t2}) from the least squares weight is identically zero. (Similar developments can be done for b_1 and b_3.)

---

[7]Throughout this discussion we must emphasize that best applies only to a class of estimates—those that are linear and unbiased. The reason for looking at just this class is straightforward. Suppose an estimate of b_2 is 17.4 regardless of the problem or the sample. The variance of that estimator is zero; it will always do as well or better than the least squares estimator on a pure variance criterion, but certainly not any criteria such as mean squared errors, which considers bias. The linear restriction is useful because it greatly simplifies the process of actually calculating b_2 and b_3 from a given data set. It also makes the problem of determining the actual distribution of the estimated coefficients much easier. If the error terms are normally distributed, least squares estimators will be best among all unbiased estimators; i.e., linearity is not necessary.

An important aspect of this development, however, is that we have accepted and used the assumptions about the error distribution and fixed X's. That is, this proof holds when $E(U_t - \overline{U}) = 0$, $E(U_t - \overline{U})^2 = \sigma^2$, and $E(U_t - \overline{U})(U_s - \overline{U}) = 0$ for all t and $s \neq t$.

The minimum variance property of OLS provides the primary justification for choosing OLS over other, perhaps computationally simpler methods. For example, in Chapter 2, we suggested an alternative estimator based simply on the average ratio of Y and X deviations about their means. That estimator is also linear and unbiased. However, the variance of the estimated parameter using this method will be greater than the variance of the OLS estimator for any sample (as long as assumptions A.1–A.4 are satisfied). Thus, for any single sample, the OLS estimator will tend to be closer to the true value than this alternative estimator.

Size of the Stochastic Component

Our interests in the model usually do not stop with the estimation of the behavioral coefficients. We would also like to know something about the distribution of the stochastic term U_t. In particular, what size is this component of the model, both absolutely and in comparison to the systematic component? The answer could come from the variance of the random component σ^2. However, since U_t is not observed, σ^2 is also unknown and we must find a way to estimate it.

Using the residuals in the estimated equation to estimate the variance for the unknown random element of the true model is a natural approach. The precise estimate used, however, allows for the fact that the e_t are themselves estimated values. In estimating these residuals we place three constraints on the sample, namely, that Eqs. (2.14)–(2.16) must be satisfied. Thus, we must allow for the "degrees of freedom" that are used in the estimation procedure. In other words, we do not have complete freedom in choosing all T values of e_t; once $T - 3$ values have been specified, the application of Eqs. (2.14)–(2.16) will supply the values for the final three residuals. Therefore, s^2, the estimator for σ^2, corrects for the number of parameters estimated:

$$s^2 = \frac{1}{T-3} \sum (e_t - \bar{e})^2 = \frac{1}{T-3} \sum e_t^2. \tag{3.10}$$

(Remember $\bar{e} = 0$.) The denominator $T - 3$ reflects the fact that three parameters (b_1, b_2, and b_3) have been estimated; in the general case, it is the number of observations minus the number of parameters estimated. By taking the expected value of $\sum e_t^2$, s^2 can be shown to be an unbiased estimator of σ^2. (For the formal proof, see Appendix 5.2.)

One very important use for this estimated value of the variance of the error term s^2 is the estimation of the variance of the coefficients. In order to estimate the variance of the OLS coefficients, we substitute s^2 for σ^2 in Eq.

(3.5a). The estimated variances of the coefficients are

$$s_{b_2}^2 = \frac{s^2}{T} \frac{1/V_2}{1-r_{23}^2} \quad \text{and} \quad s_{b_3}^2 = \frac{s^2}{T} \frac{1/V_3}{1-r_{23}^2}, \tag{3.11}$$

where $s_{b_2}^2$ and $s_{b_3}^2$ denote our estimates of the coefficients' variance. Since $E(s^2) = \sigma^2$, and the X's are fixed, this gives an unbiased estimate of the coefficients' variance. The estimated standard errors of the coefficients, symbolized as s_{b_2} and s_{b_3}, are simply the square root of these expressions.

The information about the error variance—as estimated from the sum of squared residuals—supplies us with a description of the probability function that generated the errors in the true model. However, there are two transformations of s^2 that produce more easily interpreted statistics.

The first transformation is simply the square root of s^2. This statistic s is important enough to deserve its own name: the standard error of estimate. In the first place, s is measured in the same units as Y (whereas s^2 is in the units of Y squared). The standard error of estimate is also useful because it gives some feel for the size of the dispersion when compared to tables for the normal distribution. For a normal distribution, we expect about 95% of all values to lie within plus or minus two standard deviations of the mean. Thus, if the U_t are normally distributed (as we often assume), we can expect 95% of all actual values of Y_t in our sample to be within plus or minus two standard errors of estimate away from the estimated line.[8] The estimate s thus allows an easier interpretation of the magnitude of the error terms.

The second transformation of the error variance makes a comparison with the total amount of variance in behavior $\text{var}(Y)$ existing in the sample. One may wish to compare how well an estimated model does when matched with a "naïve" guess at the behavior in question.

With no information about the underlying behavioral relationships, one guess of the value of any Y_t is the mean value of all the observed Y's. The variance of Y is then the sum of squared deviations around this guess divided by the number of observations on the behavior. The sum of squared residuals (Σe_t^2) is a measure of how far our "sophisticated" guesses, represented by $\hat{Y}_t = b_1 + b_2 X_{t2} + b_3 X_{t3}$, diverge from the actual values of Y_t. Thus, comparing the residual variance $(\Sigma e_t^2 / T)$ to the variance of Y gives some indication of the overall performance of our model relative to the simpler "model." If we look at $\Sigma e_t^2 / \Sigma (Y_t - \bar{Y})^2$, we see that this can range from zero when $\Sigma e_t^2 = 0$ to

---

[8]If we are going to use our estimated equation to forecast values of Y_t for values of X not in our original sample, the expected variance of these forecasts about the true value for Y_t is not just s^2. The variance of these forecasting errors must also take into account the expected variance of the estimated coefficients about their true values. If we denote this latter term as

$$E\left[\hat{Y}_t - (\beta_1 + \beta_2 X_{t2} + \beta_3 X_{t3})\right]^2 = E\left[(b_1 - \beta_1) + (b_2 - \beta_2)X_{t2} + (b_3 - \beta_3)X_{t3}\right]^2 = S_{\hat{Y}}^2,$$

then the variance of the forecasting error is $S_{\hat{Y}}^2 + S^2$. For a further discussion, see A. Goldberger (1964, pp. 169–170).

a maximum value of one. (Why, according to the least squares procedure, can this ratio never exceed one?) By convention, we create a new statistic, defined as,

$$R^2 = 1 - \left[\sum e_t^2 / \sum (Y_t - \bar{Y})^2 \right]. \tag{3.12}$$

This statistic, referred to simply as R-squared (or as the coefficient of determination) has the following properties: (1) when all points fall on the estimated plane so that $Y_t \equiv \hat{Y}_t$, R^2 equals one (its maximum); (2) when the mean does as well at predicting Y_t as the estimated equation, R^2 equals zero (its minimum); (3) between these two extremes, R^2 gives an ordinal measure of how well the model predicts the sample values of Y.

Another way to look at R^2 is found by transforming $\sum(Y_t - \bar{Y})^2$ as follows:

$$\sum (Y_t - \bar{Y})^2 = \sum \left[(Y_t - \hat{Y}_t) + (\hat{Y}_t - \bar{Y}) \right]^2$$

$$= \sum (Y_t - \hat{Y}_t)^2 + \sum (\hat{Y}_t - \bar{Y})^2 + 2 \sum (Y_t - \hat{Y}_t)(\hat{Y}_t - \bar{Y}).$$

Since $Y_t - \hat{Y}_t = e_t$ and $\hat{Y}_t - \bar{Y} = b_2(X_{t2} - \bar{X}_2) + b_3(X_{t3} - \bar{X}_3)$, the last summation is simply equal to $2b_2\sum(X_{t2} - \bar{X}_2)e_t + 2b_3\sum(X_{t3} - \bar{X}_3)e_t$. However, by the arithmetic properties of least squares demonstrated above, this is identically zero. Thus

$$\sum (Y_t - \bar{Y})^2 = \sum (Y_t - \hat{Y})^2 + \sum (\hat{Y}_t - \bar{Y})^2 = \sum e_t^2 + \sum (\hat{Y}_t - \bar{Y})^2. \tag{3.13}$$

This says that the variance of Y can be divided into two components. The first ($\sum e_t^2$) is the "unexplained" portion or the residual portion of the model. The second, called the "explained" portion, indicates how much better the estimated model does than using a fixed estimate of the mean would do. By rearranging (3.12), we see that

$$R^2 = \sum (\hat{Y}_t - \bar{Y})^2 / \sum (Y_t - \bar{Y})^2 \tag{3.14}$$

Thus, R^2 can be interpreted as the proportion of the variation in the sample Y_t explained by the regression equation.

Finally, if one correlates the actual and predicted values of Y, one arrives at the correlation coefficient R. Squaring this yields the same value of R^2 as found in (3.12) and (3.14).

It is important to keep in mind several aspects of the R^2 statistic when one is using it to evaluate estimated models. First, it is simply a comparison of the estimated systematic model with a very naïve model, namely the mean of the observed values of Y_t. The appropriateness of this naïve model varies considerably with the behavior being modeled and the type of data used in the study. For example, in a cross-sectional study where all observations are drawn from a relatively homogeneous population, the mean of the observed values of the dependent variable is a much stronger and informative naïve model than in a study involving a time series on a variable such as income or governmental expenditures. In the latter cases, one usually expects the variable to increase in magnitude over time, hence the mean value during the time period being studied is not very meaningful. One would never try to predict the level of Y_t in any single time period by the mean for the entire series.[9]

A second caution to be observed using R^2 as a summary statistic is that it is very sample specific. In comparing the estimated variance of the systematic component of the model with the total variance of the observed Y_t, which includes the variance of the error term, many different factors can influence the value of R^2 for a particular estimation. The sources of ambiguity in interpreting R^2 values are essentially the same as those discussed for simple correlation coefficients in Chapter 1. The variation in the systematic part of the model is a function of the variation in the observed values of the explanatory variables. If the range of values of the X's is small in a particular data set, this will reduce the variation in the systematic component of Y, and result in lower values for R^2, for a given error term.

It is also the case that all the factors that we discussed as composing the true error term will influence the R^2. Improved measurement techniques will reduce the variance of U_t and thus lead to a higher R^2. Consequently, if we are trying to compare results for a model estimated with two different data sets, the set with a better measurement of Y_t can have a higher R^2, without the characteristics of the systematic component changing. Similarly, given that another part of the error term is the effect of omitted influences that operate independently of the explanatory variables, if the variation in the values of these omitted influences is greater in one data set than another, the R^2 will be lower for that model—again without any difference in the systematic component of the behavior being studied. In summary, one must be extremely cautious in interpreting the R^2 value for an estimation and particularly in comparing R^2 values for models that have been estimated with different data sets. The value of R^2 can vary greatly without there being any differences in the systematic behavioral components.

---

[9]In many instances and for a variety of reasons, researchers may estimate a time series model where the dependent variable is the period to period change in the dependent variable (the first difference) rather than its absolute level. In such models, R^2 will generally be much lower, because the naïve model is now much different, and constitutes a much more useful comparison with the hypothesized model. The mean of the observations on $\dot{Y}_t = Y_t - Y_{t-1}$ in this first difference model is the average *change* in Y during each time period. The naïve model implied by the mean of $\dot{Y}$ is a constant time trend over the period of observation. It is completely inappropriate to say that a model with a high R^2 for one dependent variable, say the absolute level of a time series, is better in any sense than a model with a lower R^2 for a different variable, such as first differences. The usefulness of the R^2 statistics is directly related to the validity of the sample mean as an alternative model.

3.3   Distribution of b—A Monte Carlo Experiment

The fact that the estimators are random variables is easy to miss since, in most problems, we observe only a single set of estimates. This is a result of having only a single set of data for many behavioral relationships. The observation of only one set of Y_t and X_t's obscures the stochastic nature of the Y_t's, and thus of the b's. But, the true model says that each Y_t is made up of a systematic component $\beta_1 + \beta_2 X_2 + \beta_3 X_3$ and the random component U and that, if we could obtain a second data set for the same values of X, we would observe different values of Y. Any coefficients estimated with the second data set will differ from those estimated with the first set. To illustrate the assumption of fixed X's and the distributions of b_1, b_2, and b_3, we have performed a Monte Carlo experiment. This is done as follows. First, we specify a true population model that corresponds to the general model in Eq. (2.1). Here we use

$$Y_t = 15.0 + 1 X_{t2} + 2 X_{t3} + U_t. \tag{3.15}$$

Second, a set of 25 values for X_2 and X_3 are arbitrarily chosen, and for each of these "observations," the systematic component of Y calculated. Third, a set of 25 error terms U_t are drawn. Each of these U_t is taken from a distribution with a mean of zero and standard deviation of ten. Each U_t is independent of the others, so that $\sigma_t^2 = \sigma^2 = 100$ and $E(U_t U_s) = 0$. By adding the U_t to the previously calculated systematic component, a set of 25 observations on Y_t can be calculated. Fourth, using this generated sample of Y_t and X_{t2} and X_{t3}, the model's coefficients and the standard error of estimate are estimated by least squares. This set of estimates is then recorded, and the process starting at step three is repeated. Another 25 U_t are drawn and added to the same systematic component (the fixed X_t's). This gives another set of Y_t values from which another set of b's and s are estimated. This second set will be different than the first because the U_t differ and, according to (3.1) and (3.2), the estimates depend on the specific values for the U_t. In total, 400 samples are concocted in this manner; from these, the distribution of the b's can be computed, along with their mean and variance.

The distributions tend to be centered upon the true values. The means for b_1, b_2, and b_3 are 15.038, 1.003, and 1.960. Yet, for any one sample—and one is all we usually have—the values can range a fair distance away. For example, there are four negative values for b_1, and over half the estimates of b_3 are less than 1.5 or larger than 2.5. Since we usually have only one sample, this example indicates why we want the distributions bunched as close to the true values as possible, and explains our emphasis on the minimum variance properties of least squares.

We have computed the standard deviations of our 400 estimates of each coefficient. These values are 6.686, 0.148, and 0.801 for b_1, b_2, and b_3, respectively. The theoretical standard deviations (using the true value of σ^2), calculated with Eq. (3.5a), were 6.518, 0.148, and 0.786. We also estimated with Eq. (3.11) the standard error of each coefficient based upon only sample information for each replication. The averages of these estimates are 6.507, 0.148, and 0.785. These calculations substantiate the formulas derived for the characteristics of the distributions of our coefficients and demonstrate the unbiased nature of the estimates.

In discussing the variance of the estimators, we pointed out that they are determined by the number of observations, the variance of the error term, the variance in the sample values of the explanatory variables, and the correlations among the sample values of the explanatory variables. We want to use our simulations to demonstrate all of these points with the exception of the last, which is reserved for a more general discussion in the next chapter.

The formula for the variance of our estimator implies that the variance is inversely proportional to the number of observations in our sample. This attribute is quite easily demonstrated by adding additional values for X_{t2} and X_{t3}, drawing the number of additional error terms, and using the additional observations on Y_t to compute the estimated coefficients. We have done this for sample sizes of 50, 100, and 200 in addition to the previous simulation with a sample size of 25. The results of all of these simulations are shown in Tables 3.1 and 3.2. The actual distributions of the 400 coefficients for each simulation are plotted in Figs. 3.1–3.3. As expected, all the estimates are unbiased in that the means of the 400 estimates of each coefficient in each simulation are quite close to the true values. Further, the standard error of each estimate drops markedly as the sample size increases. The histograms of the distributions show this effect clearly, with the distributions for the simulation with $T = 200$ being much more peaked than in the simulation for $T = 25$. In other words, the probability of any one estimate being within a given distance of the true value is much higher for the larger sample size.

A second set of simulations takes the case where $T = 50$ and examines different values for the variance of X_3 and for the variance of the error terms σ^2. Tables 3.3 and 3.4 show the mean and standard errors for each of the

TABLE 3.1

Mean Coefficients for Different Sample Size[a]

	T			
	25	50	100	200
$\bar{b}_1$	15.038	15.275	14.939	15.013
$\bar{b}_2$	1.003	0.991	1.003	0.999
$\bar{b}_3$	1.960	1.983	1.998	2.007
$\bar{s}$	9.987	9.919	10.017	9.994

[a]Bars indicate the average values over 400 replications for each sample size. $\bar{s}$ is the mean of the standard error of the estimate computed for each replication.

TABLE 3.2

Coefficient Standard Error for Different Sample Size[a]

						T						
	25			50			100			200		
	σ_b	$\hat{\sigma}_b$	$\bar{s}_b$	σ_b	$\hat{\sigma}_b$	$\bar{s}_b$	σ_b	$\hat{\sigma}_b$	$\bar{s}_b$	σ_b	$\hat{\sigma}_b$	$\bar{s}_b$
b_1	6.518	6.686	6.507	4.756	4.742	4.715	2.846	2.837	2.851	2.030	2.178	2.029
b_2	0.148	0.148	0.148	0.108	0.112	0.107	0.073	0.075	0.073	0.053	0.059	0.053
b_3	0.786	0.801	0.785	0.537	0.531	0.532	0.362	0.355	0.363	0.260	0.267	0.260

[a] σ_b is the true standard deviation of b from Eq. (3.5a); $\hat{\sigma}_b$, the observed standard deviation of b in 400 replications; and $\bar{s}_b$ is the mean estimated standard error from Eq. (3.11).

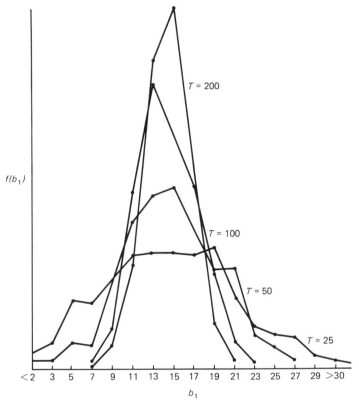

FIGURE 3.1   Distributions of simulated b_1. Horizontal axis indicates midpoint of range of values for estimated coefficient.

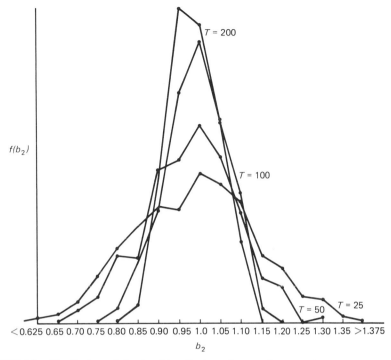

FIGURE 3.2   Distributions of simulated b_2. Horizontal axis indicates midpoint of range of values for estimated coefficient.

coefficient distributions from these simulations. Again, all our estimates are unbiased. The standard errors of the coefficients exhibit the most interesting and informative patterns. Because we have kept the correlation between X_2 and X_3 constant for all simulations, and have not altered the variance of X_2, the variance of the estimates for b_2 should change only with changes in σ, not with changes in the variance of X_3. This in fact is what we observe. For a given value for σ, the calculated and observed standard errors of b_2 are equal (within the range of sampling variations). With each doubling of σ, the average standard errors of the estimated coefficients also doubles, as expected. Also as expected, the variance of b_3 varies inversely with the variance of X_3, for given values of σ. Each quadrupling of the standard deviation of X_3 reduces the standard error of b_3 by a factor of four. When we combine differences in the variances of X_3 with changes in σ, we get very marked differences in the variance of our estimated coefficients. To get some appreciation for the magnitude of these different standard errors, consider the four different distributions for b_3 shown in Fig. 3.3. The standard deviations in these four distributions range from 0.27 to 0.80. The range when $\text{var} X_3$ and σ^2 change in the other simulations is from 0.07 to 4.4. Thus we can see that

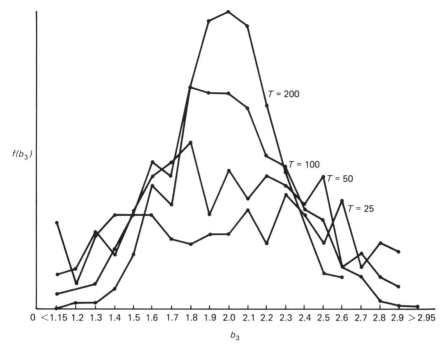

FIGURE 3.3   *Distributions of simulated b_3. Horizontal axis indicates midpoint of range of values for estimated coefficient.*

TABLE 3.3

Mean Coefficients for Variance Simulations[a]

	σ_U								
	5			10			20		
S_3	0.7	2.8	11.2	0.7	2.8	11.2	0.7	2.8	11.2
$\bar{b}_1$	14.912	14.846	15.068	14.986	15.275	15.042	14.967	15.329	15.095
$\bar{b}_2$	0.999	1.001	1.002	1.997	0.991	0.992	0.997	0.990	0.998
$\bar{b}_3$	2.073	2.019	1.994	2.137	1.983	2.005	2.076	1.977	1.993
$\bar{s}$	4.952	4.980	4.851	9.950	9.919	9.906	19.886	19.863	19.849

[a] σ_U is the standard deviation of U; S_3 is the standard deviation of X_3; and $\bar{s}$ is the mean of the standard error of the estimate computed for each replication. Sample size equals 50 and replications equal 400.

samples with small variations in the explanatory variables and large error terms will yield very imprecise estimates. One's objective in considering a

TABLE 3.4

Coefficient Standard Errors for Variance Simulations[a]

			σ_U							
			5			10			20	
	S_3	0.7	2.8	11.2	0.7	2.8	11.2	0.7	2.8	11.2
b_1	σ_{b_1}	2.377	2.377	2.377	4.754	4.756	4.754	9.508	9.508	9.508
	$\hat{\sigma}_{b_1}$	2.388	2.387	2.570	4.846	4.742	4.574	9.374	9.313	9.001
	$\bar{s}_{b_1}$	2.354	2.358	2.354	4.730	4.715	4.709	9.454	9.443	9.436
b_2	σ_{b_2}	0.054	0.054	0.054	0.108	0.168	0.108	0.216	0.216	0.216
	$\hat{\sigma}_{b_2}$	0.054	0.054	0.055	0.109	0.112	0.112	0.212	0.220	0.212
	$\bar{s}_{b_2}$	0.053	0.053	0.053	0.107	0.107	0.106	0.214	0.214	0.213
b_3	σ_{b_3}	1.073	0.268	0.067	2.146	0.537	0.134	4.292	1.073	0.268
	$\hat{\sigma}_{b_2}$	1.073	0.274	0.070	2.167	0.531	0.130	4.375	1.107	0.266
	$\bar{s}_{b_2}$	1.062	0.266	0.066	2.135	0.532	0.133	4.266	1.065	0.266

[a]σ_U is the standard deviation of U; S_3 is the standard deviation of X_3; σ_b is the true standard deviation of b from Eq. (3.5a); $\hat{\sigma}_b$ is the observed standard deviation of b in 400 replications; and $\bar{s}_b$ is the mean estimated standard error from Eq. (3.11). Sample size equals 50 and replications equal 400.

data set should be to get as large a range of values for the explanatory variables and as accurate measures for the dependent variables as possible.

3.4   Statistical Inference

Empirical social science is not concerned solely with estimating particular models and the coefficients in specific equations. The true population values are not known, and we want to use our estimates to make some statistical inferences about them. In many cases, these inferences are designed to evaluate alternative explanations of the observed phenomena, and as such constitute tests of different theories of social behavior. These inferences about the true values are based upon our estimates of the distributions from which the estimated coefficients are drawn and an assumption about the specific shape of that distribution.

A typical question in statistical inference is, What is the probability of getting a value for b that equals or exceeds our one estimate if the value of the true population coefficient is in fact some given value, such as zero? If this probability is sufficiently small, say less than 0.05 or 0.01, we may want to reject the hypothesis that the population coefficient is that given value. We can illustrate this evaluation with the help of Figure 3.4.

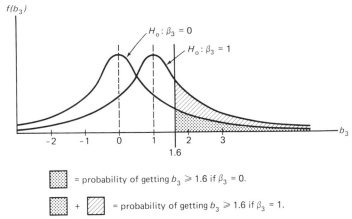

$$FIGURE\ 3.4 \quad Calculations\ of\ probabilities\ for\ alternative\ null\ hypotheses.$$

We assume that with our one data set we have obtained an estimate of 1.6 for b_3. We then show two possible distributions of estimated coefficients. If the true value of β_3 is zero, the distribution of all possible estimates, given our particular sample, would be the curve on the left (assuming our estimates are unbiased). In this situation, the probability of getting an estimate equal to 1.6 or greater is indicated by the shaded area. However, if the true value of β_3 is one, then the distribution of possible coefficients is represented by the curve on the right. In this case the probability of getting a coefficient estimate equal to or greater than 1.6 is shown by both the shaded and the lined area. As one can see, the probability of getting our estimate or a higher one is larger if the true value of β_3 is one. In general, we do not know which distribution is appropriate. What we are going to do is estimate the shaded area for a given, hypothesized distribution (the null hypothesis) and try to decide how likely our result is if the hypothesized distribution were the true one. Then based on this result, we decide whether to reject the hypothesized distribution.

The essence of the statistical tests is that if the estimated coefficient is far from the hypothesized value, we shall reject the given null hypothesis. The trick is measuring distance, which we do in probability terms.

We must first make a specific assumption about the distribution of the error terms. Without such an assumption, we cannot calculate the required probabilities. Throughout this section we shall use our final assumption, namely that U_t is drawn from a normal distribution, which has the previously assumed mean of 0 and variance of σ^2, that is,

A.5 $$\boxed{U_t \sim N\left(0, \sigma^2\right).}$$

While this assumption is not needed to obtain a coefficient estimator which is BLUE, it is necessary for common statistical inferences. Normality can be justified when the error terms are composed of many (small) influences on Y. In this case the central limit theorem implies that the error terms will be approximately normally distributed.

If $U_t \sim N(0, \sigma^2)$, and independent of U_s, then the b's, which are linear functions of U_t, are normally distributed with a variance given in Eq. (3.5a), that is, $b_2 \sim N(\beta_2, \sigma^2 / TV_2(1 - r_{23}^2))$. This implies that

$$Z_2 = (b_2 - \beta_2)/\sigma_{b_2} \quad \text{and} \quad Z_3 = (b_3 - \beta_3)/\sigma_{b_3} \qquad (3.16)$$

are distributed according to the standard normal distribution $N(0, 1)$. (This distribution is discussed in Appendix I.) Since the Z's are $N(0, 1)$, standard tables of cumulative normal distributions would yield the probability of Z being greater than any given value, or in turn would give the probability of the estimated value being more than any given distance from the true value of β, i.e., $b - \beta$. However, (3.16) depends upon σ, which is unknown.

s^2 is an unbiased estimator of σ^2 and is substituted for σ in our expression for σ_b. However, any inferences employing s^2 will not be as precise as they would be if we knew σ^2 and did not have to rely upon the random variable s^2. In order to allow for this additional imprecision, we do not use probabilities from the normal distribution. Instead, we shall rely upon the t-distribution. The definition of the t-distribution is as follows: if Z is a standard normal variable, i.e., Z is $N(0, 1)$, and if W^2 is an independently distributed chi-squared with $T - 3$ degrees of freedom, then $Z / \sqrt{W^2/(T-3)}$ is distributed according to the t-distribution with $T - 3$ degrees of freedom. We have demonstrated that $(b - \beta)/\sigma_b$ is $N(0, 1)$. It can further be shown that (Hoel, 1962, pp. 262–268)

$$W^2 = \sum e_t^2 / \sigma^2 \quad \text{is} \quad \chi_{T-3}^2.$$

For b_2 we get

$$t_{b_2} = \frac{Z}{\sqrt{W^2/(T-3)}} = \frac{(b_2 - \beta_2)/\sigma_{b_2}}{\sqrt{\sum e_t^2 / \sigma^2 (T-3)}}$$

$$= \frac{b_2 - \beta_2}{\sqrt{\sigma^2 / TV_2(1 - r_{23}^2)}} \cdot \frac{1}{\sqrt{\sum e_t^2 / \sigma^2 (T-3)}} = \frac{b_2 - \beta_2}{\sqrt{s^2 / TV_2(1 - r_{23}^2)}}$$

$$= \frac{b_2 - \beta_2}{s_{b_2}}, \qquad (3.17)$$

where $s^2 = \sum e_t^2 / (T-3)$. This variable t_{b_2} is distributed as Student's t with $T - 3$ degrees of freedom.

This conclusion permits us to test various hypotheses about the true value of β. For example, a common hypothesis is H_0: $\beta = 0$, or that X has no effect on Y. For a two-tailed test (i.e., H_1: $\beta \neq 0$), this hypothesis will be rejected at the $1 - \alpha$ level of confidence if

$$t_b = |b|/s_b > t_{\text{crit}(\alpha/2, T-3)},$$

where $t_{\text{crit}(\alpha/2, T-3)}$ is the critical value for $T - 3$ degrees of freedom and a significance level of α. The significance level α is the size of a type I error, or the probability of rejecting the null hypothesis when it is in fact true. In simple terms, the further b is from zero (i.e., the higher t_b), the less likely it is that β is really zero. The general form of the hypothesis test for H_0: $\beta = \beta^*$ is

$$t_b = |b - \beta^*|/s_b. \tag{3.18}$$

If $t_b > t_{\text{crit}(\alpha/2, T-3)}$, the null hypothesis is rejected. In other words, if the estimated value is a large distance from the hypothesized value (in terms of standard errors), it is unlikely that the null hypothesis is true. If the alternative hypothesis is more specific, such as H_1: $\beta > 0$, a one-tailed test would be appropriate, and the corresponding null hypothesis at the $1 - \alpha$ level would be rejected if $t_b > t_{\text{crit}(\alpha, T-3)}$. (See Appendix I.4.)

In terms of our earlier illustration (where $b_3 = 1.6$), if the estimated standard error of b_3 from that simulation is 0.7, then the t-statistic for the null hypothesis that $\beta_3 = 0$ is $(1.6/0.7) = 2.29$. If we are using the simulation with $T = 50$, then we have $50 - 3 = 47$ degrees of freedom, and with a confidence level of 95% for a two-tailed test, t_{crit} is 2.01, indicating that we would reject the hypothesis that $\beta_3 = 0$, at more than the 95% confidence level. If one wanted more security in rejecting the null hypothesis and set $\alpha = 1\%$, the appropriate value of t is 2.7, and one could not reject the null hypothesis that $\beta_3 = 0$, or that X_3 has no effect on Y at the 99% level. Increasing the confidence level is not without cost, however. By decreasing the probability of rejecting a true null hypothesis, we are increasing the chance of type II error —accepting a false null hypothesis.

Finally, we might reconsider the normality assumption itself. This was justified chiefly by an appeal to the central limit theorem which states that the sum of a large number of random variables will tend to be distributed normally. In making the assumption of normality, our real interest is with the distribution of the estimated coefficients and not the error term itself. Therefore, it is worth noting that if the sample with which we are making the estimates is large, the same central limit theorem implies that the coefficients will tend to be normally distributed even if the error terms are not. This obtains because the coefficients are themselves (weighted) sums of random variables (the U's), and there are no restrictions on the distribution of the random variables going into the sum in order for the central limit theorem to hold. Further, statistical tests based upon normality tend to be quite good when there are small divergences from normality (Malinvaud, 1966, pp. 251–254). Therefore, when all of the other assumptions hold, we generally do not have to be overly concerned about applying the tests suggested here.

3.5 Hypothesis Tests for Schooling/Earnings Model

The conventional hypothesis tests can be easily applied to the model relating earnings to schooling and experience discussed in Chapter 1 and estimated in Chapter 2. The null hypothesis is that earnings is unrelated to the amount of schooling an individual has. Thus, for the model

$$Y_t = \beta_1 + \beta_2 S_t + \beta_3 E_t + U_t,$$

where Y_t is the earnings of individual t, S_t the person's years of schooling, and E_t his years of experience, the null hypothesis is $H_0: \beta_2 = 0$.

A reasonable alternative hypothesis for this model is that schooling has a positive effect on earnings: $H_1: \beta_2 > 0$. This alternative hypothesis indicates that a one-tailed test is appropriate.

Before performing the test, we must choose a confidence level. A 95% confidence level ($\alpha = 0.05$) is most frequently chosen. With $\alpha = 0.05$ and 102 degrees of freedom,[10] t_{crit} is 1.66.

The estimated equation was

$$Y_t = 1045 + \underset{(151.7)}{470} \, S_t + \underset{(35.1)}{67} \, E_t,$$

where the estimated standard errors of the schooling and experience coefficients are displayed in parentheses under the coefficient values. For schooling,

$$t = 470/151.7 = 3.10 > t_{crit} = 1.66.$$

Thus, the estimated schooling coefficient is far away from zero (3.10 standard errors), implying we are quite confident that β_2 does not in fact equal zero. H_0 is rejected at the 0.05 level.[11] Similar tests can be performed for β_3.

Before leaving this example, one facet of hypothesis testing must be pointed out. We have rejected the null hypothesis that $\beta_2 = 0$ with 95% confidence in being correct; in other words, there is only a 5% probability of getting this large an estimated coefficient if $\beta_2 = 0$. This does not mean that we are 95% sure that the true value of β_2 is \$470/yr. For the true coefficient there is a range that would give an estimate this large by chance. All we have implicitly done is accept the hypothesis that $\beta_2 > 0$. However, in terms of predicting the effect of additional schooling on earnings, our best guess of this effect is \$470/yr. Sample output for this estimation from a common computer program is shown in Appendix 3.1.

3.6 Conclusion

Thoroughly understanding least squares analysis is required before good empirical work can be undertaken. Most problems involve the failure of one

---

[10]Degrees of freedom = 105 observations − 3 estimated coefficients.

[11]H_0 can be rejected at a much higher level in this case. $t_{102} = 3.09$ occurs with probability < 0.01 when β_2 actually equals zero.

or more of the assumptions made in this chapter. The ability to recognize problems and errors, both in your own work and in other people's research, begins with an understanding of these assumptions and the implications of not satisfying them. We shall now review these assumptions and point out the consequences of not meeting them.

The assumptions concern the properties of the true population model

$$Y_t = \beta_1 + \beta_2 X_{t2} + \beta_3 X_{t3} + U_t.$$

The first, and most easily satisfied, assumption is that the correlation between X_2 and X_3 be less than 1 (A.1). This requirement is necessary in order to compute the estimated coefficients, which contain the term $1 - r_{23}^2$ in the denominator. We can satisfy this condition so long as X_2 and X_3 are not linearly related to each other. Since this is an unlikely circumstance with most data sets, we can compute regression coefficients for virtually any combination of these variables.

The condition that $|r_{23}^2| < 1$ is only the beginning. It ensures only that we can estimate a set of coefficients. We cannot meaningfully interpret these estimated coefficients, however, if this is the only assumption satisfied. Since these least squares coefficients are estimates of the true coefficients and are drawn from some probability density function determined by the error term in the model, we need the remaining assumptions to be able to say something about the distribution from which these estimates are drawn and thus to obtain some information about the true values of the coefficients.

The next condition we impose is that we were working with an "experiment," albeit a natural one, in which the values of X are fixed (A.2). Thus our observed values of Y, the behavior we are explaining, represent only *one* replication out of some unspecified number of trials with the same values for the explanatory variables. With this assumption we are able to establish that the distribution from which we draw our estimates is solely a function of the distribution of the error terms in our model. Consequently, all we need to do to define the parameters of this distribution of coefficient estimates is to make some assumptions about this error term distribution. (The question of the validity of this "experiment" and of the appropriateness of generalizing from it to other situations is discussed in Chapter 1.)

The remaining conditions all refer to the parameters of the error term distribution. Once these parameters are defined, we have the necessary information to examine the distribution of estimated coefficients. The first of these assumptions is that the expected value of the error term drawn for each observation is a constant, usually assumed to be zero (A.3). In essence this is saying that the mean value of each observation's error term, taken over all implicit replications, will be the same for all observations, i.e., is the same regardless of the values taken by the explanatory variables. This assumption is necessary to obtain unbiased estimates, or to have the expected value of our estimated coefficient equal the true value. The expected value of the error terms is assumed to be zero in order for the estimate of the intercept coefficient b_1 to be unbiased. If the expected value of each error term is a constant that does not equal zero, then all the coefficients except the intercept will be unbiased, and the estimate of the intercept will be biased by an amount equal to this constant. Thus to obtain unbiased estimates of our coefficients we have only to make an assumption about the expected value of our error term.

With these three assumptions we have established that we can estimate the coefficients in our hypothesized model, that they will be drawn from a distribution that is a function of the distribution of the error terms, and that this distribution of estimates will have a mean equal to the true value. We then derived the variance of this distribution. At this point we make a fourth assumption (A.4). This assumption is that each error term is drawn randomly, or independently, from the same distribution. The random, or independent, drawing implies that $E(U_t U_{t+s}) = 0$, $s \neq 0$, meaning that there is no correlation between pairs of error terms. The requirement that all error terms come from the same distribution means that all error terms will have the same variance, i.e., $E(U^2_t) = \sigma^2$, for all t. This assumption ensures that the least squares estimators have the least variance of all linear, unbiased estimators; i.e., they are BLUE. The variance of the distribution from which we are drawing our coefficient estimates equals $(\sigma^2/T)[(1/V_X)/(1 - r^2_{23})]$.

The only thing remaining to do to make our description of the distribution complete and to meet all the requirements for classical statistical inference and hypothesis testing is to determine the actual form of the distribution of our estimates. Deriving the distribution begins by assuming that the error terms are distributed according to a normal distribution, i.e., U_t is $N(0, \sigma^2)$ (A.5). Since we have assumed that our X's are fixed, the distribution of the coefficients follows a linear transformation of the distribution of the error terms. Assuming that the error terms are normally distributed means that the estimated coefficients will be normally distributed about their true value. This assumption also means that s^2, our estimate of σ^2, can be used in conjunction with the estimated coefficients to form a t-statistic with $T - 3$ degrees of freedom. Thus we have all the necessary information and completed all the formal requirements to make inferences about the true coefficients based on our estimates of them.

In the remaining chapters, we shall discuss the effects of violating assumptions A.1–A.4. We shall also present several methods for trying to deal with circumstances where possible violations exist.

The estimation procedure we have outlined is not limited to models with two explanatory variables. The equations minimizing the sum of squared residuals, which led to the estimates for the trivariate case, can be easily expanded to take account of an arbitrary number of explanatory variables. Naturally, this leads to a more complex set of simultaneous equations and more complicated formulas for each coefficient. It is much easier at this point to develop a new notation and derive the estimates and their statistical properties with this notation than to try to expand the equations in this chapter. However, the essence of this expanded model, and the constraints placed on the error terms within this model, are exactly the same. Consequently, a thorough understanding of the trivariate case will enable you to understand and interpret larger models. We defer the presentation of the more general model until Chapter 5, after we have discussed some of the problems one usually encounters trying to estimate simple, multivariate linear models.

APPENDIX 3.1

Estimation of Schooling / Earnings Model Using SPSS Computer Program

Usually the actual computational work of statistical estimation is conducted on a high-speed computer using prepackaged computer programs. This appendix provides an example of typical output from computer estimation and relates that output to the material in the text.

The schooling-earnings model, presented in Sections 2.4 and 3.5 of the text, was estimated using SPSS (Statistical Package for the Social Sciences).[12] This particular program is used for illustrative purposes because of its wide distribution. However, the user of this program is cautioned that portions of the SPSS output may be extraneous or potentially misleading in the statistical applications considered here.

It is particularly important to note that the estimation was not done in a stepwise fashion. Stepwise regression (a method of progressively adding more exogenous variables to the model) is generally an inappropriate technique (see Section 4.5), and portions of the SPSS output pertaining to a stepwise procedure were deleted.

The computer output has been annotated to relate portions of the output to the text. Footnotes are also included to indicate portions that are not covered in the text or that are covered later in the text (Table A3.1).

REVIEW QUESTIONS

1. Given the accompanying data, calculate the least squares coefficients for the true model

$$Y_t = \beta_1 + \beta_2 X_{t2} + \beta_3 X_{t3} + U_t.$$

Y:	42	40	16	15	22	45	55	69	69	40	$\Sigma Y = 413,$	$\Sigma X_2 = 51,$	$\Sigma X_3 = 76$
X_2:	2	5	3	0	0	6	10	12	8	5	$\Sigma X_2 Y = 2732,$	$\Sigma X_3 Y = 3349$	$\Sigma Y^2 = 20501$
X_3:	20	15	3	2	10	1	1	7	12	5	$\Sigma X_2^2 = 407,$	$\Sigma X_2 X_3 = 345,$	$\Sigma X_3^2 = 958$

(a) What is the estimated variance of each coefficient?
(b) What is the R^2 for this model?
(c) What is the standard error of estimate?

[12] A description of the program can be found in Nie, N. H. *et al. SPSS: Statistical Package for the Social Sciences*, 2nd ed. New York: McGraw-Hill, 1975.

TABLE A3.1

Estimation of Schooling/Earnings Model Using SPSS Computer Program

BLACK MALE EARNINGS ₸ - BOSTON

*   M U L T I P L E   R E G R E S S I O N   * * * * * * * * * * * * * *

DEPENDENT VARIABLE..   EARNINGS

VARIABLE(S) ENTERED ON STEP NUMBER   1..   EXPERNCE
                                           SCHOOL

		ANALYSIS OF VARIANCE			
MULTIPLE R[a]	0.30067		DF[e]	SUM OF SQUARES[b]	MEAN SQUARE	F[c]
R SQUARE	0.09040[d]	REGRESSION	2.[e]	172024564.93586[f]	86012282.46793	5.06890
ADJUSTED R SQUARE[g]	0.07257	RESIDUAL	102.[h]	1730801805.17374[i]	16968645.14876	
STANDARD ERROR	4119.30154[j]					

------------- VARIABLES IN THE EQUATION -------------

VARIABLE	B[k]	BETA[l]	STD ERROR B[m]	F[n]
EXPERNCE	67.32253	0.19762	35.10072	3.679
SCHOOL	469.66440	0.31505	151.67587	9.588
(CONSTANT)	1045.44427			

[a] Correlation of $\hat{y}$ and y; also, $\sqrt{R^2}$.
[b] See Eq. (3.13).
[c] See Section 5.6.
[d] See Eq. (3.12).
[e] Value of $K-1$.
[f] Value of $K-1$.
[g] Adjusted R^2 considers loss of degrees of freedom when variables are added and is computed as

$$R^2 - \frac{(K-1)}{(T-1)}(1-R^2).$$

[h] Value of $T - K$.
[i] Value of Σe_t^2.
[j] See Eq. (3.10); $(\sqrt{s^2})$.
[k] See Eq. (2.14)–(2.16).
[l] Standardized regression coefficient; see Section 4.2.
[m] See Eq. (3.5b).
[n] $F = t^2 = (\hat{b}/s_{\hat{b}})^2$.

2.   Using the data in Problem 1, calculate the residual for each observation.
(a)   Is this the same as the true error U_t?
(b)   What is the mean of the residuals?
(c)   What is the simple correlation between the values of X_2 and the residuals?

3.   Using the data from Problem 1, test the following hypotheses:
(a) $H_0: \beta_2 = 0, H_1: \beta_2 \neq 0$;
(b) $H_0: \beta_2 = 3, H_1: \beta_2 \neq 3$;
(c) $H_0: \beta_3 = 0, H_1: \beta_3 > 0$.

4.   Begin with the least squares estimator for the parameters of a trivariate model. What is the simple correlation between b_2 and b_3?

5.   Consider estimating $Y_t = \beta_1 + \beta_2 X_{t2} + \beta_3 X_{t3} + U_t$ with least squares. Show the effect of each of the following conditions on the expected value of b_2, the variance of b_2, and statistical tests about $\beta_2 = 0$.
(a) $E(U_t) = 5$ for $t = 1, 2, \ldots, T$.
(b) $E(U_t)$ increases with the size of X_2 such that $E(U_t) = aX_{t2}$.
(c)   The correlation between X_2 and X_3 equals -1.
(d)   The correlation between X_2 and X_3 equals 0.9.
(e) U_t is not normally distributed.

6.   Consider a true model: $Y_t = \beta_1 + \beta_2 X_t + U_t$. The following estimator for β_2 is proposed:

$$b_2^* = (Y_{\max} - Y_{\min})/(X_{\max} - X_{\min}),$$

where max and min denote the largest and smallest values of X and the Y values which correspond to those X's.
(a)   Show that b_2^* is an unbiased estimator of β_2.
(b)   Using the following data, calculate $b_2(\text{OLS})$ and b_2^*.

Y_t:	10	3	21	9	18	9	7	12
X_t:	3	5	8	0	15	4	8	2

(c)   Calculate and compare $\text{var}(b_2^{\text{OLS}})$ and $\text{var}(b_2^*)$.
(d)   Which estimator would you choose in general? Why?

7.   Should one always use a BLUE estimator? In what circumstances might it be preferable to choose an estimator that was not BLUE? What criteria should be used in choosing among alternative estimators?

4 | Ordinary Least Squares in Practice

4.1 Introduction

Ordinary least squares regression analysis is a very powerful analytical tool. It is adaptable to a wide range of problems; none of the developments in the preceding chapter related to one specific problem or discipline but instead are generalizable to many applications. Further, as demonstrated before, the technique has some desirable statistical properties—at least when the set of required assumptions seems reasonable.

Yet Chapter 3 may be misleading. Certainly it is easy to program a computer to calculate the coefficient estimates and the summary statistics from the preceding chapter. In fact, computer packages to do just that are plentiful. *But the actual application of ordinary least squares is not just a technician's job.* First, the construction and interpretation of the actual model demands extensive knowledge of the relevant social science theories and of previous empirical work. Considerable expertise and experience are required to decide which variables should be included in any analysis. This topic, *model specification*, is the subject of Section 4.3 and is possibly the most important topic in this book.

The substantive interpretation of any estimated model and comparison with other work is both a necessary and a difficult part of empirical research. Useful studies may be ignored simply because the researcher could not properly interpret what had been estimated. Section 4.2 is a short discussion of the interpretation of regression coefficients and why we are concerned with them.

The last three sections present situations where modifications of the OLS model can deal with specific problems. The data at hand to test a given model may not contain enough information about behavior to allow precise estimates of these coefficients. One frequently encountered form of this problem, *multicollinearity*, again requires judgment on the part of the analyst. This problem and the choices open for its resolution are presented in Section 4.4.

75

We mentioned previously that the functional form of the model relating Y to X need not be linear. Section 4.6 presents and discusses several alternatives. Some qualitative factors—such as religion or occupation—may affect behavior but defy quantification. In this instance, a series of tests for behavioral differences in populations (as reflected in coefficient differences) through the use of qualitative variables, or dummy variables, may be called for. The range of tests for qualitative differences is presented in Section 4.7. All of these situations are dealt with through alterations of the OLS method. Other cases, where the problems are not so easily overcome and require different estimators, are the subjects of Chapters 6–10.

4.2 Interpretation of Regression Coefficients

We want to undertake a short digression to discuss the interpretation of the estimated regression coefficients. Unless one clearly understands these interpretations and the concern about getting accurate estimates of the population values, the consequences of violating the constraints placed on the error terms and other estimation problems may not be appreciated.

The coefficients in the population model represent the change in Y expected from a unit change in a particular X, the values for the other variables being held constant. This is often referred to as the "direct effect" of X on Y because the statement, "the values for the other X's held constant," excludes from consideration the possibility that changes in the X being considered may lead to changes in the other explanatory variables. In the voting example of the previous chapter, the effect of issue evaluation on 1964 presidential voting was 0.36 and that of party affiliation was 0.64. Thus the expected voting behavior of someone who favored the Democrats on all issues is 0.36 greater than someone with the same party affiliation who favors the Republicans on all issues. What we have not accounted for is the possibility that the difference in evaluations may result in altered party affiliations, which produce further differences in expected voting behavior. This consideration of direct and subsequent indirect effects will be considered in substantially more detail in Chapters 8–10. Suffice it to say here that all we are estimating with the single-equation model in Eq. (2.1) are the direct effects of each explanatory variable on the dependent variable, the values of the other explanatory variables held constant.

The direct effect represented by the coefficients in the population model describes the mean behavior for all members of the population. Thus the expected change in Y for a unit change in X, given by the appropriate parameter, describes the expected behavior of all population members, regardless of the characteristics of that member, the associated values of the other explanatory variables for that observational unit, and so forth. Because of the true coefficients' invariance among members of the population and across any sample of that population, we want to use these values, or at least our estimates of them, to discuss the substantive implications of our model.

Concentration on population coefficients has several important implications. In the first place, one must be sure that all observations do in fact represent a single population, meaning that the β's are expected to be the same for all members. One must also consider the question of whether the observations and the population they represent is the population one wants to study. One might want to consider carefully whether a voting model estimated for the 1964 presidential election describes voting behavior in all elections. These judgments, for the most part, must come from the researcher and require careful thought, experience, and modeling expertise. We shall see later in this chapter that it is possible to test for certain hypothesized differences between members of different groups to see if they do come from different populations. In fact much of social science theory testing must be concerned with exactly that question, and often it leads to further inquiry into why the populations are different. Assume the behavior of voters in the 1964 presidential election is different than their behavior in 1960 and 1968, and is different from that in the 1964 congressional elections. Is it because of the candidate differences in the 1964 presidential election? Were there fundamental changes in the American electorate in 1964? Are elections different among branches and levels of government? Or, what combination of many plausible explanations need investigation? The only good way to make these comparisons and perform the statistical tests is to estimate coefficients that best approximate the population values and whose statistical properties are least dependent upon the sample being used in the estimation. With any other estimates one does not know whether differences between studies and across data are the results of differences in the sample, e.g., the range in values of the explanatory variables and the composition and variance of the error terms, or differences in the behavior of the two populations.

The simulations described in the previous chapter illustrated the unbiased distribution of the estimated regression coefficients and their variances under different sample characteristics, such as sample size and variances of the explanatory variables and the error term. The simulations also indicate the robustness of the least square estimators to these variations in sample characteristics so long as the fundamental assumptions about the nature of the error terms are met. Regardless of how we varied the characteristics and composition of the data sets, the estimates both of the population coefficients and of the estimators' variances were unbiased. This fact is reassuring to researchers making the type of comparisons outlined above. If estimates from two different data sets of what are presumed to be the same coefficient differ, we can be confident that this difference is due either to differences in the population values (which has substantive significance) or to sampling differences associated with the fact that each estimate is an independent draw from some error probability distribution. With information about the variance and shape of this error distribution, it is possible to test the null hypothesis that the coefficient differences are simply the result of sampling differences. (The procedure for testing population differences will be dealt with in the succeeding chapter.) It is vital in these tests and for any comparisons and uses involving more than the one sample that one use estimates that are not specific to the particular sample used in the study.

A number of summary statistics vary in expected values with the particular sample. We have already discussed how R^2 can vary from sample to sample without variation in the population structure. Another often used statistic, the standardized regression coefficient, referred to as the *beta* coefficient, has the same deficiencies. The standardized coefficient is obtained by multiplying the estimated regression coefficient by the sample standard deviation of the explanatory variable and dividing by the sample standard deviation of the dependent variable, $b^* = bs_X/s_Y$. This procedure makes the resulting coefficient unit free, in the sense that a unit change in X corresponds to a change of one sample standard deviation rather than a unit of the original measurement units; and the coefficient gives the expected number of sample standard deviations change in Y expected from a one standard deviation change in X. The difficulty with this measure is that it is dependent upon the variations in X and Y observed in the particular sample. Thus samples with greater variations in one of the explanatory variables will exhibit larger standardized coefficients for that variable and smaller ones for the other variables than with smaller sample variances. Similarly, samples where the observed values of Y contain larger error terms, as measured by σ, will exhibit smaller standardized regression coefficients than samples where the error terms are smaller. These variations in coefficient estimates attributable to sample differences render comparisons between studies and data sets meaningless because one does not know whether the differences are the result of differences in the behavior of the underlying populations or simply due to differences in the samples used in the study.

It is easy to illustrate these variations in the standardized coefficients with the simulations used in Chapter 3. Table 3.3 showed the average of the 400 regression coefficients computed for each replication of the basic model under different variations in X_3 and in the error term σ. These variations are readily comparable to differences among data sets. In that table we notice that all estimates are unbiased in that the coefficient means are all centered about the true values, as one expects. We have computed the standardized regression coefficients for each replication, and Table 4.1 shows the means of the distributions of these standardized coefficients. There are striking differences in the values of these standardized coefficients among the different samples. The mean standardized value of b_2^* ranges from less than 0.50 to almost 1.0, while the range for b_3^* is from 0.06 to 0.98. The mean standardized estimate of b_2^* is nearly ten times the size of the average standardized value of b_3^* for the simulations with a small variance in X_3. Since these variations in X_3 and U_t are directly comparable to the differences one is likely to find among

different data sets, the simulations suggest that comparisons involving standardized regression coefficients are quite unreliable for measuring changes in
population behavior. It is possible, but less likely, that sample differences
could disguise true behavioral differences and result in equal standardized
coefficients.

TABLE 4.1

Mean Standardized Beta Coefficients (b*)

	σ_U								
	5.0			10.0			20.0		
S_3	0.7	2.8	11.2	0.7	2.8	11.2	0.7	2.8	11.2
$\bar{b}_2^*$	0.968	0.985	0.614	0.830	0.839	0.568	0.578	0.579	0.469
$\bar{b}_3^*$	0.101	0.399	0.980	0.089	0.327	0.921	0.061	0.232	0.750

Some way of comparing variable effects both within and across regressions
is desirable. One common measure, which generally has better properties than
standardized coefficients, is the variable's elasticity. Elasticities, used frequently in economics, are simply the percentage change in the dependent
variable that would be expected from a 1% change in an independent
variable. This measure has the advantage of being unit free. Further, its
evaluation is not as sensitive to sample definitions as the standardized
coefficients. Nevertheless, it is not immune from interpretative difficulties.
The elasticity at any point is defined as $(\partial Y/\partial X)(X/Y)$. For a linear
function $\partial Y/\partial X = \beta$, one must choose the point (Y,X) at which to evaluate
it, and the estimated elasticity will vary with the point used. (Frequently, the
point of means—$\bar{Y}$ and $\bar{X}$—is used to evaluate the elasticity, but the appropriateness of this depends upon the specific situation.)

We shall concentrate on estimating the unstandardized regression
coefficients because of their BLUE properties, their invariance under different sample characteristics, and consequently their appropriateness for comparisons, predictions, and evaluations across samples. Our discussions concentrate on how the real situations one encounters in trying to estimate actual
models with available data may result in violations of the constraints placed
on the distribution of the error terms. Different violations have different
implications for our efforts to get unbiased coefficients with the least possible
variance. We shall discuss the problems one encounters, what diagnostics
might be available to identify the problem, and finally what recourse the
researcher has to overcome the problem, if this is possible.

4.3   Model Specification

This book is really a course in application. Evaluations of applications are
based on how well reality is described, i.e., how well the estimated equations
synthesize the bits of available data and provide insights into behavior.

Probably *the most important* element in obtaining reasonable estimates of behavioral models is the statement of the model. The crucial difference between a "passing" and "failing" use of regression techniques is the development of the model—the delineation of relevant variables and the relationships among them. The refined estimates described in subsequent chapters may spell the difference between a high pass and a low pass, but they will never elevate a failing effort to a passing grade.

The general term for the description of the variables and the model is *model specification*. Because of the varied uses and applications of estimation techniques, it is not really possible to discuss in detail how one should specify a model. The specification of behavioral models relies upon the purposes of the model, the available theories of behavior, the past empirical forays into an area, and the embodied wisdom and hunches of the researcher and associates. It is not feasible for us to discuss how these elements are accumulated or combined.

As an alternative, we can discuss what happens when you miss, i.e., what happens when the model estimated differs from the true model. By following this approach, however, we can provide some general guidance on the selection of variables and models.

Throughout this book one reference point is the implicit unobservable true population model. All estimators, all statistical tests, and all future corrective actions for statistical problems are derived from a comparison with the variables, parameters, and functional form of the true model. The true model is the starting point in all of our developments and the frame of reference by which to judge results. But the exact and correct formulation is not always known. The theories of the social scientist are usually not developed to the point of giving a complete model specification.

Nor can one always expect to have the required data. In many instances data collection is separate from the research and has not been done by the researcher. Thus even when the correct model has been specified, it may not be possible to get accurate measurements for all the variables. Both of these situations, incomplete theories and incomplete data, can lead to specification errors.

The most common result of model misspecification is failure of the basic assumption that the expected value of each error term is constant. When the estimated model differs from the true model, it is very likely that the error term will be correlated with one or more of the included explanatory variables.

The cost of the failure of this assumption is reflected in biased parameter estimates. The expected value of the estimated coefficients will systematically diverge from the true population value, confusing or inhibiting attempts at substantive interpretation. The development of appropriate statistical tests is also generally impossible.

Model misspecification is easy to see in the case where the estimated model in the previous chapter omits an explanatory variable from the true model $Y_t = \beta_1 + \beta_2 X_{t2} + \beta_3 X_{t3} + \beta_4 X_{t4} + U_t$. When such a misspecification occurs, the influence attributed to the included variables is actually a combined influence of the included and excluded variables. For example, if all X variables exert a positive influence on Y_t and these variables are themselves positively correlated, the estimated coefficients for the included variables will be overstated and imply that each included variable is more important than it actually is.

The mathematics of this case is straightforward. Assume that the true model is

$$Y_t = \beta_1 + \beta_2 X_{t2} + \beta_3 X_{t3} + \beta_4 X_{t4} + U_t, \tag{4.1}$$

and that instead we estimate

$$Y_t = \beta_1 + \beta_2 X_{t2} + \beta_3 X_{t3} + U_t^* \qquad \text{where} \quad U_t^* = \beta_4 X_{t4} + U_t. \tag{4.2}$$

The least squares estimators from Chapter 2 are

$$
\begin{aligned}
b_2 &= \frac{V_3 C_{2Y} - C_{23} C_{3Y}}{V_2 V_3 - C_{23}^2} = \beta_2 + \frac{V_3 C_{2U^*} - C_{23} C_{3U^*}}{V_2 V_3 - C_{23}^2}, \\
b_3 &= \frac{V_2 C_{3Y} - C_{23} C_{2Y}}{V_2 V_3 - C_{23}^2} = \beta_3 + \frac{V_2 C_{3U^*} - C_{23} C_{2U^*}}{V_2 V_3 - C_{23}^2}.
\end{aligned}
\tag{4.3}
$$

Substituting $U_t^* = \beta_4 X_{t4} + U_t$ into the covariance expressions involving U^* gives

$$C_{2U^*} = \frac{1}{T} \sum (X_{t2} - \bar{X}_2)(U_t^* - \bar{U}^*) = \frac{1}{T} \sum (X_{t2} - \bar{X}_2)(\beta_4 X_{t4} + U_t - \beta_4 \bar{X}_4 - \bar{U})$$

$$= \frac{1}{T} \beta_4 \sum (X_{t2} - \bar{X}_2)(X_{t4} - \bar{X}_4) + \frac{1}{T} \sum (X_{t2} - \bar{X}_2)(U_t - \bar{U})$$

$$= \beta_4 C_{24} + C_{2U}$$

$$C_{3U^*} = \beta_4 C_{34} + C_{3U}.$$

Taking the expected value of b_2 and b_3, assuming fixed X and $E(U_t) = 0$, we obtain

$$E(b_2) = \beta_2 + \beta_4 \left(\frac{V_3 C_{24} - C_{23} C_{34}}{V_2 V_3 - C_{23}^2} \right) + E\left[\frac{V_3 C_{2U} - C_{23} C_{3U}}{V_2 V_3 - C_{23}^2} \right] = \beta_2 + \beta_4 b_{42}, \tag{4.4}$$

$$E(b_3) = \beta_3 + \beta_4 \left(\frac{V_2 C_{34} - C_{23} C_{24}}{V_2 V_3 - C_{23}^2} \right) + E\left[\frac{V_2 C_{3U} - C_{23} C_{2U}}{V_2 V_3 - C_{23}^2} \right] = \beta_3 + \beta_4 b_{43}, \tag{4.5}$$

where

$$b_{42} = \frac{(r_{42} - r_{32} r_{43})}{1 - r_{32}^2} \sqrt{\frac{V_4}{V_2}} \qquad \text{and} \qquad b_{43} = \frac{(r_{43} - r_{32} r_{42})}{1 - r_{32}^2} \sqrt{\frac{V_4}{V_3}}.$$

These expressions are obtained from the formula for simple correlation coefficients and appropriate division of the middle terms of Eqs. (4.4) and (4.5) by the sample variances of X_2, X_3, and X_4. The terms b_{42} and b_{43} are functions of the characteristics of the particular sample. Although X_4 is not observed and included in the data set, each observation has an implicit value for this variable associated with it. The larger the correlations between the unmeasured sample values of this excluded variable and the included variables, denoted by r_{42} and r_{43}, and the larger the variance of these implicit values for X_4, the larger the values of b_{42} and b_{43} for a given set of values for X_2 and X_3.

Equations (4.4) and (4.5) provide considerable insight into the whole question of proper model specification. The biases in the estimation with X_4 omitted are $\beta_4 b_{42}$ and $\beta_4 b_{43}$. Thus, the biases become more severe as the excluded variable becomes more important in explaining Y_t, i.e., as β_4 becomes larger in absolute magnitude. There is, however, no general effect on the magnitude of bias from changing the intercorrelations of the X's. For example, if all sample correlations (r_{32}, r_{42}, and r_{43}) are the same sign, the bias becomes larger as *sample* relationships between the included and excluded variables become stronger, i.e., as r_{42} and r_{43} increase in magnitude and as the sample variation in the excluded variable V_4 increases; yet, if any sample correlations are of opposite sign, this does not have to be true. Finally, the direction of the bias—whether b_2 and b_3 tend to over or under estimate β_2 and β_3—is solely a function of the signs of β_4 and of b_{42} and b_{43}. If both are positive or both negative, b_2 (or b_3) will be biased upward; if one is negative and one is positive, b_2 (or b_3) will be biased downward. For example, if β_4 is positive and X_4 is positively correlated with X_2 and X_3, the estimated coefficients will overestimate the true parameters; while if X_4 is negatively correlated with the included variables, b_2 and b_3 will underestimate β_2 and β_3. Note, however, that b_{42} and b_{43} do not necessarily have the same signs as r_{42} and r_{43}. Finally, the direction of bias in b_2 and b_3 does not have to be the same.

By looking at the problems of misspecified models, we get some insight into how the analyst should proceed. First, the problems of specification are related to the size of β_4. This makes a rather obvious point: proper specification is most critical in the case of very important explanatory variables. In other words, even if sunspot activity exerts some influence on voting behavior, or educational achievement, omitting a sunspot variable from these models causes serious problems only when sunspots exert a *strong* influence. A priori knowledge, based upon theory, past empirical results, hunch, etc., forms the basis for making decisions on the size of different coefficients for variables omitted from models.

The other method of reducing bias involves reducing the relationships *in the sample* between the omitted and the included variables. This is the focal point for much physical science research since laboratory experiments can be designed to reduce or eliminate the correlations with excluded variables from the experiment. Social scientists, however, do not often have the luxury of experimental design. Instead, they must rely upon "natural" experiments recorded by observers who do not know the intended use of the data.

Since the terms b_{42} and b_{43} refer to the sample used for the estimation, it is possible to reduce b_{42} and b_{43} through appropriate choice of sample. If we can find a sample where X_4 does not vary, ($V_4 = 0$), then b_{42} and b_{43} will be zero, and the bias will be removed. The educational model to be discussed in the next chapter provides a good example. The socioeconomic environment in children's homes is one of the variables expected to influence their third grade verbal achievement score. Unfortunately, data were not available to measure this variable and directly enter it in the estimated equation. Excluding this variable would seriously bias the estimated coefficients since it is undoubtedly correlated with one of the included variables, teacher characteristics—children who come from richer homes tend to go to better schools. Our solution is to estimate the achievement equation separately for children with parents from professional and from manual occupations. This sample selection then reduces the sample variance of these environmental effects. In other words, by stratifying individuals on the basis of the unmeasured variable, b_{42} and b_{43} become zero. This and other sample-related techniques of reducing specification bias will be discussed in more detail later in this chapter.

The other common sampling procedure for eliminating the biases attributable to omitted variables is to collect observations in which the excluded variable is uncorrelated with the included variables. In such a sample r_{42} and r_{43} equal zero, making b_{42} and b_{43} zero, and thus providing unbiased estimates of β_2 and β_3. The only difficulty with this procedure is that if the included explanatory variables are at all correlated, the excluded variable must be randomized with respect to *all* the exogenous variables or all the coefficients will be biased, regardless of the correlation between the excluded variable and any particular X. In the example, if r_{23} is not zero, the coefficient estimate b_2 will be biased even if r_{42} is zero because of the $r_{23}r_{43}$ term in the expression for b_{42}; similarly for b_3 if r_{43} is zero but not r_{42}. In real data sets it is hard enough to find situations where an omitted variable is uncorrelated with any included variable, let alone uncorrelated with them all.

Misspecification with additional included variables is easily considered by replacing b_{42} and b_{43} in Eqs. (4.4) and (4.5) with similar terms involving the correlations and variances of all the included and excluded variables. (This process will be greatly simplified by the developments in the next chapter.) However, the basic results for considering the bias of individual coefficients are analogous to the interpretation of Eqs. (4.4) and (4.5).

We can illustrate these results using our "laboratory" experiment, the Monte Carlo simulation. To do this, we take the simple model used in the previous simulations and add a third explanatory variable to the "true" model for Y with a coefficient of 1.0. This new model is

$$Y = 15.0 + X_2 + 2X_3 + X_4 + U. \tag{4.6}$$

X_4 is created by adding a random variable to X_3, and this random term is varied so as to obtain expected correlations between X_3 and X_4 of 0.10, 0.30, 0.50, 0.70, 0.90, and 0.94. With $T = 25$, 250 replications of the misspecified model that excludes X_4 are estimated for each of the given correlations between X_3 and X_4. Table 4.2 shows the results of this simulation. The most notable result is that, just as the mathematical explanation in Eqs. (4.4) and (4.5) predicts, the means of the 250 estimated coefficients—which represent the expected value of any single estimated coefficient—equal the true coefficient 1.0 plus $b_{42}\beta_4$ and 2.0 plus $b_{43}\beta_4$, where β_4 is 1.0. It is interesting and important to note that in this experiment the excluded variable was created with respect to only X_3 and r_{23} is close to zero. Thus b_{42} is close to zero, and the biases in the estimates of b_2 are also close to zero. This, of course, does not have to be the case, X_4 could have been related to X_2 and X_3 or r_{23} could have been nonzero, in which case both included coefficients would be severely biased.

TABLE 4.2

Mean Coefficients in the Misspecified Estimations[a]

r_{34}	b_{41}	b_{42}	b_{43}	$\bar{b}_1$	$\bar{b}_2$	$\bar{b}_3$
0.09	3.31	0.02	0.12	17.57	1.04	2.16
0.29	2.98	0.00	0.30	18.25	1.00	2.26
0.50	0.49	0.05	0.58	15.56	1.05	2.60
0.69	0.29	0.03	0.76	14.17	1.05	2.87
0.91	0.07	0.01	0.95	15.36	1.01	2.95
0.93	0.07	0.01	0.98	14.67	1.00	3.05

[a] b_{4k} is the appropriate coefficient from Eqs. (4.4) and (4.5) plus a similar coefficient for b_1. $\bar{b}_k$ is the mean of the 250 estimated coefficients.

The sociology literature contains a classic example of such a misspecification problem. William Robinson (1950) in an oft-cited article purportedly showing the pitfalls of aggregate data, uses data on individual nativitity and illiteracy from the 1930 U.S. census to compute a regression coefficient in a bivariate model stating that illiteracy is a function of being foreign born. This estimated coefficient is 0.07. In other words, a foreign-born individual would have a probability of being illiterate 0.07 higher than that of a native-born person. Robinson also estimated the simple regression coefficient in an equation that predicts the percent of a state's population that is illiterate as a function of the percent foreign born in the state. This estimated coefficient is -0.29, which is markedly different from the coefficient estimated with individual data. At the state level, it appears that foreign born are more literate than native born (because of the negative coefficient). (Robinson actually used the simple correlation coefficient, which we have transformed into the equivalent regression coefficient.)

   There are other factors besides a person's nationality contributing to the probability of being illiterate and to the literacy rates of different aggregates of people. The principal omitted factor is the quantity and quality of the individual's education or the educational services available to the different groups of individuals. These effects are excluded from Robinson's model, and they clearly have a negative effect on the probability of an individual being illiterate or on a state's illiteracy rate. Robinson's simple model then is

$$I = \beta_1 + \beta_2 F + U^*,$$

where I is percent illiterate, F is percent foreign born, and U^* is an error term containing the effect on percent illiterate of any excluded variables and different random effects. Thus the implicit value of β in the equation for the omitted schooling variable in this model is negative: the more schooling available in a state, the lower the illiteracy rate. But what are the correlations between schooling and being foreign born? At the individual level, being foreign born and the quantity and quality of the person's education are probably negatively correlated. However, at the state level in 1930, the percent of a state's population that is foreign born and the quality and availability of educational services are clearly positively correlated. Most foreign-born individuals were living in the Middle Atlantic and North Central states where the most extensive public school systems could be found. This means that b_{42} in Eq. (4.4) is positive and the misspecified estimate of β_2, the effect of percent foreign born on percent illiterate, will be underestimated, or biased downward. In this case the extent of the bias is apparently enough to give the estimate of the coefficient the wrong sign.

   To illustrate the effect of correcting the misspecification in this case, the following model explaining a state's illiteracy rate is estimated using the same 1930 census data used by Robinson, only we have included a variable measuring the quantity of schooling available to the inhabitants of the different states, along with three other minority group variables:

$$I = 86.0 \; + 0.12 \, F - 0.88 \, E + 0.10 \, B + 0.03 \, M + 0.14 \, IND,$$
$$\quad\;\; (6.23) \quad (2.82) \quad (-6.13) \quad (2.90) \quad (0.30) \quad (0.80)$$

$$(t\text{-statistics}) \qquad R^2 = 0.86, \tag{4.7}$$

where I denotes percent illiterate, F percent foreign born, E percent of a state's population aged 7–13 enrolled in school, B percent Black, M percent Mexican, and IND percent Indian.

   The exact specification of the model is not ideal as everyone is sure to note. However, by estimating a *better* specified model of state illiteracy, including variables for the amount of schooling, we have reversed the sign on the F variable so that it now conforms to our expectations. The magnitudes of the coefficients are also reasonable. One can check this last statement by calculating a predicted illiteracy rate for foreign-born individuals by assuming that they constitute 100% of the population. (The mean school attendance for foreign-born individuals to use in this calculation is 97%). Their estimated illiteracy rate with Eq. (4.7) is 12.6%, the actual rate was 10.3% and Robinson's aggregate estimate is −21.9%.

The remedy for these misspecification problems is obvious, but not necessarily easy. The excluded variable can either be included, i.e., correcting the specification as in the Robinson case, or a sample can be collected in which the covariance between the included and excluded variables is zero, either because they are uncorrelated or because the excluded variable has no variance. However, each of these solutions requires that the misspecification be recognized prior to the collection of the data. In most real world applications, the misspecification arises because researchers failed to recognize the importance of a variable, not because they were unable to obtain a measure for the excluded variable or a sample where it was uncorrelated with included variables. This will be particularly true in social science areas that do not have a well-developed a priori theory. For example, the competitive market theories of microeconomics are much more explicit about what assumptions are being made and what variables are assumed to be held constant during an analysis than are models of the educational process, of voting behavior, or of oligopolistic pricing. Consequently, in these latter and similar areas the likelihood of misspecification is increased because there is little formal theory to guide the researcher in selecting variables and ascertaining what needs to be held constant. The researcher then must be particularly careful in selecting the original variables.

One of the most important implications of both the theoretical development and the example here is that the inclusion of important variables is essential, even if one is not interested in the estimated effects of all of the variables. In order to arrive at good estimates of the parameters of interest, it may be necessary to include other variables of lesser usefulness in the given problem. Recognition of the significance of a variable in a behavioral relationship does not necessarily imply that the analyst can or wishes to interpret its coefficient, only that one wishes to avoid biasing the coefficients of real interest.

4.4 Multicollinearity

A difficulty that often exists when potential misspecifications have been taken care of by including the relevant variables is multicollinearity. This is a condition where one or more of the explanatory variables included are highly correlated in a sample of data, i.e., a high value of r_{23} in the model. Correlations among explanatory variables occur in many samples without the explanatory variables being causally related. The problems created when they are so related will be dealt with in Chapters 8 and 9. In many instances, it is just not possible to locate a data sample in which all the explanatory variables exhibit a large amount of independent variation. This will be particularly true. in small samples, such as one must deal with in a time series analysis where many variables tend to follow the same basic pattern of movement or in samples that contain highly aggregated observations such as the data on foreign born, illiteracy, and educational quantity at the state level. The question then is, What problem does multicollinearity pose for attempts to estimate the coefficients in a statistical model?

The extreme case where the explanatory variables are perfectly correlated is covered in Chapter 2. This extreme implies that $|r_{23}| = 1$ and that the estimates cannot be computed. (The case with more than two explanatory variables is covered in Chapter 5.) This seldom occurs in nature however. In the more usual case, these correlations may be high, but $|r_{23}|$ will not exactly equal 1. As the correlations increase, the value $1 - r_{23}^2$ becomes smaller. From the development of the expected value of the estimated coefficients $E(b)$ in Chapter 3, it is clear the correlation between X_2 and X_3—as long as it is not perfect—does not affect the expression for $E(b)$ because U is still uncorrelated with X_2 and X_3 so that $E(C_{2U}) = E(C_{3U}) = 0$. However, multicollinearity does influence the variance of these estimated coefficients:

$$\text{var}(b_2) = \frac{\sigma^2}{T} \frac{1}{V_2(1 - r_{23}^2)}$$

and a similar expression for $\text{var}(b_3)$. Thus the higher the value of $|r_{23}|$, the smaller the term $1 - r_{23}^2$ and the greater the variance of the estimated coefficients.

In other words the sample variance of the estimated coefficients increases as the correlation among the explanatory variables increases, giving less precise estimates of the true coefficients. This is intuitively reasonable. Each coefficient is interpreted as the independent effect of the given variable. The more two variables covary or move together in a sample, the harder it is to ascertain the independent effect of one of them, holding the other constant. The sample simply does not contain enough information about the variations in Y associated with changes in each explanatory variable for constant values of the other exogenous variables to estimate these effects accurately.

These points can be illustrated by again using the Monte Carlo simulation. All we have to do is reestimate the model shown in Eq. (4.6) and Table 4.2 including the variable X_4 in these estimations. From this we shall be able to see what happens to the estimated coefficients and their variances when the correlation between X_3 and X_4 is increased from 0.10 to 0.94. We can also get a comparison of the value of including theoretically meaningful variables in the model, even though they may be highly correlated with other explanatory variables. These simulation results are shown in Table 4.3.

TABLE 4.3

Mean Coefficients and Variances in the Multicollinearity Simulation[a]

r_{34}	$\bar{b}_1$	$\hat{\sigma}_{b_1}$	$\bar{s}_{b_1}$	$\bar{b}_2$	$\hat{\sigma}_{b_2}$	$\bar{s}_{b_2}$	$\bar{b}_3$	$\hat{\sigma}_{b_3}$	$\bar{s}_{b_3}$	$\bar{b}_4$	$\hat{\sigma}_{b_4}$	$\bar{s}_{b_4}$
0.09	14.157	6.779	6.993	1.013	0.148	0.149	2.038	0.805	0.791	1.030	0.726	0.768
0.29	15.469	7.045	6.935	0.997	0.159	0.148	1.982	0.845	0.822	0.923	0.785	0.796
0.50	15.072	6.561	6.535	1.001	0.151	0.155	2.027	0.909	0.952	0.988	0.923	0.926
0.69	13.880	7.027	6.527	1.023	0.155	0.151	2.129	1.230	1.153	0.981	1.107	1.114
0.91	15.293	6.112	6.519	1.000	0.142	0.149	2.055	2.345	2.326	0.941	2.293	2.304
0.93	14.751	6.676	6.519	0.998	0.147	0.149	2.001	3.162	3.271	1.069	3.111	3.240

[a] b_k is the observed mean estimate for the 250 trials; $\hat{\sigma}_{b_k}$ the observed standard error of the 250 estimates; and $\bar{s}_{b_k}$ the expected standard error, calculated from the formula for $E(b_k - \beta_k)^2$. The precise formula for this four-variable model is Eq. (5.16).

These results clearly show that the estimates are unbiased, even in the face of substantial multicollinearity, by properly specifying the model. The means of the 250 estimated values for β_3, which are estimates of its expected value, were quite close to 2.0 in all six simulations. This is not true in the misspecified case in Table 4.2. As predicted, the variance of the estimated coefficients does increase dramatically as the amount of correlation between the variables increases. However, this tendency does not become pronounced until the correlation exceeds 0.50. In addition the increase in variance is true only for the correlated variables. The variance of the estimate of β_2, the coefficient on the uncorrelated variable, does not change appreciably as the correlation between X_3 and X_4 increases. Thus, not only do we obtain unbiased estimates of all the coefficients in spite of the high collinearity, we can still get precise estimates of the coefficients for the relatively uncorrelated variables. However, as the correlation between X_2 and both X_3 and X_4 increases, so will the variance of b_2.

The seriousness of multicollinearity depends upon the purposes and uses of the estimated model. We can gain some understanding of the effects of multicollinearity by first considering the extreme case in the example. In that simulation, the sample correlations between the exogenous variables are $r_{34} \cong 1$ and $r_{24} = 0$. (What does this imply about r_{23}?)

If we are interested only in b_2, the estimate of β_2, we need not be very concerned about the high correlation between X_4 and X_3. When both X_4 and X_3 are included, it is difficult to disentangle their independent effects on Y, and the estimates of β_3 and β_4 will be very imprecise. Yet, including both or only one has little effect on b_2. The lack of correlation between X_3 or X_4 and X_2 implies that the variance of b_2 is not affected by r_{34}. Further, leaving out X_3 or X_4 in estimating (4.6) does not bias b_2 since b_{32} or b_{42}, the coefficients between X_2 and the excluded variable, are small. This is shown in the simulation of the misspecified model in Table 4.2.

In the above model, multicollinearity is likewise not an important concern if the only interest is predicting values of Y_t for given values of X_2, X_3, and X_4. For predicting the value of the dependent variable, imprecise coefficient estimates are less bothersome, as long as we expect the observed sample correlations r_{23}, r_{24}, and r_{34} to occur in the future (when prediction is taking place). We can then use estimates of the fully specified model or the model excluding either X_3 or X_4 with approximately equal success for prediction.

However, for different uses of the estimated model and with the relaxation of the restrictive assumptions on the simulation, the effects of multicollinearity become more severe. With the simulation, estimation of either β_3 or β_4 is inhibited by the high correlation of X_3 and X_4. We cannot simply drop out one of the variables since we would have very biased individual coefficient estimates, as indicated by the simulation in the misspecified case earlier.

Further, if the assumptions about the intercorrelations of the exogenous variables are relaxed, the problem is much more difficult. If the correlation between X_2 and the highly correlated variables X_3 and X_4 is not zero, the estimate of β_2 will be less precise, i.e., the variance of b_2 will be larger than when r_{24} and $r_{23} = 0$. Also, we are no longer indifferent to dropping out X_3 or X_4 in the estimation of b_2. Bias would again occur if that were done.

Finally, if the relationship between X_3 and X_4 changes during the prediction period, or is different in another sample being studied, the entire analysis is endangered. Particularly in the case where the model includes only X_3 or X_4 along with X_2, a change in the correlation between X_3 and X_4 implies that good estimates of both β_3 and β_4 are needed.

It is clear that multicollinearity can impair the ability to estimate a model. Yet how do we know that multicollinearity, and not something else, is causing any observed estimation difficulties? In particular, the decrease in the precision of the estimates caused by multicollinearity will generally result in very low values of the t-statistics (against a null hypothesis that the coefficient is zero). However, a low t-statistic, as we observed when they were introduced, also can result from the variable exerting no significant influence on the behavior being observed. Surely it makes no sense to blame all low t-statistics on multicollinearity since then the t-test could never tell us anything.

The distinguishing feature between multicollinearity and the problem of generally weak data is found in the relationship between the coefficient estimates. The individual coefficient estimates obtained from a given sample are not independent of each other unless all of the exogenous variables are uncorrelated. While small samples or observations of the exogenous variables with little variation may lead to imprecise estimates of individual coefficients, multicollinearity adds another dimension to the problem: high correlations among the coefficient estimates themselves. With two exogenous variables, the correlation between coefficient estimates takes on a particularly simple form. It is merely $-r_{23}$, or the correlation between the exogenous variables except with the opposite sign.[1] Thus, a positive correlation between the

---

[1] The correlation between coefficient estimates in the three-variable model can be found from extending the development in Chapter 3. Equations (3.1) and (3.2) yield expressions for $b_2 - \beta_2$ and $b_3 - \beta_3$. Taking the expected value of the product of these expressions and dividing by the square root of the variances given in Eq. (3.5a) yields the correlation between the estimates.

exogenous variables (the most common occurrence in applications) yields a negative correlation between the coefficient estimates. One implication of this is that tests of significance for individual coefficients may be quite misleading; instead, as will be discussed below, it is often necessary to test several coefficients jointly.

Identification of multicollinearity comes from experience with estimating models, and it is very difficult to outline specific criteria for labeling it severe in one case and not severe in another. As indicated in the previous section, severity is a function of the uses of the model. Beyond that it is possible to list where clues about multicollinearity can be found.

Since the interest centers upon the ability to estimate the parameters precisely, attention should first be directed toward the correlations among the coefficient estimates. In the three-variable model, this is equivalent to looking at the correlation of X's (with a sign change). In more complicated models (more exogenous variables), the relationship between the correlations of the coefficients and the exogenous variables is not so straightforward. (This relationship is discussed in Chapter 5.) Since the correlations among the X's are often more readily available, we begin with them.

The matrix of simple correlation coefficients between the exogenous variables can supply some information. If any two variables have a very high simple correlation, they are candidates for causing trouble in the estimation. High simple correlations are however neither necessary nor sufficient (except with perfect correlations) in identifying troublesome intercorrelations. As the term *multi*collinearity implies, problems can and often do arise in more than just pairs of variables. Further, it is not possible to define "high" correlations with any precision—it depends upon the specific model and sample. In our Monte Carlo experiments, "high" implies correlation of 0.7 or greater; in other samples it could be very different.

A second clue to the severity of multicollinearity comes from the stability of estimated coefficients derived from slightly different model specifications or samples. When multicollinearity is severe, parameter estimates are based upon very little information; i.e., the "independent" variation of each variable upon which they are based is small when the explanatory variables are highly intercorrelated. In such a case, the coefficients become very sensitive to slight modifications of the model specification or of the sample used for estimation. Since many uses of estimated models delve into uncharted areas where there is little guidance to the precise definitions of variables, it is often appropriate to experiment in the estimation of models by varying the definition of essentially the same conceptual variable. If this is done and large changes in coefficient estimates (particularly of the other coefficients in the model) result, it may be taken as prima facie evidence that serious multicollinearity is present. Similarly, if slight changes in the sample used for estimation yield large changes in estimated parameters, multicollinearity may well be present and serious.

Finally, a variety of diagnostic statistics have been suggested for multicollinearity. They generally suffer from an inability to distinguish between serious and not so serious multicollinearity, but they can provide further supportive evidence. One such statistic, suggested by Farrar and Glauber (1967), is the determinant of the simple correlation matrix for the exogenous variables.[2] This is a handy statistic since it is possible to derive both upper and lower bounds for it, corresponding to no multicollinearity and to perfect collinearity. When there are no intercorrelations among the exogenous variables, the determinant is one. This upper limit then corresponds to no problem. At the other extreme, if there is a linear relationship among any set of explanatory variables, the determinant of the correlation matrix will be zero. Between the two extremes, the determinant is a continuous function that uniformly decreases with higher intercorrelations. By observing the value of this determinant (which is calculated in many computer programs for regression analysis), some feel for the severity of multicollinearity can be observed. There is still, however, a problem of judging what is a "low" value and what is not. But it can give another useful clue in addition to looking at different specifications of the model.

A situation that is often observed when the exogenous variables are highly correlated is the following: individual t-tests on the coefficients indicate that the null hypothesis that each is zero cannot be rejected, but the model has a good fit for the observed values of Y, indicating that the variables included in the analysis do account for the systematic variations in Y. In these cases one cannot rely upon the statistical tests developed in Chapter 3 to provide much information about proper specification. The tests are still appropriate in the sense that the assumptions underlying them are unaltered by considerable multicollinearity. However, the tests are not very discriminating between true and false hypotheses. (Chapter 5 presents a way to test hypotheses about parts of or the whole systematic component rather than individual coefficients.) This again reflects the high correlation between coefficient estimates when there is substantial multicollinearity. It is common with multicollinearity to have t-tests on two individual coefficients fail (i.e., inability to reject the null hypothesis of no effect) and still find that the variables are jointly significant. (Testing two or more coefficients jointly is accomplished with an F-test, described in Chapter 5.)

Serious multicollinearity—i.e., the inability to obtain precise enough coefficient estimates because of intercorrelations in the exogenous variables— is an unenviable position. The sample of data used for estimation does not

[2]The determinant of a matrix is discussed in Appendix II and its role in estimation in Chapter 5. In the case of more than two explanatory variables, the value of the determinant of the correlation matrix of the exogenous variables replaces the expression for D in the expressions in Chapter 2. Just as D approaches 0 as r_{23} approaches 1, in the larger model the value of the determinant approaches 0 as the multiple correlation between each explanatory variable and all remaining exogenous variables approaches 1.

contain enough information about the independent effects of each of the exogenous variables on the endogenous variable. When multicollinearity is particularly harmful for the specific analysis, the remedy is generally of little consolation to the analyst. In simplest terms more information is needed.

Obtaining more information is usually not easy. In order to get the best possible estimates (smallest variance) of the individual parameters in the model, one should use all of the available data during the original estimation process. This leaves no reserve data to be called out when more information is needed to combat multicollinearity. Clearly if more data are available, they should be used.

But what if no more observations can be readily found to use in breaking the multicollinearity among the explanatory variables? One alternative is to introduce information about the parameters themselves. This information may come from coefficients estimated in other samples, from theoretical developments (say of the relationship between two parameters in the model), or from strong a priori beliefs about a given parameter. The essence of this notion is easily seen: if two variables are highly intercorrelated, the sample does not provide enough information to estimate the coefficients accurately. However, if information from outside the sample is available, many times it can be used to improve the estimates.

The exact estimation procedures for this purpose are beyond the scope of this book. They require pinpointing the exact location of the damaging intercorrelations, a task that can be quite difficult in complex multivariate models.[3] They also require some adjustment in the estimation procedure to incorporate this information.[4] However, even after these tasks are accomplished, this method of combating multicollinearity can be quite hazardous since one seldom has reliable information about the precise parameter values to use.

Thus, we are really left with the following summary: multicollinearity can be a serious problem that impedes estimation. When it is present to a high degree, one must either live with imprecise estimates or find more information. The prospects for readily finding more information are often bleak.

One postscript of our discussion is required. Multicollinearity is a problem of not enough information. When it is a serious problem, individual coefficient estimates are very frail, i.e. they are very sensitive to the selection of the sample and to the precise model specification. Adding or subtracting one observation from the sample used for estimation can often have dramatic effects on the magnitude, and even the sign, of different coefficient estimates. One implication of the multicollinearity problem is never to reduce the

---

[3] Some suggestions of tests for pinpointing the sources of multicollinearity can be found in Farrar and Glauber (1967).

[4] Methods of incorporating this information can be found in Goldberger (1964, pp. 255–266); Theil (1971, pp. 282–315); Johnston (1972, pp. 155–159). A discussion of one method—merging cross-sectional and time series information—can be found in Meyer and Kuh (1957).

sample; a lack of information is not solved by throwing out information. A given observation or set of observations may contain redundant information about the effects of two variables, i.e., the variables in those observations may exhibit the same intercorrelations as observed in other observations, but that is no reason to eliminate the observation.

4.5 Model Specification and Multicollinearity in Practice

Multicollinearity can have a powerful effect upon model specification and, particularly, on statistical tests of model specification. When we discussed model specification in terms of an omitted variable, there were two possible situations: data on the omitted variable exist but were ignored, and data on the omitted variable do not exist. In the first case, the standard t-test of the null hypothesis $\beta = 0$ is a test of the specification that includes X, and performance of that statistical test provides information about appropriate model specification. But multicollinearity confounds this test and weakens the ability to judge among model specifications. Since multicollinearity reduces the precision of the estimates (increases their variance), it becomes difficult to develop tests that are good at distinguishing between alternative values of a parameter and alternative specifications of the model.

Likewise we must be very careful about the model specification. If minor changes in the model specification or the definition of variables yield large changes in the estimated coefficients, the model should be treated with some caution. In particular, the precise estimates of any given specification may represent an artifact more of the sample than of the true underlying structure. They may rely heavily upon one or two data points that exhibit a slightly different pattern of intercorrelations but which are not necessarily representative of the population. In other words, multicollinearity reduces our confidence in any particular point estimates of parameters. This lowered confidence is usually, but not always, reflected in the estimated coefficient variances. Moreover, since the estimates become very sensitive to sample and specification, the results that are obtained from experimentation with a variety of specifications and variable definitions are quite suspect by themselves. They require more than the usual amount of verification from other samples of data.

We have concentrated our discussion on the problems associated with omitting an important variable from the analysis. The reverse case is also true: there are costs associated with including irrelevent variables. The expected value of the estimated coefficient for such a variable will of course be zero, and the other coefficients will remain unbiased, so that faulty conclusions are not expected. However, the effects on the single estimate of each coefficient obtained from one data set may not be trivial because the variance of all estimates will increase. The formula for the variance of each estimated coefficient does not take into account that one of the included variables has no influence on Y; it includes only terms involving the variances and covariances of the explanatory variables. For example, in the model examined in Chapter 2, the variance of b_2 is a function of the variance of X_2 and of the correlation between X_2 and X_3. In that case if X_3 does not influence Y and should be omitted from the model, including it only increases the variance of the estimate of β_2. Thus the cost of including extraneous variables in the estimation is reflected in higher variances for the estimates of the coefficients for the variables that belong in the equation. In equations with more than two explanatory variables, irrelevant variables have the same effect because they generally increase the collinearity within the set of included variables, which reduces the size of the determinant and increases the variance of each estimated coefficient. The implication of this discussion is that you do not want to include unnecessary variables, particularly if they are collinear with other variables, just as you do not want to omit necessary ones. The question is how to tell the two apart. The only certain answer is with the theory used to construct the model in the first place. This, and possibly previous empirical findings, are the only sure way to make such decisions.

In some instances researchers will use the statistical tests described in Chapter 3 to decide whether a variable is extraneous or not. If the t-statistic falls below the critical value for a specified confidence level, say 0.1, the researcher will decide to accept the null hypothesis that the true coefficient for that variable is zero, and reestimate the equation with that variable omitted. This process runs the very considerable risk of biasing the remaining coefficients because the statistical tests used are not set up to test the null hypothesis implicit in this decision process. The null hypothesis the researcher is actually using is that the true coefficient is not zero, H_0: $\beta \neq 0$; but the t-statistic is testing the null hypothesis that β equals zero. Not being able to reject the null hypothesis that $\beta = 0$ is not equivalent to rejecting the null hypothesis that $\beta \neq 0$.

Nature, as the designer of social scientists' experiments, is particularly perverse on this problem. Low t-statistics can result either from the true coefficient being close to zero or because the estimated coefficient has a high variance, possibly caused by multicollinearity. If one could be sure that the true influence of a variable is small and that the low t-statistic is the result of the true coefficient being close to zero, the amount of bias from omitting this variable would be relatively small. The gain in precision by reducing the variance of the remaining coefficients could offset this small bias. However, large gains in precision are possible only when highly collinear variables are omitted. Unfortunately, this collinearity increases the variance of the estimated coefficients and implies that one cannot be confident that the true value of the coefficient is close to zero. The result of using t-tests to justify omitting variables may lead to the exclusion of collinear but substantively important variables. Thus there is considerable risk of biased coefficients if one adopts this strategy. We return to our previous comment that in the face of multicollinearity the researcher must be more cautious in evaluating and interpreting the results and must provide much more information about the behavior being modeled. This information can come only from theoretical considerations and previous empirical work.

A common estimation procedure known as stepwise regression is particularly vulnerable to the problems of specification and multicollinearity just described. In stepwise regression, the researcher specifies only the dependent variable and a list of possible explanatory variables rather than the exact model to be estimated. The program doing the regression[5] then successively selects variables for inclusion in the equation on the basis of which one will yield the greatest increase in R^2. In some cases cutoffs are established in terms of the number of variables to be included or the minimum change in R^2 required for inclusion of the next variable.

Stepwise regression represents a series of ordinary least squares estimates where the number of variables is progressively increased. At any stage, the coefficient estimates, estimated standard errors, R^2, etc. arrived at through a stepwise procedure will be identical to the estimates obtained from a simple OLS regression that includes the same variables. Thus, the issue is not the numerical coefficient estimates. Instead, it is whether the additional information generated in intermediate stages of the stepwise process is useful in interpreting (or constructing) the model itself or in ascertaining anything about relationships between individual independent variables and the dependent variable.

Two common justifications for the use of stepwise regression are that such a procedure is useful in determining the "most important" variables in explaining the behavior in question and that, because there is uncertainty about just which variables should be in the equation, this procedure allows the data "to tell the best model." Let us consider these in order.

R^2 was interpreted as the amount of dependent variable variation explained by the exogenous variables. Thus, it seems logical that the variables that "explain the most" are the "most important" in determining the behavior. However, R^2 is a sample specific statistic. As such it is determined not only by the strength of the relationship (the β_k) but also by the intercorrelations among the exogenous variables and by the variance in each of the exogenous variables. These last two terms are dependent upon the specific characteristics of the sample and, thus, cannot be easily generalized to the entire population under consideration. Further, since the procedure operates on increments to R^2, or changes in explained variation, the amount of

---

[5]Some stepwise programs start with a full set of variables identified in the equation and eliminate variables on the basis of the smallest reduction in R^2. This is referred to as stepwise deletion. Other programs combine these two methods.

variation attributable to any variable is dependent upon the order in which it is entered, i.e., on the set of other variables that are already in the equation (entered in an earlier step) and on the set not yet entered.

Consider a simple example where two exogenous variables each have a strong influence on the dependent variable (i.e., a large value of β_k) but which are highly correlated with each other in the sample. A stepwise procedure would select one of the variables for inclusion but might neglect the second because it would add little to R^2. The individual parameter estimate for the included variable would be biased (as discussed previously), and the procedure would be misleading if we interpreted the stepwise regression as indicating that the included variable was important and the excluded variable was unimportant. The individual coefficients at any stage in the procedure are biased in just the way discussed under the heading of model specification.

Further, there is little assurance that the final model—the model selected at the end of the entire stepwise procedure—bears any relationship to the underlying population model. First, a variable entered at an early stage may have no influence on the dependent variable ($\beta_k = 0$) but may be correlated in the sample with several other variables that do influence the dependent variable. The stepwise procedure may include this variable because it "proxies" several other variables—variables that do have a significant relationship with the dependent variable. Because of the level of intercorrelation, the true variables may never be included. Second, it is possible that some important variables are not included (and neither are any proxies for them). A set of variables might be skipped in the search process if their effects are "offsetting"; i.e., in the case of two variables, if both have similar effects on Y_t (in terms of β_k) but are negatively correlated in the sample, or if each has an opposing effect on Y_t but they are positively correlated in the sample, the estimated importance of either one taken separately will be understated.

The point of this discussion is simple. Stepwise regression appears to promise something that it cannot deliver. It is not possible to use stepwise regression to give both the model and the parameter estimates. Nor is it possible to use either the order of entry into a stepwise procedure or the parameter estimates of intermediate stages to make inferences about the importance of particular variables (except in the context of one specific sample). To the extent that the purpose of estimation is to make inferences about population relationships on the basis of sample information, a stepwise procedure can be very misleading.

4.6 Functional Forms

The previous discussion of model specification has centered exclusively on which variables should be included. Specifying the relationship among them is also very important. Choosing the correct *functional form* of the model is often done with even less guidance than choosing the variables of the model.

Theory rarely provides more than a few very general conditions on the form.

The linear form, which we have relied upon until now, has two principal justifications. It has been well analyzed so that the statistical methods and properties are understood; and many nonlinear functional forms look linear over small ranges. Nevertheless, the linear form has both conceptual and empirical drawbacks. Fortunately, a considerable variety of functional forms can be estimated with only minor modifications of the previous developments of least squares techniques.

A linear relationship between the exogenous and endogenous variables is not required for the previous developments to hold. The least squares estimator is still appropriate when the dependent variable is linearly related to a *known function* of the X's and U's. In other words, if transformations of the X's can be substituted into a relationship that is then linear, least squares methods are still appropriate. As an example, assume that the true relationship is

$$Y_t = \beta_1 + \beta_2 X_t^2 + U_t.$$

This is a nonlinear relationship between Y and X, but, by defining $Z_t = X_t^2$ and regressing Y_t on Z_t, we can use least squares. In other words, we have substituted a known function of X_t and can treat the relationship as linear in terms of the substituted variable. This clearly opens up least squares to a great many nonlinear relationships. Relationships such as these that are nonlinear in terms of the variables but linear in terms of the parameters can be easily handled.

Below we shall display a few of the more common nonlinear functions that can be treated with the least squares technique. Some of these deal with the more difficult problem of nonlinearity in terms of the parameters. Nonlinearity in the parameters is not as easily dealt with and is often intractable. This section sketches some of the more common forms where estimation with least squares is tractable. In the course of this discussion, we shall point out the most important attributes of these functions. All of these different functional forms require that either or both the independent and dependent variables be transformed in some way, such as taking the logarithm or the reciprocal of each observation of the variable. In other words, the variable included in the regression equation is the log, the reciprocal, or the square of the original variable. Models where the dependent variable is transformed are those that are originally nonlinear in the parameters, but are linear in the parameters after transformation.

One of the most commonly used functional forms other than the linear relationship is the log–log form. This model is estimated by taking the natural logarithm of all variables (except the constant term). This is shown in

$$Y^* = \beta_1 + \beta_2 X_2^* + \beta_3 X_3^* + U^* \text{ or } \log Y = \beta_1 + \beta_2 \log X_2 + \beta_3 \log X_3 + \log U,$$

$$(4.8)$$

where the asterisks indicate the natural log of the original variable. Taking the antilog of each term permits us to show the hypothesized nonlinear relationship between Y and each X. This true relationship is shown in Fig. 4.1 (assuming that variable X_3 is held constant) and in

$$Y = \beta_1^* X_2^{\beta_2} X_3^{\beta_3} U \quad \text{where} \quad \beta_1^* = e^{\beta_1} \quad \text{and} \quad U = e^{U^*}. \tag{4.9}$$

There are several important implications in this model, differentiating it from a linear relationship. The log–log model hypothesizes that the effect on Y of any given change in X varies with the magnitude of X. In fact, Eq. (4.9) indicates a constant relationship between percentage changes in X and percentage changes in Y, for small changes. Economists refer to this as a constant elasticity, and verbally it means that for every *percent* change in X_2, Y changes by β_2 *percent*. By way of comparison, the linear relationship implies that for each *unit* change in X_2, Y changes by β_2 units. (Please note that if the variables themselves are percents, such as the percent of the population that is black, we are *not* talking about a 0.01 change in the proportion measured by the variable, e.g., going from 15 to 16% black.)

Another difference between the log–log and the linear forms is that the change in Y associated with a change in X_2 varies with the magnitudes of both X_2 *and* the other explanatory variables in the log–log model. In the linear model, this marginal effect of any given variable is always equal to that variable's coefficient and, thus, is independent of the values of the variable itself or of the other explanatory variables. The expression for the marginal change in Y associated with a small change in X_2 in Eq. (4.9) is

$$\partial Y / \partial X_2 = \beta_2 \beta_1^* X_2^{\beta_2 - 1} X_3^{\beta_3} = \beta_2 Y / X_2$$

This marginal relationship can be seen to depend upon the values for the other explanatory variables and the value of X_2 itself. Whether this marginal effect (holding the other variables at a constant level) increases or decreases as X_2 increases depends upon the size of β_2. As indicated in Fig. 4.1, if β_2 is less than one, this marginal effect decreases as X_2 increases. If β_2 is greater

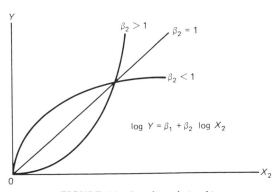

FIGURE 4.1   Log–log relationships.

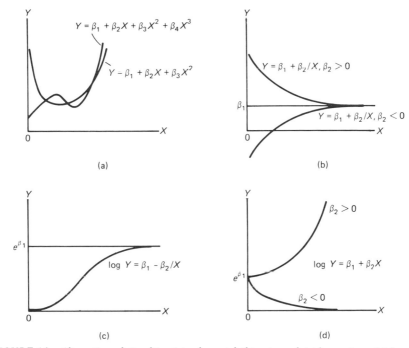

FIGURE 4.2   *Alternative relationships: (a) polynomial, (b) reciprocal. (c) log–reciprocal, (d) semilog.*

than one, the opposite is true and the marginal effect increases as X_2 increases in size. In addition, the larger the terms β_1^* and $X_3^{\beta_3}$, the larger the change in Y for a given change in X_2.

The final difference is in the assumptions about the error term. As we emphasized in Chapter 3, our estimation model assumes the error term in the model being estimated has an expected value of zero. Consequently, the log of the error term hypothesized in Eq. (4.9) must have an expected value of zero. The log–log model also hypothesizes that the effect of *the error term is multiplicative*, as are the effects of the explanatory variables.[6]

Other transformations can be used to alter the form of the relationship between the dependent and explanatory variables. Each one requires a different interpretation of the estimated coefficients and implies different assumptions about the error term. Several of these possible transformations are shown in Fig. 4.2a–d along with the specific equations.

Some of these transformations have more variables in the estimating equation than in the conceptual model. For example, a general quadratic

---

[6]In the log–log model, the error term is the ratio of the observed value to the systematic component $Y/\beta_1^* X_2^{\beta_2} X_3^{\beta_3} = U$, rather than their difference, as in the linear model, $Y - (\beta_1 + \beta_2 X_2 + \beta_3 X_3) = U$. This means the magnitude of the error term is proportional to the size of the systematic component.

relationship between Y and a single exogenous variable such as

$$Y_t = \beta_1 + \beta_2 X_t + \beta_3 X_t^2 + U_t$$

actually requires estimating a model with two right-hand side variables (X and X^2) even though there is only one conceptual variable. Further, even though X and X^2 are related to each other, they are not linearly related. Therefore, they can both be included in the same model and the parameters can be estimated. Depending upon the variance in X, there may be multicollinearity problems—even though the collinearity is not perfect.

The important points to keep in mind with all these transformations, however, are the assumptions being made about the implicit error term in the model relating Y to the explanatory variables. The assumptions of course are the ones discussed in Chapter 3, and they refer to the transformed error term in the equation being estimated.

One note of caution should be introduced. Some transformations of variables are undefined for certain variable values. For example, logarithms are undefined for values less than or equal to zero. To handle these situations, it is sometimes necessary to do a preliminary transformation of the variables; such procedures should be considered carefully however since they might make interpretation difficult or might make assumptions about the error terms untenable.

Choice of functional form is often difficult. Each of the different functional forms displayed has implications both for the relationship between the independent variables and the dependent variables and for the characteristics of the error term. Before using a specific functional form in an analysis one should be satisfied that the following questions have been answered: What is the expected change in Y for a one unit change in X_2? Does the effect of a change in X_2 depend upon the magnitude of X_2? Upon the magnitude of X_3? Must any special assumptions be made about the distribution of the error term?

In practice the responses to these questions and the judgment between two forms—say linear and log–log—represents a blending of a priori views about the variables to be included, the magnitude of effects related to the included variables, the precision of coefficient estimates, and the goodness of fit of the overall equation. Nothing precise can be said about any of these factors except perhaps the last one.

The expected value of the residual variance in a correctly specified model can be shown to be smaller than that in an incorrectly specified model.[7] This result gives some guidance to model selection, but it is not foolproof. In particular, just because the expected value is smaller, the residual variance for any particular sample may not be smaller. This property holds only over a large number of samples, and it could fail for many individual samples.

---

[7]This theorem is proved in Theil (1971, p. 543).

Further, it does not necessarily hold when the dependent variable is transformed and different across models.

While the residual variance theorem does not strictly hold across transformations of the dependent variable, it may still be useful to compare the residual variances (or equivalently the standard errors of estimate) across different specifications. Importantly, however, any residual variance comparisons must be made in the same units. One cannot compare the residual variance in logarithms to the residual variance in normal or untransformed units. Instead, residuals must be calculated in the natural units for study. In other words, if one compares a log–log and a linear model, the residuals must be put in the same units—usually those of the linear model. This involves taking the antilog of the predicted values in the log–log model, calculating the squared difference between actual and predicted, and comparing the resultant variance figure with that from the linear model. (In this case, can one just use the antilogs of the observed residuals in the log–log model as a starting point in calculating the residual variance? Why not?) One bad, but often used, variation of the residual variance test is the comparison of values of R^2 between say a log–log and a linear model. This is clearly inappropriate since the variances of the dependent variables in each equation and the residuals from each estimation are not commensurable. (Write out the formula for the variance for Y and $Y^* = \log Y$ and the residual from each regression to demonstrate this.)

4.7 Dummy Explanatory Variables

With model misspecification, the error term is correlated with the set of explanatory variables and biased parameter estimates are obtained. A related problem occurs when several potentially different populations have been combined into one sample. In other words, our observations come from groups that could display somewhat different behavioral patterns. For example, in the educational production function to be estimated in Chapter 5, both males and females are combined into one sample to estimate the achievement relationships. But, if girls learn faster than boys, or test better than boys at that age, girls will have higher achievements than boys, everything else being equal. If this is true, and the student's sex is in any way correlated in the sample with the other explanatory variables, such as females having more verbal teachers, combining girls and boys into the same sample would place us in precisely the same position as we were in with a misspecified model.

When faced with the possibility of behavioral differences among different subpopulations of the sample, there are two basic alternatives: stratify the sample into the different subsamples relating to each group or specify the behavioral differences within one overall model. Stratification, while an obvious way to deal with the problem, has some serious drawbacks. In order to obtain the most precise parameter estimates possible, we want to use the largest possible sample; stratification implies reducing the sample size. In fact in some cases where there are very few original observations, stratification may make estimation totally impractical.

Thus, in an attempt to avoid stratification, we shall outline methods of including behavioral differences directly into an overall model. This problem is subsumed into the general topic of qualitative explanatory variables, or dummy variables. Throughout the discussion we shall emphasize tests for partial behavioral differences among subpopulations. The development begins with the simplest way of including qualitative differences and progresses to a test of completely different behavior among subpopulations. In the last case—when there is completely different behavior—there are no gains from pooling the observations into one sample for estimation, and stratification is a necessity. (Chapter 5 discusses stratification in more detail and the problems posed by qualitative dependent variables are reserved for Chapter 7.)

In many cases where observations from several potentially different samples have been combined into one data set there is no way to "measure" the difference in the populations. When looking at a time series of a process with seasonal aspects, one is hard pressed to measure "winter," or the relative value of "fall" compared to "spring." Or in using a cross-sectional sample of counties to test hypotheses about voting behavior, it is not possible to attach a value to "Southerness" or "Northerness," even though you may believe it to be an important influence. Or when analyzing individual behavior, who can measure "femaleness" or "Catholicism"?

Dummy variables are a way to introduce this type of information into the quantitative statistical models we have been considering. A dummy variable is simply a binary variable that takes the value one if the observation comes from the population with the qualitative factor being considered, and is zero for all other observations. For example, in the model of educational achievement, the hypothesis that females are different in terms of third grade verbal achievement is incorporated by introducing a sex dummy variable. This variable is one for all girls in the sample and zero for all boys. In the case where we have more than two distinct groups, we need only add additional dummy variables to represent the population completely. The party identification variable could be restructured as four separate dummy variables: one each for strong Republicans, weak Republicans, weak Democrats, and strong Democrats.

There are several ways in which we can incorporate these population differences into statistical models. The appropriate way will depend upon the hypotheses about what effect coming from one group rather than another has on the model. The simplest hypothesis, and the one used in the education example, is that there is a difference only in the constant term for the two groups.

In two dimensions, we can show graphically the effect of the qualitative difference in constant terms for populations A and B. Let

$$Y_t = \beta_1 + \beta_2 X_{t2} + \beta_3 X_{t3} + U_t, \tag{4.10}$$

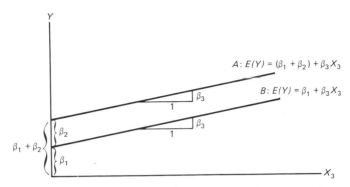

FIGURE 4.3   Bivariate relationship with intercept dummy variable.

where $X_{t2} = 1$ for observations in population A and $X_{t2} = 0$ for observations not in population A. Line A in Fig. 4.3 represents the systematic behavior for members of group A, e.g., female; and line B applies to those not in A, e.g., male.

The introduction of a dummy constant term variable into the model includes some assumptions about behavior. In particular, it assumes the other behavioral relationships in the model are the same across observations. Thus, the effect of teachers on boys and girls is the same, i.e., they have the same teacher coefficients. This is seen graphically in Fig. 4.3 by the fact that the two lines are parallel to each other—they have the same slope as given by β_3.

The use of dummy variables admits to a lack of knowledge and/or data. It says that we cannot explicitly model a behavioral difference. We do not know the underlying cause of the differences in the populations, or we cannot break out separate elements of this different behavior. For example, knowing that girls achieve more in school does not help policy makers very much unless they know why girls score better. We cannot transform all boys into girls; and, even if we could, we probably would not get the predicted gain in achievement. Most likely this difference arises from a set of attitudinal differences which, if they could be adequately measured, might yield some interesting policy implications. As it stands, the dummy variable representation does not supply such information. The moral is that, if possible, quantifying the underlying behavioral relationships is generally superior to the dummy variable formulation.

This absence of knowledge does not imply leaving out hypothesized qualitative aspects. For example, if Fig. 4.4 represents data generated by the model in Eq. (4.10) (dashed lines for true model), an estimated line constraining the groups to the same intercept would overestimate the true effect of X_3. Note that a sample with small values of X_3 for members of group A and large values of X_3 for those in group B would possibly lead to an estimate of β_3 that was negative or at least badly underestimated.

Until now, we have considered only a binary relationship, i.e., one condition that either holds or does not. However, this is not the only type of relationship possible. For example, in a seasonal relationship there are four

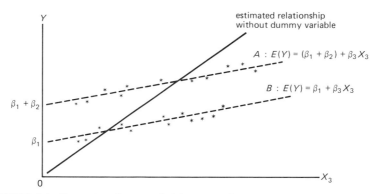

FIGURE 4.4   Estimation of misspecified bivariate relationship excluding dummy variable.

possible "states" or conditions. Multidimensional qualitative factors can be easily included in a dummy variable format. The approach is generalized by defining a series of dichotomous relationships. For example, in the seasonal case, define $X_{t2} = 1$ if t is spring and $X_{t2} = 0$ otherwise; define $X_{t3} = 1$ if t is summer and $X_{t3} = 0$ otherwise; and $X_{t4} = 1$ if t is fall and $X_{t4} = 0$ otherwise. It is important to note that the generalization relies on dichotomizing the attribute; it is not developed by letting spring equal one, summer equal two, and so forth. The latter technique would force a rigid relationship between the seasons. But, is fall really three times spring? The same analysis could be done for a set of regional variables, $X_{t2} = 1$ for the Northeast, $X_{t3} = 1$ for the South, $X_{t4} = 1$ for the Midwest, etc.

A technical constraint is present in the use of those dummy variables. It is not possible to estimate a model that contains both an intercept and a complete set of dummy variables. Thus, it is not possible to have a separate variable for each of the four seasons or all regions plus an intercept. Such an estimation violates the assumption that no exact linear relationship exists among the explanatory variables. This is easily seen with a strictly dichotomous variable like sex. Let X_2 equal one for females and zero otherwise; let X_3 equal one for males and zero otherwise. Since individuals are either males or females, $X_2 + X_3 = 1$, the constant term, implying a linear relationship among X_2, X_3, and the constant term. The same result holds for qualitative variables with more attributes. If we consider a model with four regional dummy variables and try to include all four variables plus a constant term in the equation $X_2 + X_3 + X_4 + X_5 = 1$, and we have the same problem. In practice, this means that one of the attribute variables should be omitted $(X_2, X_3, X_4,$ or $X_5)$ or the intercept should be eliminated (X_1). It does not matter which is done since the net intercept terms for each case will be identical in the two "different" estimated models.

We can extend this analysis to consider several qualitative factors within the same equation. It is important in this case to exclude one out of each set of the dummy variables representing each qualitative factor. This will prevent the existence of linear relationships among the X's and permit estimation of the coefficients in the model. In this case, the constant term in the regression is interpreted as the intercept for the observations falling into the excluded category for each factor. Each dummy variable then represents the addition to the constant term for the observations satisfying the attribute associated with each dummy variable.

A short schooling/earnings example using dummy variables may help clarify their interpretation. Let us suppose as before that a person's income depends upon his experience and level of schooling. Define Y_t to be annual income in dollars, and the following dummy variables: $E_1 = 1$ for people whose experience is between 10–30 years, $E_2 = 1$ for people whose experience is greater than 30 years, $S_1 = 1$ for people with only a high school degree, $S_2 = 1$ for people with both a high school and a college degree, but no advanced degree, and $S_3 = 1$ for people with an advanced degree. The equation

$$Y_t = \$4212 + \underset{(2.0)}{586\,E_1} + \underset{(2.6)}{817\,E_2} + \underset{(1.5)}{1950\,S_1} + \underset{(3.5)}{5325\,S_2} + \underset{(4.5)}{7405\,S_3} \qquad (4.11)$$

is an estimate of an earnings relationship for Boston black males with different income, experience, and educational characteristics. Calculated t-statistics for the null hypothesis that $\beta = 0$ are shown in parentheses below each coefficient.

This estimated equation gives the predicted incomes of individuals in each of the 12 possible experience and education categories. By selecting the right combination of coefficients, we can estimate the expected income of people in any of the classifications. For example, the expected income of a person with less than 10 years experience and with less than a high school education is $4212. This is just the intercept term because this individual defines the standard case and is measured by zeros for all of the dummy variables. The expected income of a person with less than 10 years experience but with a high school degree is $6162 (4212 + 1950). And a person with more than 30 years experience and with an advanced degree has an expected income of $12,434 (4212 + 817 + 7405).

The coefficients on the education variables measure the expected monetary returns for additional years of schooling. High school graduates are expected to earn $1950 more each year than people who do not finish, college graduates annually earn $3375 more than high school graduates on average, and an advanced degree is expected to be worth $2080 (above a college degree) annually. The relatively low t-statistic for the coefficient of S_1 indicates that we cannot reject the null hypothesis that high school graduates earn more than nongraduates with much confidence. However, since one presumably has fairly strong beliefs that high school graduates do earn more than dropouts, we would be making a mistake to delete S_1 from the model. It is also the case that so long as we believe that high school graduates earn more than nongraduates, this estimate of a $1950 difference is the best estimate of that difference, at least until one examines additional data.[8]

Finally, we might reconsider the specification of the model. In particular, including continuous education and experience variables, such as years of schooling or precise years of experience, would provide more information. This alternative specification—which is possible when we have an idea about how to quantify a variable—provides an estimate of the value of an additional year of schooling, a value we cannot estimate with this hypothetical model. The cost of changing the model specification is that we must make stronger assumptions about the model's functional form.

Slope Dummy Variables

As an alternative or additional hypothesis about the way our different populations behave, we may hypothesize that the influence of one or more of the independent variables is different for the members of each group, in other words, that one of the β_k (for k greater than 1) is different for each group. This hypothesis can be easily tested with the inclusion of what is called a slope dummy variable. This new variable is created by multiplying the dummy variable used in the previous discussion times the explanatory variables hypothesized to have different effects. The regression is then run using the original variables plus the new slope dummies. For example, if the hypothesized model is

$$Y = \beta_1 + \beta_2 X_2 + \beta_3 X_3 + U \tag{4.12}$$

and we have two groups in our sample used to estimate it, say males and females, we may hypothesize that the effect of X_3 on Y may be more for females than for males. This latter hypothesis may be tested by creating a new variable D that equals one if observation t is a female and zero if t is a male, and estimating the model

$$Y = \beta_1 + \beta_2 X_2 + \beta_3 X_3 + \beta_4 DX_3 + U = \beta_1 + \beta_2 X_2 + (\beta_3 + \beta_4 D)X_3 + U. \tag{4.13}$$

The implication of this model is that for males the estimated coefficient on X_3 is $b_3 + b_4 \cdot 0 = b_3$ since $D = 0$, while for females, the coefficient for X_3 is $b_3 + b_4$ since $D = 1$. We can use the standard error for the coefficient b_4 to test the null hypothesis that the effect of X_3 is not different between the two groups the same way we use the t-statistic to test the null hypothesis that the variable X has no influence on Y. If b_4 divided by its standard error exceeds the predetermined confidence level, then we can reject this null hypothesis that the effect of X_3 is the same in both groups. Figure 4.5 illustrates the effect of a slope dummy variable for a simple two-variable model,

$$Y = \beta_1 + \beta_2 X_2 + \beta_3 DX_2 + U.$$

---

[8]When several dummy variables are used to measure a single conceptual variable, such as schooling, the proper significance test is not whether an individual coefficient is zero (as performed by the t-test), but whether the coefficients in all dummy variables are zero. This test is discussed in Chapter 5.

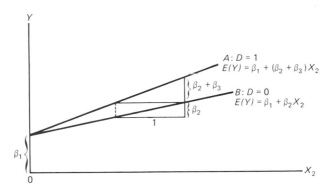

FIGURE 4.5   Bivariate relationship with slope dummy variable.

We can illustrate the use of a slope dummy variable with our voting equation from the previous chapter. The equation at the end of Chapter 2 implied that the influence of party affiliation on voting is the same for all people classed as voters and members of one of the two major parties, regardless of any other considerations. In the context of this model, it may very well be the case that the influence of party affiliation on voting is greater for people who do not have clear issue preferences during that election. Thus we are postulating two populations of voters: those who have a preferred party on the basis of issue positions (and we use one set of weights to combine evaluations and party affiliations into an expected vote decision) and another group of voters who have no preference between the parties on the issues (and we give a different weight to party affiliation in reaching a voting decision). Presumably, the importance of party affiliation among this second set of voters will be greater than among the first set.

The precise equation embodying this proposition is

$$V = \beta_1 + \beta_2 E + \beta_3 P + \beta_4 DP + U, \tag{4.14}$$

where D is a dummy variable defined as one for those people who are indifferent between the parties on the various issues, measured as those for whom $E = 0.50$. The explicit voting equation for nonindifferent voters is $\beta_1 + \beta_2 E + \beta_3 P + U$, and for indifferent voters is $(\beta_1 + 0.5\beta_2) + (\beta_3 + \beta_4)P + U$. Estimating this model gives

$$V = 0.074 + 0.39E + 0.61P + 0.12DP.$$

This implies that the party coefficient for indifferent voters is 0.12 higher (0.73 versus 0.61) than for nonindifferent voters.

The examples so far have assumed either an intercept dummy or a slope dummy. In some cases researchers may want or need to include both, depending upon the hypothesized differences in the populations measured by the dummy variable. This is quite simply accomplished by specifying a model that includes both D, the intercept dummy variable, and DX, the appropriate slope dummy. In terms of the model in Eq. (4.13), this is

$$Y = \beta_1 + \beta_5 D + \beta_2 X_2 + \beta_3 X_3 + \beta_4 DX_3 + U$$

so that the intercepts are $\beta_1 + \beta_5$ and β_1 for members of groups A and B, respectively.

There are other techniques and questions associated with the estimation of models over aggregations of different populations. They are extensions of this basic framework to models with more explanatory variables, where most of the additional explanatory variables come from the appropriate slope and intercept dummy variables to model the presence of the different populations. We are deferring discussion of these techniques until the end of the next chapter where we have developed the general multivariate estimation model and its associated statistical tests in more detail.

REVIEW QUESTIONS

1.   Consider the following model:

$$Y_t = 21.7 + 0.83 X_{t2} + 1.07 X_{t3}, \qquad R^2 = 0.34 \qquad \text{(standard errors in parentheses).}$$
$$\qquad\quad (0.38) \qquad (0.78)$$

$T = 25$

(a)   Test the following hypotheses: H_0: $\beta_2 = 0$, H_0: $\beta_3 = 0$.

(b)   The government proposes a policy that will change X_2 by 12; what effect on Y would be predicted?

(c)   Is your answer to (b) affected by the level of the variable X_3?

2.   Two different models are estimated:

$$\text{I.} \quad Y_t = 20.7 + 14.2 X_{t2} + 0.086 X_{t3}, \qquad R^2 = 0.63, \quad \hat{\sigma} = 38.3;$$
$$\qquad\qquad (9.30) \qquad\quad (.098)$$

$$\text{II.} \quad Y_t = 18.6 + 17.2 X_{t2}, \qquad R^2 = 0.58, \quad \hat{\sigma} = 37.9.$$
$$\qquad\qquad (6.12)$$

Which model would you choose? Why?

3.   Can the following models be estimated with OLS? If so, how would you estimate them?

(a) $Y_t = \beta_1 X_{t2}^{\beta_2} X_{t3}^{\beta_3} + U_t$.

(b) $Y_t = \beta_1 e^{-X_{t2}\beta_2 - X_{t3}\beta_3} + U_t$.

(c) $Y_t = (1 + e^{-\beta_1 X_{t2} - \beta_2 X_{t3}})^{-1}$.

(d) $Y_t = \beta_1 + \beta_2 X_{t2} + \beta_3 X_{t3} + \beta_4 X_{t2} X_{t3} + U_t$.

(e) $Y_t = \beta_2 X_{t2} + \beta_3 X_{t3} + U_t$.

4. For each of the models in Problem 3, determine the effect on Y_t of a unit change in X_{t2}, holding X_{t3} constant (i.e., find $\partial Y / \partial X_2$).

5 | Multivariate Estimation in Matrix Form

5.1 Introduction

Development of the ordinary least squares estimation model in Chapter 2 concentrated on the trivariate case, with one dependent or endogenous variable and two explanatory variables. This limitation was adopted strictly for expositional reasons, although several mentions were made that the model could be generalized and extended to a larger set of explanatory variables. This larger model and the development of its associated statistics could be presented by expanding the equations of Chapter 2 to include more terms. This approach leads to many more simultaneous linear equations of the form shown in Eqs. (2.11)–(2.13), which become difficult to solve with the algebraic methods used in Chapter 2. It is much easier to use the notation and operations of matrix algebra to accomplish this development. Matrix algebra has been developed, among other things, to represent and manipulate large-scale linear equation systems with which we are dealing. Appendix II presents a summary and review of the matrix algebra notations and operations relevant to our discussions in this and subsequent chapters. You may find it useful to review that appendix at this point and to refer to it during the subsequent discussions.

In addition to developing this expanded version of the multivariate estimation model, we shall examine some additional statistical tests concerning the presence of more than one population in a data set. These tests and their application are different from those in the previous chapter because they require equations with considerably more than the three or four variables we have dealt with so far. The discussion of these models and the derivation of the appropriate statistics are greatly facilitated by the matrix notation used hereafter.

5.2   The Least Squares Estimators

The development of the estimators for the coefficient values in the expanded model is done in two parts. First, the estimators are derived using the standard scalar notation of Chapter 2. Then the equivalent estimators are developed in matrix terms. This approach should provide a bridge between Chapter 2 and the subsequent discussions.

As we enter into the multivariate realm, notation becomes more important. Here we shall introduce the notation to be used for the remainder of this book. The true population model is written as

$$Y_t = \beta_1 + \beta_2 X_{t2} + \beta_3 X_{t3} + \cdots + \beta_K X_{tK} + U_t \qquad \text{for} \quad t = 1, \ldots, T, \quad (5.1)$$

or compactly,

$$Y_t = \beta_1 + \underbrace{\sum_{k=2}^{K} \beta_k X_{tk}}_{\text{systematic}} + \underbrace{U_t}_{\text{stochastic}},$$

where Y_t is the dependent, or endogeneous, variable for observation t; X_{tk} the tth observation on the kth independent or exogenous variable; β_1 is a constant, or intercept; β_k the coefficient for the kth variable (a constant slope for all T observations of the given variable X_k), and U_t is the stochastic element or random error for observation t.

As in the model in Chapter 2, for each observation, the value of the dependent variable, Y_t is determined by adding a systematic component ($\beta_1 + \sum_{k=2}^{K} \beta_k X_{tk}$) and a random component (U_t). This model depicts the true or population behavorial relationships that generate the outcome represented by Y_t. The systematic part contains all the hypotheses about variables that influence the behavior being explained. The random part (U_t) again represents several factors in addition to purely random behavior. These factors are the small measurement errors that occur randomly in observing Y_t, the relatively insignificant variables excluded from the model which may influence each Y_t in a very small and nonsystematic fashion, and any misspecifications of the proper form of the function relating Y_t to the values of X.

Several aspects of the basic population model are worth emphasizing before proceeding. The coefficients of the system (the β_k) are constant for the whole population; they do not change for different cases or observations. Moreover, they represent the independent effect of each exogenous or independent variable on the endogenous or dependent variable when all the other variables are held constant. Thus, β_2 is the amount of change in Y that corresponds to a unit change in X_2 with "other things equal," i.e., with X_3, $X_4, \ldots, X_K$ held constant.[1] Take the estimated model of presidential voting in

---

[1]Note that the subscript for observation t is dropped. This will be done when the context causes no confusion.

Eq. (2.20). The coefficient of 0.36 on the issue preference variable implies we can expect the person who favors the Democrats to have a 0.36 higher probability of voting for Johnson, regardless of party affiliation.

Our interests are mainly in understanding the systematic portion of the population model. Thus, from our observations of past behavior we will attempt to deduce the values of the β_k. Again, as in Chapter 2, it would be very simple to do this if it were not for the stochastic element U_t. If U_t were always zero, we would only have to observe K different (independent) cases of the behavior in question. Then the K equations (corresponding to observations) could be solved simultaneously for the K unknowns β_k. However, the introduction of a stochastic term U_t precludes this. The answer to the problem again lies in inferential statistics. By collecting many observations of past behavior, we can estimate the values of the parameters. It is then possible to derive the same desirable estimators of β_k as we did in Chapter 2.

In the multivariate case, the estimates of the β_k are found in a manner similar to the procedure used in the bivariate case. In the multivariate case we have observed a sample of values for Y and X_2 through X_K for a total of T observations[2] $(T > K)$. The criterion again calls for minimizing the sum of squared residuals or squared differences between actual values of Y and estimated values of Y. We denote the estimated value of Y_t by $\hat{Y}_t$, and the residual as $e_t = Y_t - \hat{Y}_t = Y_t - b_1 - \Sigma b_k X_{tk}$. The problem is to minimize the sum of squared residuals (SSE) where

$$SSE = \sum_{t=1}^{T} e_t^2 = \sum_{t=1}^{T} \left(Y_t - \hat{Y}_t \right)^2 = \sum_{t=1}^{T} \left(Y_t - b_1 - \sum_{k=2}^{K} b_k X_{tk} \right)^2. \quad (5.2)$$

At this point there is some confusion about terminology, and we shall digress. In the behavioral model of Eq. (5.1), Y and the X's are the variables. These variables are constructed to represent behavior and to measure the difference in our observations. However, in the problem of finding estimators for the β_k from a sample of Y's and X's, these observations are no longer variable. We mentioned this in Chapter 2, but it is worth emphasizing again that after collecting the data we cannot vary Y and X_k to minimize SSE. In minimizing SSE, the coefficient estimates (b_k) become the "variables" that must be adjusted to find the minimum; and the values for Y and X_k, and their covariances, are treated as constants. There are an infinite number of b_k that could be used as estimates of the β_k (i.e., any set of K real numbers is a possible estimate). We shall use calculus to obtain the best possible estimates of the b_k, according to the least squares criterion. The problem then is to minimize SSE with respect to the estimates of b_k, which are the variables for the minimization.

---

[2]The reasons for requiring more observations than parameters to be estimated will become apparent soon. However, consider the nonstochastic case ($U_t = 0$): there only $T = K$ is necessary in order to find a unique solution to the set of linear equations used to find the β_k.

The minimum value of SSE is found by setting the first partial derivative of SSE with respect to each b_k equal to zero, written as $\partial SSE/\partial b_k = 0$. This operation yields K linear equations, which can (generally) be solved uniquely for the K unknowns $b_1, b_2, \ldots, b_k$.

In order to find $\partial SSE/\partial b_k$, we use Eq. (5.2) which expresses SSE in terms of the b_k. From calculus, the partial derivative of the sum of squared errors (SSE) with respect to any coefficient estimate b_k is found as follows:

$$\frac{\partial SSE}{\partial b_k} = \frac{\partial \left(\sum e_t^2 \right)}{\partial b_k} = \sum_{t=1}^{T} \left(\frac{\partial e_t^2}{\partial b_k} \right)$$

and

$$\frac{\partial \left(e_t^2 \right)}{\partial b_k} = \frac{\partial \left(e_t^2 \right)}{\partial e_t} \frac{de_t}{db_k} = 2e_t \frac{de_t}{db_k},$$

according to the rules of calculus. Then from the definition of e_t, for any observation,

$$2e_t \frac{de_t}{db_k} = 2e_t(-X_{tk}) = -2\left(Y_t - \hat{Y}_t \right)(X_{tk})$$

$$= -2\left(Y_t - b_1 - \sum_{j=2}^{K} b_j X_{tj} \right)(X_{tk}),$$

$$\frac{\partial SSE}{\partial b_k} = \sum_{t=1}^{T} \frac{\partial e_t^2}{\partial b_k} = -2 \sum_{t=1}^{T} \left[\left(Y_t - b_1 - \sum_{j=2}^{K} b_j X_{tj} \right)(X_{tk}) \right]$$

$$= -2\left(\sum_t Y_t X_{tk} - b_1 \sum_t X_{tk} - \sum_{j=2}^{K} b_j \sum_t X_{tj} X_{tk} \right).$$

This expression holds for[3] $k = 2, \ldots, K$. For $k = 1$, the constant term

$$\frac{\partial e_t^2}{\partial b_1} = 2e_t \frac{de_t}{db_1} = 2e_t(-1) = -2\left(Y_t - b_1 - \sum_{j=2}^{K} b_j X_{tj} \right)$$

so that

$$\frac{\partial SSE}{\partial b_1} = \frac{\partial \sum e_t^2}{\partial b_1} = -2 \sum_t \left(Y_t - b_1 - \sum_{j=2}^{K} b_j X_{tj} \right).$$

The set of estimates of the b_k minimizing the sum of squared errors SSE is found by solving the following K simultaneous equations for the K unknowns

[3]The development so far simply extends the discussion in Chapter 2 to an arbitrary number of explanatory variables.

b_1 through b_K:

$$\frac{\partial SSE}{\partial b_1} = 0 = -2\left(\sum_{t=1}^{T} Y_t - \sum_{t=1}^{T} b_1 - b_2 \sum_{t=1}^{T} X_{t2} - \cdots - b_k \sum_{t=1}^{T} X_{tk} - \cdots \right.$$

$$\left. - b_K \sum_{t=1}^{T} X_{tK} \right)$$

$$\frac{\partial SSE}{\partial b_2} = 0 = -2\left(\sum Y_t X_{t2} - b_1 \sum X_{t2} - b_2 \sum X_{t2}^2 - \cdots - b_k \sum X_{tk} X_{t2} - \cdots \right.$$

$$\vdots \qquad \left. - b_K \sum X_{tK} X_{t2} \right)$$

$$\frac{\partial SSE}{\partial b_k} = 0 = -2\left(\sum Y_t X_{tk} - b_1 \sum X_{tk} - b_2 \sum X_{t2} X_{tk} - \cdots - b_k \sum X_{tk}^2 - \cdots \right.$$

$$\vdots \qquad \left. - b_K \sum X_{tK} X_{tk} \right)$$

$$\frac{\partial SSE}{\partial b_K} = 0 = -2\left(\sum Y_t X_{tK} - b_1 \sum X_{tK} - b_2 \sum X_{t2} X_{tK} - \cdots - b_k \sum X_{tk} X_{tK} - \cdots \right.$$

$$\left. - b_K \sum X_{tK}^2 \right)$$

Rearranging these equations, we have the following K nonhomogeneous linear equations, called the normal equations:

$$\sum Y_t = Tb_1 + b_2 \sum X_{t2} + \cdots + b_k \sum X_{tk} + \cdots + b_K \sum X_{tK} \qquad \text{for } b_1, \quad (5.3)$$

$$\sum_{t=1}^{T} Y_t X_{tk} = b_1 \sum X_{tk} + b_2 \sum X_{t2} X_{tk} + \cdots + b_k \sum X_{tk}^2 + \cdots + b_K \sum X_{tK} X_{tk}$$

$$\text{for } b_2, \ldots, b_K. \quad (5.4)$$

Expressions (5.3) and (5.4) are completely analogous to Eqs. (2.11)–(2.13), only expanded to include more X's and b's. The solution to these K equations is best left to the computer. Anybody who has ever solved systems of three or four equations and three or four unknowns should appreciate the difficulty as K becomes large.

5.3 Least Squares in Matrix Notation

The solution of these equations and the properties of the final estimates are, however, better seen by putting the problem in terms of matrix algebra.

In addition, computers, which are particularly good at doing a sequence of simple repetitive operations, can be organized to deal with arrays or matrices. Our discussion then will be compatible with existing regression programs.

The model expressed in Eq. (5.1) can be summarized in matrix form as

$$Y = X\beta + U \tag{5.5}$$

where Y is a $T \times 1$ column vector of observations on the dependent variable, X is a $T \times K$ matrix of observations on the independent or exogenous variables, β is a $K \times 1$ column vector of the true coefficients, and U a $T \times 1$ vector of the true error terms:

$$
Y = \begin{bmatrix} Y_1 \\ Y_2 \\ \vdots \\ Y_t \\ \vdots \\ Y_T \end{bmatrix}, \quad
X = \begin{bmatrix}
1 & X_{1,2} & \cdots & X_{1,k} & \cdots & X_{1,K} \\
1 & X_{2,2} & \cdots & X_{2,k} & \cdots & X_{2,K} \\
\vdots & \vdots & & \vdots & & \vdots \\
1 & X_{t,2} & \cdots & X_{t,k} & \cdots & X_{t,K} \\
\vdots & \vdots & & \vdots & & \vdots \\
1 & X_{T,2} & \cdots & X_{T,k} & \cdots & X_{T,K}
\end{bmatrix},
$$

$$
\beta = \begin{bmatrix} \beta_1 \\ \beta_2 \\ \vdots \\ \beta_k \\ \vdots \\ \beta_K \end{bmatrix}, \quad
U = \begin{bmatrix} U_1 \\ U_2 \\ \vdots \\ U_t \\ \vdots \\ U_T \end{bmatrix}.
$$

A row of ones has been inserted in the explanatory variable matrix to permit estimation of the constant term.

The equation

$$Y = Xb + e \tag{5.6}$$

shows our estimation relationship for the T observations in matrix form, where b is the vector of our estimates of the true coefficients and e is the vector of observed residuals or errors in our estimation. These replace β and U, respectively, in Eq. (5.5).

Our objective here is to obtain a set of estimates, the b_k, for the true coefficients, the β_k. In addition we need to obtain information about the distributions from which these b_k are drawn so that we may make some inferences about the true coefficients, the β_k.

The criterion used in deriving these estimates of the coefficients is again the least squares principle. We have shown in Chapter 3 that this criterion yields estimates of the coefficients with very desirable properties, e.g., unbiasedness, least variance, etc. We shall do the same for the general multivariate case.

The sum of squared errors in vector terms is

$$\sum_{t=1}^{T} e_t^2 = e'e = (Y-Xb)'(Y-Xb) = Y'Y - 2b'X'Y + b'X'Xb \qquad (5.7)$$

since $Y'Xb$ is a scalar and equals $b'X'Y$. We need to choose a set of b_k minimizing this sum. We can accomplish this by setting the derivative of $e'e$ with respect to each b_k equal to zero.

This system of equations for all the b_k was shown in Eqs. (5.3) and (5.4) where $X_{t1} = 1$ for all t. This system of equations can be expressed compactly in matrix form as

$$\frac{\partial e'e}{\partial b} = -2X'Y + 2X'Xb = 0 \qquad \text{or} \qquad X'Xb = X'Y. \qquad (5.8)$$

It is easy to complete the multiplications in Eq. (5.8) to see that it gives the K separate equations shown in Eqs. (5.3) and (5.4). These are generally referred to as the first order conditions.

At this point we impose the first of five important conditions on the data, namely

A.1
$$\boxed{\text{rank of } X = K.}$$

When this condition holds, $(X'X)^{-1}$ will exist (be nonsingular)[4] and Eq. (5.8) can be rewritten as

$$b = (X'X)^{-1}X'Y, \qquad (5.9)$$

where b is then a $K \times 1$ column vector of estimated values for β. Thus, (5.9) represents a set of equations whose solution vector b yields the minimum sum of squared errors.

In arriving at the estimated coefficients that minimize $\sum_{t=1}^{T} e_t^2$, the only assumption needed is that the rank of X is K. More importantly, this condition implies that there are more observations than parameters to be estimated ($T > K$). This also implies that no independent or exogenous variable is a linear combination of the other exogenous variables. This condition on X, that rank $X = K$, has considerable intuitive meaning. If we imagine that regression analysis is associating movements in the various exogenous variables with movements in the endogenous variable, it is immediately obvious that one cannot distinguish between the effects of two variables that are linearly associated with each other. Chapter 4 showed that inclusion of a complete set of dummy variables in the equation resulted in an

---

[4]From Appendix II, an inverse of a matrix exists if and only if it has full rank, i.e., a $K \times K$ matrix must have rank K. If X is $T \times K$ with $T > K$ and rank K, $X'X$ has rank K. This assumption implies that no variable in X_k is a linear combination of the other X's. This is analogous to the restriction in Chapter 2 that $|r_{23}| < 1$.

exact linear relationship among the explanatory variables, preventing estimation of the model. This is a case where the rank of X is less than K. When the rank of X is less than K, $(X'X)^{-1}$ does not exist and it is impossible to estimate b according to (5.9).

5.4   Properties of Least Squares

Chapter 3 showed that our emphasis on the least squares estimators arises because of their known statistical properties—they have the smallest variance of any unbiased, linear estimator (BLUE)—as long as the error terms satisfy certain conditions. We now want to develop these properties for the general multivariate estimator. The discussion will parallel that of Chapter 3.

We can establish the mean and variance for the estimated coefficients by substituting $Y = X\beta + U$ from Eq. (5.5), for Y in Eq. (5.9). This gives

$$b = (X'X)^{-1}X'(X\beta + U) = \beta + (X'X)^{-1}X'U \qquad (5.10)$$

since $(X'X)^{-1}X'X = I$. This indicates that the estimated coefficients are a function of the true coefficients (β) and the true error term (U). To establish the distribution of these estimated coefficients, the simplest development assumes that

A.2 | The X's are fixed for repeated trials. |

In other words, we are saying that implicitly we could repeat the "experiment" used to generate our observed values of Y with the same values of the explanatory variables. Repeating the experiment or fixing the X's thus implies gathering another sample with a different set of U's and therefore different observed Y's. The fact that the estimated coefficients are themselves random variables can then be seen from the fact that in drawing another sample, the X's remain fixed but the U's are different. This clearly produces a different set of b's. The distribution of these estimates will depend upon the characteristics of X and the distributions of the true error terms as we shall show next. (The simulations in Chapter 3 illustrate the distributional nature of our coefficients and how these distributions vary with different sample properties.)

Unbiasedness

The first parameter describing the distribution of coefficients will be their means. Taking the expected value of both sides of Eq. (5.10) gives

$$E(b) = E(\beta) + E\big[(X'X)^{-1}X'U\big].$$

Since β and X are assumed to be fixed, we have

$$E(b) = \beta + (X'X)^{-1}X'E(U). \qquad (5.11)$$

We now impose the important condition that

A.3

$$E(U)=0.$$

This condition allows us to rewrite Eq. (5.11) as $E(b)=\beta$, showing that our estimated coefficients are unbiased. The statement $E(U)=0$ means $E(U_1)= E(U_2)= \cdots = E(U_t)= \cdots = E(U_T)=0$. In other words, the error terms associated with each observation over all replications of the experiment have a zero mean, and this holds for all observations. (See Chapter 3 for a more detailed discussion of this point.) This condition means that the values of U_t are distributed independently of the sample values of each X, $E(U_t|X_{tk})= E(U_t)=0$.

The constraint that the mean of the unknown errors is zero is actually stronger than needed. $E(U)=0$ is needed only to show that the intercept b_1 is unbiased. All of the other estimates, b_2 through b_K, will be unbiased as long as the expected value of each error is a constant, i.e., $E(U)=C$. This is shown for the three-variable case in Chapter 3, and its extension applies here. Regressing $\tilde{Y}=(Y-\bar{Y})$ and $\tilde{X}=(X-\bar{X})$ according to (5.9) and taking the expected value of the estimators yields

$$E(b^*)=\beta^* + (\tilde{X}'\tilde{X})^{-1}\tilde{X}'E(U-\bar{U}), \qquad (5.12)$$

where b^* and β^* are identical to b and β except that the intercepts b_1 and β_1 are not included.[5] As long as the expected value of each error term U_t equals the mean of all error terms, $E(U_1)= E(U_2)= \cdots = E(U_t)= \cdots = E(U_t)= \bar{U}$, then b^* will be unbiased.[6]

Arithmetic Properties

As with the previous model, we unfortunately cannot use the residuals from our estimated equation to test assumption A.3. In particular, $\Sigma e_t = \bar{e}=0$ or the mean observed residual is zero, regardless of the actual distribution of the observed error U. It is also true that the correlation between e and each X is zero. These arithmetic properties are easy to see from Eq. (5.8):

$$0=X'Y-X'Xb=X'(Y-Xb)=X'(Y-\hat{Y})=X'e.$$

The first row of X' is a vector of 1's. Premultiplying the vector e by such a vector gives $\Sigma e_t=0=\bar{e}$. The remaining rows of X' are the T observations on

---

[5]The fact that β^* is identical to the 2 through K elements in b is sometimes used to reduce the dimension of the problem. By using deviations from the mean, $(XX')^{-1}$ becomes of order $K-1$ instead of K. The intercept can then be found from $b_1= \bar{Y}-b_2\bar{X}_2-\cdots -b_K\bar{X}_K$. This reduction in the dimension of the problem was more significant in the days of hand computation than it is now.

[6]Getting this formulation made use of the fact that $\tilde{Y}=\tilde{X}\beta^*+(U-\bar{U})$. The student should be able to verify that this is true.

each explanatory variable. Thus each multiplication gives $\Sigma X_{tk} e_t$ or $\text{cov}(X_k e)$ $= 0$.

These arithmetic properties are important because we assume that $E(U) = 0$ or at least that the expected value of U_t is the same for all observations, meaning that X and U are independent, in order to demonstrate that b is unbiased. But, by the above demonstration, these assumptions are *untestable* from our observations of e.

Variance of *b*

The preceding discussion showed that the least squares estimators are unbiased under a set of simple assumptions. In other words, if we can draw repeated samples of Y (and therefore U) for the same X's, the mean estimate of b would tend to equal the true and unknown population values β. However, it is not altogether encouraging to know that the estimates are good if we replicate the estimates many times. Most social science events cannot be repeatedly sampled. It is very difficult in practice to restart the economy at 1945 levels and conditions or completely replicate a presidential election in order to gather a new sample of error terms for the same values of X. Given this, we are very interested in the dispersion or variance of the estimator. Further, we are interested in how the variance of the least squares estimator compares with the variance of other possible estimators. We will be able to show that least squares estimators are best (minimum variance) among all linear unbiased estimators *if* a further assumption about the distribution of U can be made. This latter demonstration is often called the Gauss–Markov theorem.

The variance of the estimator for each parameter can be written in a general way as $E[(b - \beta)(b - \beta)']$, where

$$E[(b-\beta)(b-\beta)']$$

$$= \begin{bmatrix} E(b_1-\beta_1)^2 & E[(b_1-\beta_1)(b_2-\beta_2)] & \cdots & E[(b_1-\beta_1)(b_K-\beta_K)] \\ E[(b_2-\beta_2)(b_1-\beta_1)] & E(b_2-\beta_2)^2 & \cdots & E[(b_2-\beta_2)(b_K-\beta_K)] \\ \vdots & & & \vdots \\ E[(b_K-\beta_K)(b_1-\beta_1)] & & \cdots & E(b_K-\beta_K)^2 \end{bmatrix}.$$

$$(5.13)$$

The variance of each parameter estimator is found on the main diagonal, i.e., the variance of b_k is the $(k\text{th}, k\text{th})$ element of the variance–covariance matrix of the coefficients. The off-diagonal terms represent the covariance between any two coefficients.

Rearranging (5.10) slightly, we have

$$b - \beta = (X'X)^{-1}X'U. \tag{5.14}$$

Then

$$(b - \beta)(b - \beta)' = (X'X)^{-1}X'UU'X(X'X)^{-1}. \qquad (5.15)$$

However, before taking the expected value of this expression, we make the same two assumptions about the distribution of U as we did in Chapter 3. First, we assume that each U_t is drawn from a distribution with the same σ^2. This requirement is commonly called homoskedasticity. Secondly, we assume that each U_t is independent of the other U_{t+s}, $s \neq 0$, or *no serial correlation*. If we examine

$$E(UU') = \begin{bmatrix} E(U_1^2) & E(U_1U_2) & \cdots & E(U_1U_T) \\ E(U_2U_1) & E(U_2^2) & \cdots & E(U_2U_T) \\ \vdots & \vdots & & \vdots \\ E(U_TU_1) & E(U_TU_2) & \cdots & E(U_T^2) \end{bmatrix},$$

we see that the diagonal terms are the variance of the error term in each observation and the off-diagonal terms are the covariances between the stochastic terms in any two observations [using the previous assumption that $E(U_t) = 0$]. We are imposing the conditions that

$$E(UU') = \begin{bmatrix} \sigma^2 & 0 & \cdots & 0 \\ 0 & \sigma^2 & \cdots & 0 \\ \vdots & \vdots & \vdots & \vdots \\ 0 & 0 & \cdots & \sigma^2 \end{bmatrix}.$$

Thus our fourth condition can be summarized by the matrix equation

A.4 $\boxed{E(UU') = \sigma^2 I_T}$, where I_T is an identity matrix of order T.

Taking the expected value of (5.15),

$$E[(b - \beta)(b - \beta)'] = (X'X)^{-1}X'E(UU')X(X'X)^{-1} = \sigma^2(X'X)^{-1}. \quad (5.16)$$

This follows by sequential application of the facts that the X's are fixed, assumption A.4 that $E(UU') = \sigma^2 I$, that $XI = X$, and that σ^2 is a scalar. This expression means that the variance of the kth coefficient equals the kth diagonal element of $(X'X)^{-1}$ times the variance of the true error terms. The covariance between any two coefficients will be σ^2 times the appropriate off-diagonal element of $(X'X)^{-1}$. [As an exercise, compute $X'X$ and $(X'X)^{-1}$ for the three variable model of Chapter 2. You should then be able to see the correspondence between the elements of $(X'X)^{-1}$ and the formulas for the variance of b_2 and b_3 in that model.]

Best

We have shown that the least squares procedure yields unbiased predictors of the coefficients β, that these are linear in terms of Y, and that they have a variance–covariance matrix equal to $\sigma^2(X'X)^{-1}$. We can now show that these estimated coefficients have the least variance among all possible linear unbiased estimators. This proof is shown in Appendix 5.1 and will only be sketched here to illustrate the point. We can define an arbitrary linear estimator of β as $b^{\#} = CY$, where C is a $K \times T$ nonstochastic matrix. We can further define C to be the least squares weights plus an additional matrix D, $C = (X'X)^{-1}X' + D$, where D is a $K \times T$ matrix, without any loss of generality. In the formal demonstration, we then show that under the conditions that $b^{\#}$ be unbiased, its variance–covariance matrix is

$$E(b^{\#} - \beta)(b^{\#} - \beta)' = \sigma^2\left[(X'X)^{-1} + DD'\right]. \tag{5.17}$$

The first term in this expression $\sigma^2(X'X)^{-1}$ is the matrix of variances and covariances of the least squares estimator. Since we are concerned with the variances of the estimators, we are interested in the size of the main diagonal elements. The diagonal elements of DD' must all be positive since each is a sum of T squared numbers. Thus, the variance of $b^{\#}$ is greater than var(b), unless $D \equiv 0$. When $D \equiv 0$, we are left with the least squares estimators. Consequently, the least squares estimators will have the least variance of any unbiased $[E(b^{\#}) = \beta]$, linear $[b^{\#} = CY]$ estimator. In effect, this is saying that the variance of the least squares estimate of β is less than that of any other estimator that is a linear function of the Y vector and that also gives unbiased estimates of the true parameters.

The above proof shows that the variance of each parameter estimate will have a smaller variance than any other linear, unbiased estimator. In fact, a stronger result has also been established. This stronger result, known as the Gauss–Markov theorem, is that any linear combination of the individual parameter estimates obtained by OLS will have a smaller variance than the same linear combination of alternative parameter estimates. This is an important result if one considers predictions of Y based upon the parameter estimates. This theorem says that $\hat{Y} = Xb$ will have a smaller variance than the prediction based upon alternative linear, unbiased estimators of β.

5.5   Distributional Aspects of the Error Term

An important part of the estimation procedure is finding out something about the distribution of U. If our assumptions hold, our real interest centers upon σ^2, the variance of U, because the variance of the parameter estimates was shown to be a function of σ^2. Also, we are often interested in knowing how large the stochastic element is relative to the systematic component.

Since U is not observed, σ^2 will not be observed. However, as in Chapter 3, σ^2 can be estimated from the observed residuals e_t. An unbiased estimator of σ^2 is

$$s^2 = \sum e_t^2 / (T - K). \tag{5.18}$$

(A proof of this unbiasedness is found in Appendix 5.2.) This estimator is simply the sample residual variance with an adjustment for the degrees of freedom used in estimating the K parameters in the model. Before proceeding to the uses of the estimated variance of U, it should be emphasized that s^2 is itself an estimator and a random variable that is dependent upon U. The estimator s^2 can also be seen to be unbiased in the simulations in Chapter 3. The estimated value for s^2 is substituted for σ^2 in Eq. (5.16) to estimate the variance–covariance matrix for b.

As we discussed in Chapter 3, s^2 is not so useful a measure for summary purposes as some which have been developed. A useful transformation is simply the square root of s^2. This measure s, or the standard error of the estimate, has the same units as Y (where s^2 has the units of Y^2), which allows some interpretation of the size of the error term. If we are further willing to make some assumptions about the actual distribution from which U_t is drawn, we can use s to estimate the probability of getting an error term greater than a certain size.[7]

The second widely used statistic is the coefficient of determination or R^2, defined in Chapter 3 as

$$R^2 = 1 - \frac{\sum e_t^2}{\sum (Y_t - \overline{Y})^2} = \frac{\sum (\hat{Y}_t - \overline{Y})^2}{\sum (Y_t - \overline{Y})^2}.$$

This is the most common summary measure used. If all observations fall precisely on the plane estimated, then $\sum e_t^2 = 0$ and $R^2 = 1$. On the other hand, if Y is unrelated to all of the X_K, the least squares estimates would be $b_1 = \overline{Y}$ and $b_2 = \cdots = b_K = 0$. This case of no relationship yields $e_t = Y_t - \overline{Y}$ or $\sum e_t^2 = \sum (Y_t - \overline{Y})^2$ and $R^2 = 0$.[8] Thus, R^2 ranges between zero and one.

The coefficient of determination, R^2 can be interpreted in a number of ways. First, if we consider a naïve estimate of any behavioral variable to be its mean, which is our best guess at the population mean, then R^2 gives a comparison of how well our "sophisticated" estimate relying on the values of a set of different variables X_k does in comparison to the mean. In a crude sense, divergence from an R^2 of zero gives the gains in explaining the total

---

[7]For normal variates, 68% will fall within ± 1 standard deviation from the mean; 95% within ± 2 standard deviations; and 99% within ± 3 standard deviations.

[8]This section assumes that the estimated model includes a constant term. In such a case, $\overline{e} = 0$, and R^2 cannot be negative. Without an intercept, $\overline{e}$ is not necessarily 0, and R^2, computed as above, can be negative.

behavior of Y attributable to our model. Alternatively, if we look at $e'e$, we see from (5.7) that with $e = Y - \hat{Y} = Y - Xb$,

$$e'e = Y'Y + \hat{Y}'\hat{Y} - 2\hat{Y}'Y = Y'Y + \hat{Y}'\hat{Y} - 2b'X'(Xb + e)$$

$$= Y'Y + \hat{Y}'\hat{Y} - 2b'X'Xb - 2b'X'e = Y'Y - \hat{Y}'\hat{Y}, \tag{5.19}$$

since $\hat{Y}'\hat{Y} = b'X'Xb$ and $X'e = 0$ by the arithmetic properties of the residuals. If we subtract $(\Sigma Y_t)^2 / T$ (or $T\bar{Y}^2$) from both sides and rely upon the fact that $\bar{Y} = \bar{\hat{Y}}$ because $\bar{e} = 0$, we see that

$$\text{var}(Y) = \text{var}(\hat{Y}) + \text{var}(e).$$

In other words, the variance of the dependent variable can be decomposed into an explained variance [var($\hat{Y}$)] and an unexplained variance [var(e)], just as in the trivariate case. Further,

$$R^2 = 1 - \frac{\text{var}(e)}{\text{var}(Y)} = \frac{\text{var}(\hat{Y})}{\text{var}(Y)}$$

Thus, R^2 can again be interpreted as the percentage of the variance of Y explained by the regression. These are the same results and interpretations we attained in Chapter 3. These two summary statistics—the standard error of estimate and R^2—are usually displayed with an estimated equation.

5.6   Statistical Inference

The task of model building usually does not end with estimation of a set of parameters. Since they are estimates that are a function of the sample U's, the true population values are not known. The next step is to make some inferences about the true population values.

Throughout this section we shall use our final assumption, namely that U_t is drawn from a normal distribution. While this assumption was not needed to obtain an estimator which is BLUE, it is needed to make common statistical inferences. If U is $N(0, \sigma^2 I)$, then b, a linear function of U, is distributed as $N[\beta, \sigma^2 (X'X)^{-1}]$. (The student should be able to demonstrate this.) This implies that

$$Z_k = (b_k - \beta_k) / \sigma \sqrt{m_{kk}} \tag{5.20}$$

is $N(0, 1)$, where m_{kk} is the (kth, kth) element of $(X'X)^{-1}$. Since Z is $N(0, 1)$, it would be possible to find the probability of the estimated value being more than any given distance from the true value β_k. However, (5.20) depends upon σ, which is unknown.

We can use s, our estimate for σ, to estimate the coefficient variance–covariance matrix. But any inferences here will not be as precise as in the case of using (5.20) because both b_k and s are random variables. In order to allow for this, we shall not use probabilities from the normal distribution. Instead, we will rely upon the t-distribution. The definition of the t-distribution is that, if Z is a standard normal variable, $Z \sim N(0,1)$, and if W^2 is an independently distributed χ^2 with $T-K$ degrees of freedom, then $Z/\sqrt{W^2/(T-K)}$ is distributed according to the t-distribution with $T-K$ degrees of freedom. We have demonstrated that b_k is normally distributed about β_k with a variance equal to

$$\sigma^2(X'X)_{kk}^{-1} \quad \text{or} \quad \sigma^2 m_{kk}.$$

Thus

$$(b_k - \beta_k)/\sigma\sqrt{m_{kk}} \quad \text{is} \quad N(0,1).$$

It can be shown (Hoel, 1962, pp. 266–268), that

$$W^2 = \Sigma e_t^2/\sigma^2 \quad \text{is} \quad \chi^2_{T-K}.$$

From this,

$$\frac{Z}{\sqrt{W^2/(T-K)}} = \frac{(b_k - \beta_k)/\sigma\sqrt{m_{kk}}}{\sqrt{\Sigma e_t^2/\sigma^2(T-K)}} = \frac{b_k - \beta_k}{\sigma\sqrt{m_{kk}}} \frac{1}{\sigma^{-1}\sqrt{\Sigma e_t^2/(T-K)}}$$

$$= \frac{b_k - \beta_k}{s\sqrt{m_{kk}}} = t_{T-K} \qquad (5.21)$$

is distributed as Student's t with $T-K$ degrees of freedom.[9]

This conclusion permits us to test various hypotheses about the true value of β. For example, a common hypothesis is H_0: $\beta_k = 0$, or that X_k has no effect on Y. This hypothesis will be rejected if

$$|b_k|/s\sqrt{m_{kk}} > t_{\text{crit}(\alpha/2, T-K)},$$

where $t_{\text{crit}(\alpha/2, T-K)}$ is the critical value for $T-K$ degrees of freedom and significance level α.[10]

---

[9]The independence of Z and W^2 follows from showing that the covariance of e and b is zero. With normal variables, covariance of zero implies independence. The demonstrations of zero covariance are

$$E[e(b-\beta)'] = E\left[(I - X(X'X)^{-1}X')UU'X(X'X)^{-1}\right]$$

$$= \sigma^2\left[X(X'X)^{-1} - X(X'X)^{-1}X'X(X'X)^{-1}\right]$$

$$= \sigma^2\left[X(X'X)^{-1} - X(X'X)^{-1}\right] = 0$$

from $e = Y - Xb = Y - X(X'X)^{-1}X'Y = X\beta + U - X(X'X)^{-1}X'(X\beta + U) = (I - X(X'X)^{-1}X')U$ and $b - \beta = (X'X)^{-1}X'U$.

[10]Perfectly analogous to this hypothesis test is forming the $100(1-\alpha)$ confidence interval of $b_k \pm t_{\text{crit}(\alpha/2, T-K)}s\sqrt{m_{kk}}$.

The general form of the hypothesis test for H_0: $\beta_k = \beta_k^*$ and H_1: $\beta_k \neq \beta_k^*$ is

$$t_{T-K} = |b_k - \beta_k^*| / s\sqrt{m_{kk}}$$

If $t_{T-K} > t_{\text{crit}(\alpha/2, T-K)}$, the null hypothesis is rejected. In other words, if the estimated value is a large distance from the hypothesized value (in terms of standard errors), it is unlikely that the null hypothesis is true.

One important special application of t-statistics is the test for equality of two substantively comparable coefficients in the same model.[11] The null hypothesis of the coefficient equality can be stated:

$$H_0: \quad \beta_i - \beta_j = 0.$$

The appropriate test statistic is then

$$t = |(b_i - b_j) - 0| / s_{(b_i - b_j)}.$$

The variance of the difference of two random variables is (see Appendix I):

$$\text{var}(b_i - b_j) = \text{var}(b_i) + \text{var}(b_j) - 2\,\text{cov}(b_i, b_j).$$

The variance of b_i and b_j are s^2 times the appropriate diagonal terms of $(X'X)^{-1}$, (m_{ii} and m_{jj}); the covariance term is s^2 times the appropriate off-diagonal element in $(X'X)^{-1}$, (m_{ij}). Thus the standard error of the difference in two coefficients is $s\sqrt{m_{ii} + m_{jj} - 2m_{ij}}$.

Hypothesis Tests for Sets of Coefficients

In many instances, we want to test hypotheses about a set of the estimated coefficients, not just a single coefficient as was done previously. We may have several variables measuring different aspects of the same effect and want to test whether this effect, rather than merely the individual aspects of it, has any statistically significant influence on the dependent variable Y. For example, in many social models an important influence is hypothesized to be socioeconomic status. This general concept is then assumed to be related to, but not strictly measured by, people's income, education, and possibly occupation. Testing the null hypothesis that *SES* has no influence on behavior is then the test of whether the coefficients on all three variables are zero.

A second situation where we need to test hypotheses about a set of coefficients was mentioned briefly at the end of Chapter 4. When a model is being estimated with data that may contain observations from more than one population, the coefficients on some or all of the variables will differ for each population. We discussed this possibility as it applied to the intercept and/or just one of the explanatory variables in the preceding chapter. That test of population differences can be extended to cover additional explanatory variables by including slope dummies for as many variables as one needs to model the hypothesized differences in populations. The test for whether the

---

[11]To be substantively comparable, the coefficients must be in the same units.

populations are different is now the test of whether all the slope dummies are zero. If this hypothesis is rejected, we are rejecting the notion that the populations are similar.

The appropriate statistical test for a set of parameters, e.g., whether $\beta_2 = \beta_3 = \cdots = \beta_k = 0$, is not the simple test for *each* estimated coefficient taken separately, e.g., that $\beta_2 = 0$, that $\beta_3 = 0$, that $\beta_k = 0$, and so on. Unless the off-diagonal terms of $(X'X)^{-1}$ are zero, which occurs only if the covariances among the explanatory variables equal zero, the distributions of the estimated coefficients are not independent. Consequently, the proper statistical test is the probability of the joint outcome that *all* estimates exceed some critical value. Specifically, what is the probability that $b_2 > b_2^*$ *and* $b_3 > b_3^*, \ldots,$ and $b_k > b_k^*$, where the asterisked terms indicate the appropriate critical values for the test being used. This test is not based on the t-test for the individual coefficients. We shall develop a procedure that is appropriate for the joint hypothesis about sets of parameters.

We arbitrarily place the variables whose coefficients are to be included in the test at the end of the equation and shall denote this set of coefficients and variables as β_2 and X_2, respectively. Thus the model we want to estimate is

$$Y = X_1 \beta_1 + X_2 \beta_2 + U,$$

where β_1 is the $H \times 1$ vector of coefficients not being examined, X_1 is the $T \times H$ matrix of observations on the associated explanatory variables, β_2 the $(K - H) \times 1$ vector of coefficients we are testing, and X_2 is the $T \times (K - H)$ matrix of observations of the explanatory variables associated with β_2. Our null hypothesis is that each of the elements of β_2 is zero: $\beta_{H+1} = \beta_{H+2} = \cdots = \beta_K = 0$. If, based on the statistic we develop, the null hypothesis is rejected, the conclusion would be that the factor represented by the $K - H$ variables in X_2 does influence Y. In the socioeconomic status example, X_2 would be the explanatory variables income, education, and occupation (which itself might be several dummy variables). The test of whether *SES* is an important influence is the null hypothesis that all elements of $\beta_2 = 0$, rather than the test of each individual coefficient. The estimated effect of *SES* however is measured by the individual coefficients, and one must specify income and education levels and an occupation for two different *SES* levels in order to measure the "effect" of *SES*. This effect is computed by summing the product of each coefficient and variable difference, e.g., $b_{H+1} \Delta X_{H+1} + b_{H+2} \Delta X_{H+2} + \cdots + b_K \Delta X_K$, and is the difference in Y attributable to the differences in *SES* measured by $\Delta X_{H+1}, \Delta X_{H+2}, \ldots,$ and ΔX_K.

The correct statistical test is based on the increase in the sum of squared residuals resulting from the deletion of the K-H variables in X_2. Intuitively, any time we delete variables from a regression, the sum of squared residuals increases if for no other reason than there are fewer terms (estimated coefficients) to use in minimizing the sum of squared errors (an increase in the degrees of freedom). The statistical question is whether the increase in the residual sum of squares from deleting X_2 is primarily due to the decrease in the number of coefficients being estimated or to the removal of a causally important set of influences. The greater the increase, the more likely we are to have excluded an important set of variables. We shall denote the residuals from the estimation with X_2 excluded as e_1 and the residuals from the full equation including both X_1 and X_2 as e. We are thus concerned with the magnitude of the scalar term $e_1'e_1 - e'e$. The precise derivation of how this term is incorporated into the statistical test is done by Fisher (1970) and only summarized here. The core of the argument is that b_2 from the full regression is a vector of normally distributed random variables with an expected value of β_2 and a variance–covariance matrix given by σ^2 times the appropriate submatrix of $(X'X)^{-1}$ [given as $(X'X)_{22}^{-1}$], or $b_2 \sim N[\beta_2, \sigma^2(X'X)_{22}^{-1}]$. Since we are testing the null hypothesis that $\beta_2 = 0$, we need to know the probability of getting the estimated values for b_2 by chance if the true mean of each element is zero. The derivation in Fisher shows that with the distribution of b_2 being multivariate normal and the assumption that $\beta_2 = 0$, $(e_1'e_1 - e'e)/\sigma^2$ is distributed as a χ^2 with $K - H$ degrees of freedom. Thus we need to ascertain the probability of obtaining a χ^2_{K-H} value this large by chance (e.g., is the probability less than 0.05 or 0.01?) before we can decide to reject the null hypothesis. Again we do not know σ^2. However, we have previously used the fact that $e'e/\sigma^2$ is distributed as a χ^2_{T-K} and that the residuals e are independent of b (see p. 123). We can then use these two χ^2 distributions and the definition of the F-distribution (Appendix I) to show that

$$\frac{(e_1'e_1 - e'e)}{\sigma^2(K-H)} \frac{1}{e'e/\sigma^2(T-K)} = \frac{(e_1'e_1 - e'e)/(K-H)}{e'e/(T-K)} = F_{T-K}^{K-H}. \quad (5.22)$$

The test for whether a set of coefficients equals zero is then simply a comparison of the reduction in the sum of squared errors obtained by including these variables with the sum of squared errors in the complete model with both the numerator and denominator adjusted by their appropriate degrees of freedom.

An important special application of this test occurs when we test the complete set of explanatory variables except the constant term. This is equivalent to setting $H = 1$, and the null hypothesis is $\beta_2 = \beta_3 = \cdots = \beta_K = 0$. We are, in effect, testing the complete model against an alternative one that predicts each observation of Y to be its sample mean, which implies that all the variance in Y is accounted for by the error term. (The reader should be able to see why the coefficient for the constant term, if all other variables are omitted, will be the sample mean of Y.) If the computed value of the F-statistic exceeds the established critical value, we would reject the naïve model of no systematic explanation for Y in favor of the hypothesized model. The computation for this test is quite simple. Because X_1 is only a constant term whose coefficient will be $\bar{Y}$ for the estimation with the other variables excluded, $e_1 = Y - \bar{Y}$ and $e_1'e_1$ is T times the variance of Y, $e_1'e_1 = T\text{var}(Y)$.

The *F*-statistic for the test of the entire model specification is the "explained" variance divided by $K-1$, divided by the unexplained variance divided by $T-K$. Most regression programs include this *F*-statistic as well as the residual sum of squares as part of their output. We can also express this *F*-value in terms of R^2, as[12]

$$F_{T-K}^{K-1} = \frac{R^2/(K-1)}{(1-R^2)/(T-K)}.$$

The second important application of this statistical test is the situation where the researcher suspects the data contain observations from more than one population and the coefficients on some of the explanatory variables differ between the populations. This situation can be modeled by including slope dummy variables for each of the variables suspected of having different coefficients and a dummy intercept for each subpopulation. We shall denote the set of suspect variables as X_2 through X_M in our model and include $DX_2, DX_3, \ldots, DX_M$, and D in our estimation, where D is the dummy variable indicating those observations that are members of the second subpopulation (additional dummy variables and a parallel specification can be used to model additional subgroups). This model is shown as

$$Y = \beta_1 + \beta_2 X_2 + \beta_3 X_3 \cdots + \beta_M X_M + \cdots + \beta_K X_K$$
$$+ \beta_1^* D + \beta_2^* DX_2 + \cdots + \beta_M^* DX_M + U. \qquad (5.23)$$

We have arbitrarily set up the model so that the $M-1$ variables being tested with slope dummies are variables $X_2, \ldots, X_M$ in the original model, and that these and the intercept dummies are placed at the end of the model as variables $X_{K+1}, \ldots, X_{K+M}$. When we estimate this model, the estimated coefficients $b_1^*, \ldots, b_M^*$ test whether the intercept and the effects of variables $X_2, \ldots, X_M$ are different in group B. The *t*-statistics computed for each coefficient test the null hypothesis that the effects of each variable, taken separately, are the same in each group. It is also possible to test the single null hypothesis that $\beta_1^* = \beta_2^* = \cdots = \beta_M^* = 0$. The implication of this hypothesis is that there is no difference in the effects of this set of variables in the second group, i.e., no differences in the two populations. This hypothesis is appropriately tested by the *F*-test. This is done by computing the sum of squared residuals from two separate regressions, one with the dummy variables included, $(e'e)_{K+M}$, and one with variables excluded, $(e_1'e_1)_K$, where $(e'e)_{K+M}$ is the residual sum of squares with all the dummy variables

---

[12]To see this step,

$$F = \frac{(e_1'e_1 - e'e)/(K-1)}{e'e/(T-K)} = \left(\frac{e_1'e_1}{e'e} - \frac{e'e}{e'e}\right)\frac{T-K}{K-1} = \left[\frac{(Y-\bar{Y})'(Y-\bar{Y})}{e'e} - 1\right]\frac{(T-K)}{(K-1)}$$

$$= \left(\frac{1}{1-R^2} - 1\right)\left(\frac{T-K}{K-1}\right) = \frac{R^2}{1-R^2}\frac{T-K}{K-1} = \frac{R^2/(K-1)}{(1-R^2)/(T-K)}.$$

included and $(e'_1 e_1)_K$ is the value with the set of M dummy variables excluded. Then the ratio

$$\frac{[(e'_1 e_1)_K - (e'e)_{K+M}]/M}{(e'e)_{K+M}/(T-K-M)} = F_{T-K-M}^M \tag{5.24}$$

is distributed as an F-statistic with M and $T-K-M$ degrees of freedom. A high value of F would lead to a rejection of the hypothesis that $\beta_1^* = \beta_2^* = \cdots = \beta_M^* = 0$. Intuitively, this is saying that if the inclusion of the variables producing different coefficients in group B reduced the sum of squared errors by a large amount, then Y is the product of a different behavioral model in group B.

At this point we should return to the introductory discussion of this section. When we suspect that there may be some behavioral differences among different subpopulations, we can attempt to include these in an overall model through the use of dummy variables. This can be looked upon as an alternative to stratifying the observations into separate data sets for each population and estimating the model separately for each group. But what are the gains of this procedure, especially in the more general form of Eq. (5.23) where there can be significant behavioral differences?

The gains of modeling the differences in a pooled sample over stratification come in the coefficients that are restricted to be the same in each of the subpopulations, i.e., b_{M+1} through b_K. These parameter estimates rely upon the total information from the pooled sample and will thus be more efficient —have smaller variances—than the same coefficients estimated from stratified samples.

The ultimate extension of the tests for coefficient equality comes when all coefficients are allowed to vary among subpopulations, i.e., when $M = K$ in Eq. (5.23). If the F-test in Eq. (5.24) indicates with considerable confidence that the coefficients in the subpopulations are different (the F-statistic is large) when all coefficients are allowed to vary, pooling of observations is not appropriate. That indicates that the observations come from entirely different populations, and sample stratification is appropriate.

The test in Eq. (5.24) when a complete set of dummy variables is included ($M = K$) is equivalent to a test commonly referred to as a Chow test, which accomplishes the same purpose. The initial formulation of the Chow test suggested running three separate regressions to get the information needed for Eq. (5.24). If we estimate the basic model separately for each population and once for the entire sample with no dummy variables included, $(e'e)_{2K}$ is simply the sum of squared residuals from the regressions on the two subpopulations added together, $(e'_1 e_1)_K$ is the sum of squared errors from the pooled regression, and the ratio in Eq. (5.24) is distributed as an F-statistic with K and $T-2K$ degrees of freedom.

In using either form of this complete coefficient test we must be especially careful about the statistical test that we are applying. The null hypothesis is that all coefficients of subsample differences are zero. It is thus a good test of when pooling of observations is not appropriate. However, we are often interested in just the opposite situation—when pooling is appropriate. Given the situation with two samples that we suspect are different but which we would like to pool for efficiency reasons, the appropriate null hypothesis (the one we would like to reject with some confidence) is that the coefficients are different from zero. (The alternative hypothesis that would be accepted with a rejection of the null hypothesis is that the observations come from the same population.) Since our statistical tests are all designed to test simple hypotheses—that the parameter equals a given value—we cannot test this complex hypothesis. The implication is that the hypothesis test specified in Eq. (5.24) is too rigid if we are really interested in the possibility of pooling, but it is not possible to specify a better test with any precision. In applying this test to see if we could possibly pool, a lower level of acceptance, say the 25% level, is probably more appropriate than the standard 5% level. Even this must be evaluated in the context of the uses of the model, however.

5.7 Multivariate Education Example

Estimation of a model of educational achievement provides an illustration of the statistical tests discussed in this chapter. A model of educational performance is important to school administrators. Knowledge of the effects of teachers and specific teacher attributes on educational output is necessary to improve the efficiency of school operations. Yet, education is not solely a function of schools; it is also a function of family and friends. Therefore, even though the entire interest may be in the inputs of schools and teachers to the educational process, models of education must include other inputs to satisfy an implicit *ceteris paribus* assumption. The problem is to relate test scores to a complete set of explanatory variables designed to measure the influence of family, friends, and school characteristics.

Educational inputs of the family have been shown to be positively related to the socioeconomic status of the family. This relationship derives from attitudes, inherited capabilities, and direct assistance through examples of language structure, problem solving, etc. Thus, measures of family status such as educational levels, occupation, income, and family size provide proxies for the educational inputs of the family.

Schools are hypothesized to contribute several inputs. School administrators pay for teaching experience and advanced education—suggesting one set of hypotheses about teacher effects. Also, common beliefs indicate that such attributes as class size, teacher intelligence, knowledge of subject, and so forth should be important.

The need for a multivariate analysis should be clear. It is necessary to include measures of the students' backgrounds even though we are primarily concerned with estimating the effects of teachers and teacher characteristics on learning. Because of the nonrandom assignment of pupils to schools and teachers, the background characteristics will be positively correlated with the measures of the influences we are trying to study. Exclusion of the background variables would result in a misspecified model and biased coefficients. One way of removing this possible bias is by including an extensive set of variables in the model to represent both school and teacher attributes and child background effects. The second way is to prevent some of the background influences from varying in the sample by an appropriate stratification. We shall follow both strategies.

The second reason a multivariate analysis is demanded is that it requires more than one variable to assess teacher attributes. Thus simply to raise the question of "Do teachers count?" requires that we include several variables assessing teacher characteristics and perform the joint test on the set of coefficients.

A specific multivariate test of the above model of the educational process was performed on a sample of 515 third graders in a California school system. These children are all from families with fathers in blue-collar occupations, thus the largest family educational inputs measured by broad socioeconomic status are held constant by sample design. The multiple linear regression equation estimated by applying the least squares criterion is

$$A_3 = 20.8 + 2.81\,F - 6.38\,R + 0.79\,A_1 - 0.07\,D + 0.09\,T_3 - 0.57\,Y_3 + 0.06\,T_2 - 0.68\,Y_2,$$
$$\quad\;\;(1.2)\quad\;\;(2.3)\quad\;\;(.04)\cdot\quad\;(0.03)\quad\;(0.04)\quad\;(0.39)\quad\;(0.03)\quad\;(0.24)$$

$$R^2 = 0.51, \qquad SEE = 13.5. \tag{5.25}$$

These results indicate that the third grade reading achievement test score of any individual student (A_3) is a function of sex (F), grade repeats (R), initial achievement level (A_1), percent of class time devoted to discipline (D), teacher verbal facility (T_3 and T_2), and years since latest educational experience (Y_3 and Y_2) for third and second grade teachers.[13] Subscripts refer to the appropriate grade level. (Note: the standard error of each coefficient is displayed immediately beneath the coefficient; SEE is the standard error of estimate for the entire equation.)

Equation (5.25) indicates that, after allowing for differences in sex, grade repeats, and initial achievement levels, two aspects of both the third and second grade teachers affect educational outcomes. The higher a teacher's score on a verbal facility test for teachers (T), the higher is the students' achievement. A one-point increase in third grade teacher verbal score, *ceteris paribus*, yields a 0.09 point increase in student achievement. Also, more recent educational experience is related with higher achievement. Since we explicitly introduced initial achievement as an explanatory variable, the coefficients estimate the achievement gains associated with more verbally

---

[13] Sex and grade repeats are dummy variables. F equals one if the individual is a girl; zero if a boy. R equals one if the student repeated a grade, zero otherwise.

facile teachers regardless of entering achievement; in other words, the effect of increasing the third grade teacher's verbal facility, holding first grade achievement constant.

By comparing the standard errors for each coefficient (in parentheses) with its coefficient, it is seen that most of the coefficients are statistically different from zero at the 5% level with a two-tailed test. (Remember that, for H_0: $\beta = 0$, $t = b/s_b$, and that $|t| \geqslant 1.96$ implies that the null hypothesis H_0 is rejected at the 5% level for a two-tailed test.) As indicated by the R^2 value, slightly over half of the variance in achievement is explained by the model.

With these data, the standard hypotheses about experience and advanced degrees were also tested, and the effects of these variables were found to be insignificant at the 5% level. (The t-ratio for H_0: $\beta = 0$ was consistently very low, i.e., much less than 1.96. Thus, it was not possible to reject the null hypothesis of no effect.) Since these are the factors purchased by schools, it is apparent that schools are being run inefficiently in this area. It was not possible to test hypotheses about class size adequately because there was very little variance in class size within the sample. We can also test the hypothesis, Do teachers count? This hypothesis implies that the coefficients T_2, T_3, Y_2, and Y_3 are all zero. The appropriate test in this instance is an F-test on the coefficients of these four variables. When accomplished according to Eq. (5.22), we find that $F_{506}^4 = 5.68$. This large value is sufficient to reject the "no effect" null hypothesis at the 0.01 level. In other words, we are very confident that differences in these teacher attributes lead to significant differences in student achievement.

5.8 Multicollinearity

Within the general multivariate regression model, we return briefly to the problem of multicollinearity. High correlations among the exogenous variables lead to imprecise coefficient estimates and to high correlations among the estimated coefficients. These results of multicollinearity can seriously handicap our ability to make inferences about individual coefficients.

In the simple model with two exogenous variables discussed in Chapter 4, there is an exact (negative) correspondence between the correlation of the X's and the correlation of the estimated coefficients. With more than two exogenous variables, the relationship between the correlations in the X's and the correlations of the coefficients is considerably more complicated. A beginning place in analyzing estimation results for the possibility of serious multicollinearity is examination of the simple correlations among estimated coefficients. The simple correlation coefficients can be found from the estimated variance–covariance matrix of the coefficients $s^2(X'X)^{-1}$. The simple correlation between any two coefficients, say b_i and b_j, is

$$r_{b_i b_j} = s^2 m_{ij} / \sqrt{s^2 m_{ii} s^2 m_{jj}} = m_{ij} / \sqrt{m_{ii} m_{jj}} \, ,$$

where m_{ij} is the (i,j) element of $(X'X)^{-1}$. If there is the suspicion that multicollinearity is a problem, the simple correlations among estimated coefficients can be easily calculated from the variance–covariance matrix for the coefficients. Examination of these correlations is not conclusive since the relationship between more than two coefficients may be important. Nevertheless, a check of these correlations may provide evidence as to whether imprecise coefficient estimates are the result of multicollinearity.

If multicollinearity is a serious problem, it becomes difficult to test hypotheses about individual coefficients. Instead, it is often necessary to resort to joint hypothesis tests, i.e., F-tests on sets of coefficients. It is very possible with multicollinearity to not reject a null hypothesis of $\beta = 0$ for each parameter in the collinear set but to reject the null hypothesis that all parameters in the set are zero. In other words, there could be low t-statistics for each parameter yet a high F-statistic for the set.

This can be shown diagrammatically by looking at confidence intervals for two coefficients. For an individual parameter, a $1 - \alpha$ confidence interval for β_k is simply

$$\beta_k = b_k \pm t_{\text{crit}(\alpha/2, T - K)} s_{b_k}$$

where $t_{\text{crit}(\alpha/2, T - K)}$ is the value of the t-variable for probability $\alpha/2$ and $T - K$ degrees of freedom. This corresponds to the probability statement

$$\Pr\left(b_k - t_{\text{crit}(\alpha/2, T - K)} s_{b_k} < \beta_k < b_k + t_{\text{crit}(\alpha/2, T - K)} s_{b_k}\right) = 1 - \alpha.$$

The confidence interval for a single parameter can be generalized to handle several coefficients.[14] With two variables, the confidence interval translates into a joint confidence region that is an ellipse. The confidence ellipse indicates a probability statement about joint values of parameters. Figure 5.1 displays a confidence ellipse for two parameters β_i and β_j. The true parameter values will fall within the shaded area with probability $1 - \alpha$.

The effect of multicollinearity is to elongate and tilt the confidence ellipse. With high positive correlation between two X's, the coefficients become negatively correlated, and the confidence ellipse slopes downward as in Fig. 5.2. Figure 5.2 also displays the confidence intervals for each individual parameter.

The important point of Fig. 5.2 is that the confidence interval on each parameter separately includes the value zero. Thus, we would not reject at the $1 - \alpha$ level the hypothesis that each parameter equals zero. However, the joint confidence region does not include zero. Therefore, we would reject at the $1 - \alpha$ level the hypotheses that *both* parameters equal zero. The confidence

---

[14]The generalized confidence interval for a set of coefficients, β_2, is

$$(\beta_2 - b_2)'\left[(X'X)_{22}^{-1}\right]^{-1}(\beta_2 - b_2) \leqslant s^2(K - H)F_\alpha$$

where $(X'X)_{22}^{-1}$ is the appropriate portion of $(X'X)^{-1}$, and $K - H$ is the number of parameters in the vector β_2.

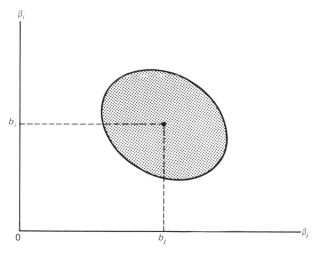

FIGURE 5.1   Joint confidence ellipse.

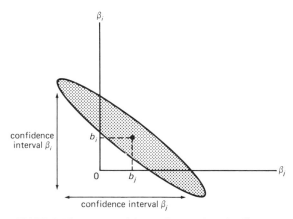

FIGURE 5.2   Joint confidence ellipse with multicollinearity.

region does cross both the β_i and β_j axes. Therefore, it is conceivable that *one* of the parameters is zero. However, the coefficients are so intertwined that it is not possible to speak about one parameter at a time when there is serious multicollinearity.

5.9   Conclusion

We want to reiterate the comments made at the end of Chapter 3 about the importance of thoroughly understanding least squares analysis. Proper analysis requires several assumptions about the structure of the model and data.

These conditions must be met in order to make any substantive interpretations about behavior from the estimated coefficients. The assumptions concern the structure of the true population model, the equation

$$Y = X\beta + U, \qquad (5.26)$$

and the process that generated the data used to estimate the model.

The first and most easily satisfied of the assumptions is that X is a $T \times K$ matrix of explanatory variables, with $T > K$ and rank $X = K$. This assumption is required in order to invert the cross-products matrix, or to ensure that $X'X$ is not singular. If this assumption cannot be met, then it is impossible to compute the estimated coefficients, just as it is impossible to compute a value for any expression divided by zero. The rank of X will equal K as long as there are more observations than there are variables $(T > K)$ and as long as no one variable is a *linear* function of the other explanatory variables. Fortunately, variables are seldom a strictly linear function of each other in nature. Practically speaking, this means that we can obtain estimates for the coefficients in almost any model for which we can collect data. The most common cause of singularity in $X'X$ is the creation of linearly dependent variables by the analyst, such as with the inclusion of too many dummy variables. The first check when a singular matrix is discovered thus is a bit of introspection by the analyst.

Assuming that the problem is not created by the analyst, one thing is clear: we cannot even obtain estimates for the coefficients in our hypothesized model. Consequently, there is little point in trying to debate this assumption or finesse our way around it. There are only two possible options. The first is to obtain additional data that have enough independent variation in the explanatory variables to ensure that X will have sufficient rank. The second alternative is to try to locate the offending variable or variables and delete them from the analysis. This however could lead to biased estimates of the remaining coefficients. Yet if these are not important coefficients, in the sense that one is trying to obtain accurate estimates of their value, but interest is centered on other variables in the model, and these linearly related variables have been included only to obtain the proper specification of the model, the consequences of deleting one of these variables may not be serious. This fortuitous circumstance occurs in some instances, and the offending variables can usually be located and manipulated so as to prevent singularity and permit the coefficients to be estimated.

This assumption is only the beginning however. Having rank $X = K$ ensures only that we can estimate a set of coefficients. The difficulty comes however in trying to interpret these estimated coefficients meaningfully if this is the only assumption satisfied. Since the least squares coefficients are estimates of the true coefficients and are drawn from some probability density function determined by the error term in the model, we need the remaining assumptions to be able to say something about the distribution from which these estimates are drawn and thus obtain some information about the true values of the coefficients.

The next assumption is that we are working with fixed values of X. Thus our observed values of Y, the behavior we are explaining, represent only *one* replication out of some unspecified number of trials with the same values for the explanatory variables. With this assumption we are able to establish that the distribution from which we drew the estimates is solely a function of the distribution of the error terms in the model. Consequently, all we needed to do to define the parameters of this distribution of coefficient estimates was to make some assumptions about the error term distribution.

We should examine the representativeness of this particular experiment or data set to the broader class of behavior we are examining before making any general behavioral interpretations based on our inferences about the true coefficients. We may have adequately described the 1964 presidential election, or the educational process of Southern California third graders, and our inferences about the true coefficients for these populations may be correct in that if we could run each of these experiments a large number of times, as we did with the Monte Carlo example, the distribution of estimated coefficients would conform to our inferences. However, if these experiments are not typical of presidential elections or elections in general, or if the school environment and learning process of these particular third graders is not representative of all students, then it becomes difficult to use our inferences about the true coefficients in our models to draw conclusions about the more general classes of voting or learning behavior.

We can modify the assumption of fixed X and examine the consequences of permitting the explanatory variables to be stochastic and likely to vary each time we replicate the experiment. This is saying that each of the T values of X_k are drawn from a distribution with some mean and variance. Once we do this, we must alter our derivations of the expected value and variance of our estimated coefficients. In Eq. (5.11), the expected value of b is given as

$$E(b) = \beta + E[(X'X)^{-1}X'U].$$

If we no longer assume fixed X, then we cannot pass the expected value operator through the first part of this expression to get $E(U)$. If, however, we assume that each X_k and U are independent, then this equation becomes merely

$$E(b) = \beta + E[(X'X)^{-1}X']E(U) = \beta \qquad (5.27)$$

since we may still assume that $E(U) = 0$. Thus even though the X's are no longer fixed, as long as they are independent of the error term, we still get unbiased estimates of the coefficients. If X and U are not independent but are likely to covary, then we can no longer contend that our estimates are unbiased.

Relaxing our assumption of fixed X means that our discussion of the variance of the estimated coefficients must also be altered. Again we cannot pass the expected value operator through the expression given in Eq. (5.15):

$$E(b-\beta)(b-\beta)' = E[(X'X)^{-1}X'UU'X(X'X)^{-1}].$$

This now expresses the variance–covariance of our estimates as a multivariate joint distribution of U and X. It is obviously a complicated matter to try to sort out the precise parameters of this multivariate distribution. An alternative is to make the variance–covariance matrix conditional on the values of X. Once we do this we obtain the same expression for the variance–covariance matrix, $E(b-\beta)(b-\beta)' = \sigma^2 E(X'X)^{-1}$, that we obtained in the fixed regressor case, only now this expression is conditional upon X and contains the expected value of $(X'X)^{-1}$. Goldberger has an extended discussion of this derivation and argues that since we have not placed any restrictions upon our values of X, these results apply for all possible values and thus are not very restrictive (Goldberger, 1964, p. 268). The consequence of these results is that as long as we can argue that U and each X_k are distributed independently from each other, we can achieve the same results as we did in the fixed X case, namely that our estimates are unbiased and have a variance–covariance equal to $\sigma^2(X'X)^{-1}$. The important point here is to make sure that U and each X_k are independent.

The first two assumptions, that rank $X = K$ and that the X's could be fixed, were used to make sure that we could obtain estimates for the coefficients in our model and to specify that the distribution of estimated coefficients (from which we are drawing only one value) is solely a function of the distribution of the stochastic component in the behavior we are modeling. The remaining assumptions all refer to the parameters of this error term distribution. Once these parameters are defined, we have the necessary information to talk about the distribution of estimated coefficients. The first of these assumptions is that the expected value of the error term drawn for each observation is a constant, usually assumed to be zero. In essence this is saying that the mean value of each observation's error term, taken over all implicit replications, will be the same for all observations, i.e., is the same regardless of the values taken by the explanatory variables, and, one hopes, is zero. This assumption is necessary to obtain unbiased estimates, or to have the expected value of our estimated coefficient equal the true value. The expected value of the error terms is assumed to be zero in order that the estimate of the intercept coefficient b_1 be unbiased. If the expected value of each error term is a constant that does not equal zero, then all the coefficients except the intercept will be unbiased and the estimate of the intercept will be biased by an amount equal to this constant. Thus to obtain unbiased estimates of our coefficients we only have to make an assumption about the expected value of our error term, in addition to the assumptions of a fixed X having a rank equal to K.

With these three assumptions we have established that we can get estimates of the coefficients in our hypothesized model, that they will be drawn from a distribution that is a function of the distribution of the error terms, and that this distribution of estimates will have a mean equal to the true value. However, we have not established the variance of this distribution yet. This is the purpose of the fourth assumption. This assumption assumes that each error term is drawn randomly, or independently, from the same distribution and that this distribution has a variance equal to σ^2. The random, or independent, drawing implies that $E(U_t U_{t+s}) = 0$, $s \neq 0$, meaning that there is no correlation between pairs of error terms. The requirement that all error terms come from the same distribution means that all error terms will have the same variance $E(U_t^2) = \sigma^2$. Once we make this additional assumption, we can show that the variance of the distributions from which we are drawing our estimates equals $\sigma^2(X'X)^{-1}$.

The only remaining thing to do to make our description of the distribution complete and to meet all the requirements for classical statistical inference and hypothesis testing is to determine the actual form of the distribution of the estimators. This is accomplished by assuming that the error terms are distributed according to a normal distribution, i.e., U is $N(0, \sigma^2 I)$. Since we have assumed that the X's are fixed, this means that the distribution of the coefficients follows a linear transformation of the distribution of the error terms. Assuming that the error terms are normally distributed implies that the estimated coefficients will be normally distributed about their true value. Substitution of s^2 for the unknown term σ^2 results in the ratio $(b_k - \beta_k)/s_{b_k}$ being distributed as t-statistic, with $T - K$ degrees of freedom. Thus we have all the necessary information and completed all the formal requirements to make inferences about the true coefficients based on the estimates of them.

APPENDIX 5.1

Proof of Best

To show that the least squares estimators have the smallest variance among all linear unbiased estimators, we shall proceed to develop an arbitrary alternative estimator. The alternative linear estimator was written as

$$b^\# = CY = \left[(X'X)^{-1}X' + D \right] Y.$$

Consequently for $b^\#$ to be unbiased,

$$E(b^\#) = E\left[(X'X)^{-1}X' + D \right]\left[X\beta + U \right]$$

$$= E\left[(X'X)^{-1}X'X\beta + (X'X)^{-1}X'U + DX\beta + DU \right]$$

$$= E(\beta) + (X'X)^{-1}X'E(U) + E(DX\beta) + DE(U) = \beta + DX\beta = (I + DX)\beta$$

since $E(U) = 0$ and $DX\beta$ is nonstochastic. Thus for $b^\#$ to be unbiased, we must impose the condition that $DX = 0$. With this condition, we can express $b^\#$ as $b^\# = [(X'X)^{-1}X' + D][X\beta + U] = \beta + (X'X)^{-1}X'U + DU.$

Thus

$$(b^* - \beta) = \left[(X'X)^{-1}X' + D\right]U,$$

$$E\left[(b^* - \beta)(b^* - \beta)'\right] = E\left[(X'X)^{-1}X' + D\right]UU'\left[D' + X(X'X)^{-1}\right]$$

$$= \sigma^2\left[(X'X)^{-1}X' + D\right]\left[D' + X(X'X)^{-1}\right]$$

since $E(UU') = \sigma^2 I$

$$= \sigma^2\left[(X'X)^{-1}X'D' + DX(X'X)^{-1} + (X'X)^{-1} + DD'\right]$$

$$= \sigma^2\left[(X'X)^{-1} + DD'\right]$$

since $DX = X'D' = 0$. This is then the expression shown in Eq. (5.17).

APPENDIX 5.2

Proof of Unbiasedness of the Estimator for σ^2

We shall show that with K variables, $E(\Sigma e_t^2) = (T - K)\sigma^2$. We start with $\Sigma e_t^2 = e'e$, and

$$e = Y - \hat{Y} = Y - Xb = X\beta + U - Xb = X(\beta - b) + U = -X(X'X)^{-1}X'U + U$$

$$= \left[I - X(X'X)^{-1}X'\right]U.$$

Thus

$$e'e = U'\left[I - X(X'X)^{-1}X'\right]\left[I - X(X'X)^{-1}X'\right]U = U'\left[I - X(X'X)^{-1}X'\right]U$$

$$= U'U - U'QU,$$

where Q is the $T \times T$ matrix $X(X'X)^{-1}X'$. Taking expected values,

$$E(e'e) = E(U'U) - E(U'QU)$$

$$= E\sum U_t^2 - E\sum Q_{tt}U_t^2 - 2E\sum\sum Q_{ts}U_tU_s = T\sigma^2 - \sigma^2\sum Q_{tt}$$

since $E(UU') = \sigma^2 I$. The sum of the main diagonal elements of a square matrix is called its trace,

$$\sum_{t=1}^{T} Q_{tt} = \mathrm{tr}(Q) = \mathrm{tr}\left[X(X'X)^{-1}X'\right].$$

One of the properties of the trace operator is that $\mathrm{tr}(AB) = \mathrm{tr}(BA)$. Thus letting

$$A = X(X'X)^{-1} \quad \text{and} \quad B = X', \ \mathrm{tr}\left[X(X'X)^{-1}X'\right] = \mathrm{tr}\left[X'X(X'X)^{-1}\right] = \mathrm{tr}\, I_K = K.$$

Thus $E(e'e) = T\sigma^2 - K\sigma^2 = \sigma^2(T - K)$ and $E[\Sigma e_t^2/(T - K)] = \sigma^2$, our unbiased estimator of σ^2.

REVIEW QUESTIONS

1. Given the following matrix representation of data:

$$Y = \begin{bmatrix} Y_1 \\ Y_2 \\ \cdot \\ \cdot \\ \cdot \\ Y_T \end{bmatrix}, \qquad X = \begin{bmatrix} 1 & X_{12} & X_{13} \\ 1 & X_{22} & X_{23} \\ \cdot & \cdot & \cdot \\ \cdot & \cdot & \cdot \\ \cdot & \cdot & \cdot \\ 1 & X_{T2} & X_{T3} \end{bmatrix}.$$

(a) Calculate the elements of $X'X$.
(b) Calculate $(X'X)^{-1}$.
(c) Calculate $X'Y$ and $(X'X)^{-1}X'Y$. How do these elements compare to the coefficient estimates in Chapter 2?

2. Given that

$$y = Y - \bar{Y} - \begin{bmatrix} y_1 \\ y_2 \\ \cdot \\ \cdot \\ \cdot \\ y_T \end{bmatrix}, \qquad x = X - \bar{X} = \begin{bmatrix} x_{12} & x_{13} \\ x_{22} & x_{23} \\ \cdot & \cdot \\ \cdot & \cdot \\ x_{T2} & x_{T3} \end{bmatrix}.$$

Calculate the elements of $(x'x)^{-1}x'y$ and compare these to the estimates in Problem 1.

3. For $x = X - \bar{X}$ and $y = Y - \bar{Y}$, a sample of data is found such that X_2 and X_3 are uncorrelated. This implies that

$$x'x = \begin{bmatrix} x_2^2 & 0 \\ 0 & x_3^2 \end{bmatrix}.$$

(a) Find $(x'x)^{-1}x'y$ and compare these results to those in Problem 2 and to bivariate regressions of Y on X_2 and Y on X_3.
(b) Calculate the variance–covariance matrix for the estimated coefficients and compare this to the results in Chapter 3.

4. The following models have been estimated for subgroups of the total population:

group I: $Y_t = 103.7 + 3.75X_{t2} + 10.51X_{t3}$, $R^2 = 0.48$, $SSE = 7262$, $T_1 = 105$;
              (1.30)   (3.85)

group II: $Y_t = 95.6 + 2.90X_{t2} + 13.3X_{t3}$, $R^2 = 0.41$, $SSE = 1131$, $T_2 = 65$;
              (1.97)   (6.80)

pooled: $Y_t = 98.3 + 3.60X_{t2} + 12.78X_{t3}$, $R^2 = 0.38$, $SSE = 8451$, $T = 170$.
              (1.13)   (4.20)

(a) Test whether the coefficients in the models are different.
(b) Should the observations be pooled?

5. The following model is estimated:

$$Y_t = -34.6 + 3.38X_{t2} + 1.97X_{t3}, \qquad R^2 = 0.38, \quad T = 43.$$

The estimated variance–covariance matrix for the coefficients $s^2(X'X)^{-1}$ is

$$\begin{bmatrix} 256 & -83.2 & -53.1 \\ -83.2 & 1.96 & -1.04 \\ -53.1 & -1.04 & 1.00 \end{bmatrix}.$$

Test the hypotheses:

(a) $H_0: \beta_2 = 0$, $H_1: \beta_2 \neq 0$;

(b) $H_0: \beta_3 = 1$, $H_1: \beta_3 \neq 1$;

(c) $H_0: \beta_2 = \beta_3$, $H_1: \beta_2 \neq \beta_3$;

(d) $H_0: \beta_2 = \beta_3 = 0$, H_1: the estimated model is correct.

6. You are given the data matrices

$$Y = \begin{bmatrix} 10 \\ 13 \\ 7 \\ 5 \\ 2 \\ 6 \end{bmatrix}, \quad X = \begin{bmatrix} 1 & 1 & 2 \\ 1 & 1 & 4 \\ 1 & 1 & 4 \\ 1 & 0 & 1 \\ 1 & 0 & 2 \\ 1 & 0 & 5 \end{bmatrix}, \quad \tilde{X} = \begin{bmatrix} 0 & 1 & 2 \\ 0 & 1 & 4 \\ 0 & 1 & 4 \\ 1 & 0 & 1 \\ 1 & 0 & 2 \\ 1 & 0 & 5 \end{bmatrix}.$$

(a) Calculate the least squares estimator for $Y = X\beta + U$ and $Y = \tilde{X}\tilde{\beta} + U$.

(b) Compare the estimates of b_1 and b_2 from the two different models.

(c) What does this imply for the way dummy variables are defined?

7. You wish to test the hypothesis that state expenditures on education are related to income levels in the states. At the same time you are worried that state education levels and state size also influence expenditures.

(a) How would you proceed in the analysis?

(b) How would you test the additional hypothesis that the proportion of people in the state with a college education has a greater effect on expenditures that does the average level of income?

(c) How would you test the hypothesis that Southern states spend less on education, other things being equal?

(d) Use data from the *U.S. Statistical Abstract* to perform these tests.

6 | Generalized Least Squares

6.1 Introduction

We now want to discuss the second important assumption made in Chapters 2 and 5, namely, that all error terms not only have a zero expected value (or at least are distributed independently of each explanatory variable) but are independently drawn from the same distribution. This assumption is $E(UU') = \sigma^2 I$ or $E(U_t^2) = \sigma^2$ for all t and $E(U_t U_s) = 0$ for all $t \neq s$. In other words, each error term, observed across all possible replications, has the same variance, regardless of the observation, and there are no correlations among the error terms associated with different observations. This chapter presents circumstances that raise doubts about the validity of these assumptions, the consequences of erroneously assuming they are met, and possible diagnoses and remedies when one suspects trouble.

The important points to remember are that we are still concerned with how these problems distort any inferences about the true, or population, parameters underlying the model and any estimates of the magnitudes of the influences present in the model. To anticipate the results of this chapter slightly, the difficulties with assumptions about the variances and covariances of the error terms do not affect the unbiasedness of our estimators but do increase their variance. Further, standard statistical tests will be inaccurate since the estimated variance will be biased. Thus the OLS estimator is unbiased, but it is no longer the best linear unbiased estimator. It is possible to generalize the least squares estimator to take into account the expected variance and covariance of the error terms, and thus to recapture this desirable property. This chapter develops both of these points and discusses how one goes about diagnosing the problem and applying the corrective generalized least squares estimator.

6.2   Heteroskedasticity and Autocorrelation

The terms used to describe situations where the previous assumptions about the variances and covariances of the error terms do not hold are heteroskedasticity and autocorrelation. Heteroskedasticity indicates that the variances of the error terms are not equal for each observation. (The previous assumption is that they are homoskedastic, or have equal variance.) Autocorrelation or serial correlation means the error terms are not drawn independently, but are correlated across observations. In the purely heteroskedastic case, $E(UU')$ is still diagonal, but the elements on the diagonal are not equal. In the autocorrelated case, $E(UU')$ has nonzero off-diagonal elements, indicating that pairs of error terms are correlated. The heteroskedastic case is more frequently associated with cross-sectional analysis where the observations are drawn from different units, such as firms, counties, or voting districts, and even people in some instances. Autocorrelation is most often a time series or longitudinal problem where data are collected from the same observational unit at successive points in time.

Heteroskedasticity is illustrated in Fig. 6.1. In this diagram, the true population relationship of the bivariate model is plotted in the X, Y plane. The third dimension shows the probability density function for the error term where σ_t^2 is an increasing function of X. In the simple case discussed in Chapters 2, 3, and 5, the density function for each value of X would be identical. With σ_t^2 positively related to X, the density function for larger X's has larger tails, making large absolute values of the error terms more likely when X is large.

If we think back to the discussion about the composition of the individual error terms and the various factors contributing to their variance, we can see what circumstances may lead to violations of the basic assumption. The first contributor to these errors is the difficulty in accurately measuring the behavior being modeled. If it is easier to measure this behavior for some

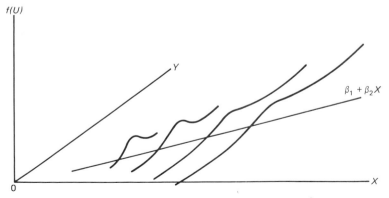

FIGURE 6.1   *Error distribution with heteroskedastic X's* $[\sigma_t^2 = f(X_t)]$.

observations, if some countries or firms keep more accurate records, or if some people give more accurate responses to questions, the result is a smaller variance in the error term for those units. Such problems are more likely in cross sections where data are collected for different units.

The most common illustration of heteroskedasticity is with aggregate data where the dependent variable is a proportion or a mean value for the people in the observational unit, such as with aggregate voting studies or average educational performance. The accuracy of the dependent variable will be a function of the number of individuals in the aggregate. Assume that all individuals behave according to the same population model $Y_i = X_i\beta + U_i$, where $E(U_i^2) = \sigma^2$ for all individuals. If there are N_t people in the tth aggregate unit, then the variance of the average of their behavior will be distributed about the true mean with a variance equal to σ^2 divided by the number of people in the aggregate, σ^2/N_t. (This argument is simply the statistical property that the variance of a sample mean about the true population mean is the population variance divided by the number of units in the sample.) In effect, then, observations for more populous units are presumably more accurate and should exhibit less variation about the true value than data drawn from smaller units. This clearly leads to different values of the error term variance for each observation, the heteroskedastic problem.

There are other reasons why the error term variance might not be equal across observations drawn from a cross-sectional sample. Consider a decision process, say a production decision by a firm or a budget decision by a political jurisdiction. If decisions tend to deviate from expectations in equal percentage terms, the error variance would be proportional to the size of the decision unit; e.g., large firms or jurisdictions would likely have larger absolute error terms. In other situations, such as with underdeveloped areas, there may be greater difficulty getting accurate measures of the necessary data for some observations. For example, census taking efforts may be less accurate (although not systematically biased) in some areas rather than others. These problems, as well as others one may encounter, will lead to different variances for each error term.

In the time series model, measurement errors may lead to autocorrelation if the same difficulties or problems are encountered during several successive time periods. If some component of the observed behavior is misinterpreted or not included in the measurements for several successive time periods or if the same measurement problems are encountered in successive periods, positive covariances among the error terms for successive observations will result.

The second cause for the stochastic error term is the presence of small, excluded effects that are presumed to be systematically unrelated to the explanatory variables. Although these effects may be small and independent of the explanatory variables, and hence do not violate the assumption of independence between X and U, they may still influence the variances and covariances among the error terms. The most obvious occurrence of such a problem is with time series data where the observations refer to the same unit. Any omitted variables are clearly present in all observations, with similar values from one observation to the next, unless they are time independent. If, as is likely, these omitted variables follow any pattern from observation to observation and cause any systematic variation in the behavior being modeled, it will result in error terms that are correlated from one time period to the next, even though they may vary independently of the explanatory variables. This is a clear case of autocorrelation.

The problems created by omitted variables and serial correlation in cross-sectional work are not so obvious. Again, assuming that the error terms represent a series of small omitted variables that are independent of the exogenous variables, there may be some geographical correlations of these omitted influences such that the covariances of errors across observations are not zero. This possibility, while perhaps real, is difficult to treat in a general manner but will lead to the same problems as serial correlation in time series analysis. It is reasonable to assume that the likelihood and magnitude of the serial correlation is less for cross-sectional data than for time series data, however. It is more conceivable that omitted variables differ in magnitude for cross-sectional analysis such that concentration upon heteroskedasticity with cross sections has a higher potential payoff and thus is a better general strategy.

Misspecification of functional form may also lead to the autocorrelation problem. If a linear form is being used to approximate a nonlinear relationship, there will be ranges for the explanatory variables where the estimated model consistently under- or overestimates the true values. This means that the error terms for observations with values for the explanatory variables in each of these ranges will be systematically correlated. If we are dealing with a time series model, where the explanatory variables are generally increasing over time, then successive error terms will be correlated. If the ordering of the observations does not produce a systematic ordering of the explanatory variables, the implicit covariances will not be ordered in any way, but there will be nonzero off-diagonal elements throughout $E(UU')$.

The decision to consider whether these problems exist in fact, must come from the researcher's knowledge about the data collection methods and the likelihood of systematic problems. If there are strong statistical grounds for believing a problem exists, such as when using aggregate data collected from units of vastly different sizes, one may presume a priori that the problem exists. In other circumstances, the presence of a problem is less obvious a priori.

All of these problems and situations may make the assumption that $E(UU') = \sigma^2 I$ untenable. The important questions are, What are the consequences of violating these assumptions? How do they affect the ability to estimate the true values and to make appropriate inferences about the true model?

6.3 Formal Statement of the Problem

Because of its more generalized format, we shall discuss the problems outlined above in the general matrix fashion developed in the previous chapter. The basis for our discussion is the variance–covariance matrix for the individual error terms. We write this generalized variance–covariance matrix as

$$E(UU') = \Sigma_U. \tag{6.1}$$

Σ_U is a $T \times T$ matrix with each element being the expected variance or covariance of the appropriate error terms. For example, σ_{tt} is the variance of the error term associated with the tth observation and σ_{ts} is the covariance of the error terms associated with the tth and sth observations.

As long as we can maintain the assumption that $E(U) = 0$, it is easy to see that our ordinary least squares estimates are still unbiased:

$$E(b) = \beta + E(X'X)^{-1}X'U = \beta + (X'X)^{-1}X'E(U) = \beta. \tag{6.2}$$

Thus we can see that the heteroskedastic and autocorrelated problems do not affect the unbiased property of our OLS estimates.[1]

What the problems do create however are inefficient estimators and biased estimates of the coefficient standard errors if the ordinary least squares model is used inappropriately. This property is easy to see by noting that with the assumption of fixed X,

$$E(b - \beta)(b - \beta)' = E(X'X)^{-1}X'UU'X(X'X)^{-1}$$

$$= (X'X)^{-1}X'E(UU')X(X'X)^{-1}$$

$$= (X'X)^{-1}X'\Sigma_U X(X'X)^{-1}. \tag{6.3}$$

It should be clear that, unless $\Sigma_U = \sigma^2 I$, Eq. (6.3) does not reduce to the previous expression for the variance of the estimated coefficients.

The most serious consequence of this problem is that our least squares estimator, shown in Eq. (6.2), although unbiased, is not the best, or least variance, linear unbiased estimator. In Appendix 6.1, we show this property in a proof which parallels the proof of best in Appendix 5.1. This is a fairly

---

[1]There are certain specialized cases where the estimates could be biased. If one of the explanatory variables is a lagged value of the endogenous variable, e.g., Y_{t-1}, the explanatory variables will not be independent of the error terms when there are autocorrelated errors. This is discussed in Section 6.10.

serious consequence. One of the considerations that makes the least squares estimator so desirable is its attribute of being the best linear unbiased estimator. When the error matrix is not $\sigma^2 I$, OLS no longer makes the most efficient use of the information in the data for estimating β. Given only one sample with which to estimate β, the property of best is quite desirable. This consequence, by itself, should cause us to reconsider the least squares estimator in cases where either heteroskedasticity or autocorrelation is suspected.

The second consequence of either of these problems is that OLS no longer accurately estimates the variance of the estimator with the formula developed in Chapters 3 and 5. In the homoskedastic and nonautocorrelated cases, the variance–covariance matrix for the individual coefficients is

$$\Sigma_b = \sigma^2 (X'X)^{-1}.$$

However with the problems encountered in this chapter, this is no longer an accurate estimate of the variances and covariances because

$$\sigma^2 (X'X)^{-1} \neq (X'X)^{-1} X' \Sigma_U X (X'X)^{-1}.$$

We can summarize the difficulties at this point quite simply. Heteroskedasticity and autocorrelation violate the assumption that $E(UU') = \sigma^2 I$. Consequently, OLS gives unbiased estimates, but these estimates are not the best, or most efficient, linear unbiased estimators. Secondly, the OLS formula for computing the variances of the estimated coefficients is wrong if the error terms are heteroskedastic or autocorrelated, and thus standard statistical tests are wrong. (Note that biased variance estimation is not a problem with OLS per se; it is simply that the OLS formulas based upon an incorrect error matrix are wrong.)

6.4   Generalized Least Squares

Fortunately, there is a way around these problems, provided one is willing to do some additional work and to supply additional information about the problem. This method is referred to as generalized least squares (GLS) or Aitken estimators. The key to applying this method is a knowledge of the error term variance–covariance matrix, Σ_U. (As will be discussed later, an estimate of Σ_U can often be used.) We shall use a different expression for Σ_U to simplify our presentation. Let $\Sigma_U = \sigma^2 \Omega$, where σ^2 is a scalar, and Ω summarizes the pattern of unequal variances and covariances. (In the previous case, $\Omega = I$.)

The generalized least squares estimator[2] is

$$b = (X'\Omega^{-1}X)^{-1} X'\Omega^{-1}Y = \beta + (X'\Omega^{-1}X)^{-1} X'\Omega^{-1}U. \tag{6.4}$$

[2]This is also called an Aitken estimator for an early development of it by A. C. Aitken.

It is easy to see that the ordinary least squares estimator of Chapters 2, 3, and 5 is simply a special case of this estimator with $\Omega = I$. In this case, Eq. (6.4) reduces to the estimator in Eq. (5.10), and ordinary least squares is a specific application of the generalized least squares estimator.

Statistical Properties

The statistical properties of the generalized least squares estimates for the heteroskedastic and autocorrelated cases are quite similar to those of ordinary least squares in Chapters 3 and 5. In addition to the previous assumption that X and U are independent, we shall now assume that Ω is fixed and does not vary from replication to replication. From this assumption, and Eq. (6.4) we can see that the generalized least squares estimator is unbiased:

$$E(b) = \beta + (X'\Omega^{-1}X)^{-1}X'\Omega^{-1}E(U) = \beta. \tag{6.5}$$

Thus it is a linear unbiased estimator.

The coefficient variance–covariance matrix is

$$E(b-\beta)(b-\beta)' = E(X'\Omega^{-1}X)^{-1}X'\Omega^{-1}UU'\Omega^{-1}X(X'\Omega^{-1}X)^{-1}$$

$$= (X'\Omega^{-1}X)^{-1}X'\Omega^{-1}E(UU')\Omega^{-1}X(X'\Omega^{-1}X)^{-1}$$

$$= \sigma^2(X'\Omega^{-1}X)^{-1}. \tag{6.6}$$

Again, in the standard case where $\Omega = I = \Omega^{-1}$, the variance of the GLS estimators is identical to that of the OLS estimators—further demonstrating that OLS is a special case of GLS.

To estimate the variances and covariances of the estimated coefficients we need an estimate for σ^2. As in the OLS model, this estimate is obtained from the residuals of the estimated equation. Only now this estimated equation is

$$\hat{Y} = Xb = X(X'\Omega^{-1}X)^{-1}X'\Omega^{-1}Y = X\beta + X(X'\Omega^{-1}X)^{-1}X'\Omega^{-1}U.$$

The residuals are

$$e = Y - \hat{Y} = \left[I - X(X'\Omega^{-1}X)^{-1}X'\Omega^{-1} \right]U.$$

The appendix to this chapter shows that an unbiased estimator for σ^2 is

$$s^2 = e'\Omega^{-1}e/(T-K).$$

Thus the estimate for the variance of b_k is

$$s_{b_k}^2 = s^2(X'\Omega^{-1}X)^{-1}_{kk}. \tag{6.7}$$

The most important aspect of the generalized estimator however is the property that it is the best (minimum variance) linear unbiased estimator. This is demonstrated in the appendix to this chapter. The proof begins by noting that the generalized least squares estimator is an alternative linear unbiased estimator to OLS. Thus another way to write the GLS model is $(X'X)^{-1}X' + D$, which is the expression for any other alternative linear estimator (see the discussion of best in Chapters 3 and 5). The criteria of unbiasedness implies a constraint on D such that $DX = 0$. The proof in the appendix demonstrates that choosing D to minimize the variance of the estimator, subject to the constraint that $DX = 0$, yields

$$D = (X'\Omega^{-1}X)^{-1}X'\Omega^{-1} - (X'X)^{-1}X',$$

which results in the generalized least squares estimator.

The discussion of the problems of heteroskedasticity and autocorrelation can be easily summarized at this point. These violations of the OLS assumption about the structure of the variances and covariances of the implicit error terms are contained in the matrix $E(UU') = \sigma^2\Omega$, $\Omega \neq I$. The consequences of this condition are inefficient but unbiased estimates of the equation's parameters and erroneous estimates of the variances of these estimates. These consequences can be remedied however by using the generalized least squares estimator shown in Eq. (6.4). This GLS estimator is linear, unbiased, and best among the set of unbiased linear estimators. The true variances and covariances of these estimates are given by Eq. (6.6), and their estimated variance by Eq. (6.7).

The proof that generalized least squares estimators are BLUE relies upon knowing the true error variance–covariance matrix $\Sigma_U = \sigma^2\Omega$. Clearly, there are few actual estimation situations where the true Ω is known. However, there are many situations where it is possible to use an estimate of Ω to improve the estimates. This chapter discusses various ways of estimating Ω and using such estimates in the GLS format. In many circumstances, it is possible to use the residuals in an OLS estimation to estimate Ω. The main requirement for doing this is that we must be able to specify the structure generating the error terms, such as the factors that cause their variances to differ or the patterns of correlations among successive terms. The complete matrix of residuals ee' from an OLS estimation is singular and therefore cannot itself be used to estimate Σ_U. The next sections describe plausible structures for the error terms and how these can be used to estimate Ω.

When an estimated error matrix $\hat{\Omega}$ is used in GLS estimation, the coefficient estimates can be shown to be asymptotically best; i.e., as the sample gets very large, these estimators are more efficient than any other linear unbiased estimator. The properties of GLS estimators in small samples are generally unknown; however, there is some evidence that they perform better than OLS estimators when $\hat{\Omega}$ is reasonably well estimated and when divergences from an error matrix of $\sigma^2 I$ are significant.

6.5 Generalized Least Squares and Examples
of Heteroskedasticity and Autocorrelation

The best way to visualize the generalized least squares estimator is for us to demonstrate its application to two simple cases. The first is the hetero-skedastic situation, where the error terms for each observation are drawn from distributions with different variances, but independently of the other error terms. Thus we write $E(UU')$ as

$$E(UU')=\Sigma_U=\sigma^2\begin{bmatrix} h_1 & 0 & 0 & \cdots & 0 \\ 0 & h_2 & 0 & \cdots & 0 \\ 0 & 0 & h_3 & \cdots & 0 \\ \vdots & \vdots & \vdots & \vdots & \vdots \\ 0 & 0 & 0 & \cdots & h_T \end{bmatrix}=\sigma^2\Omega,$$

where $E(U_t^2)=\sigma_t^2=\sigma^2 h_t$, where each h_t is an observation specific scalar. Then

$$\Omega^{-1}=\begin{bmatrix} 1/h_1 & 0 & 0 & \cdots & 0 \\ 0 & 1/h_2 & 0 & \cdots & 0 \\ 0 & 0 & 1/h_3 & \cdots & 0 \\ \vdots & \vdots & \vdots & \vdots & \vdots \\ 0 & 0 & 0 & \cdots & 1/h_T \end{bmatrix}.$$

Consider the expressions $X'\Omega^{-1}X$ and $X'\Omega^{-1}Y$. We can write their individual elements as $\Sigma_{t=1}^T(X_{ti}X_{tj}/h_t)$ and $\Sigma_{t=1}^T(X_{ti}Y_t/h_t)$, respectively. Thus in creating the sum of cross products of the variables, each observation is weighted in inverse proportion to the variance of the error term associated with that observation $(1/h_t)$. This should make intuitive sense in that we want to give more weight to those observations that have the least error and less weight to those with larger error terms.

The simplest form of the autocorrelated case is where we assume that the error term in period t, U_t, is a function of the error term in period $t-1$ plus a random component. We write this as

$$U_t=\rho U_{t-1}+\varepsilon_t. \tag{6.8}$$

where $E(\varepsilon\varepsilon')=\sigma^2 I$, $|\rho|<1$, and ε_t is independent of U_{t-1}. This expression is called a first order autoregressive process (because each error term is related only to the immediately preceding one). ρ is called the serial correlation coefficient. By extending Eq. (6.8) backward, we can write each U_t as a function of only previous values of ε_t:

$$U_t = \rho^2 U_{t-2} + \rho\varepsilon_{t-1} + \varepsilon_t = \rho^3 U_{t-3} + \rho^2\varepsilon_{t-2} + \rho\varepsilon_{t-1} + \varepsilon_t = \sum_{i=0}^{t} \rho^i \varepsilon_{t-i}.$$

From this expression,[3] we can write the variances and covariances for each element in Σ_U,

$$E(U_t^2) = E(\varepsilon_t^2) + \rho^2 E(\varepsilon_{t-1}^2) + \rho^4 E(\varepsilon_{t-2}^2) + \cdots$$

$$= \sigma^2(1 + \rho^2 + \rho^4 + \cdots) = \frac{\sigma^2}{1-\rho^2} = \sigma_U^2,$$

$$E(U_t U_{t-s}) = \sigma_U^2 \rho^{|s|}.$$

These expressions for $E(U_t^2)$ and $E(U_t U_{t-s})$ imply that the expected correlation between U_t and U_{t-1} is ρ, a fact used subsequently in applying the GLS estimator. Consequently,

$$E(UU') = \sigma_U^2 \begin{bmatrix} 1 & \rho & \rho^2 & & \rho^{T-1} \\ \rho & 1 & \rho & & \rho^{T-2} \\ \vdots & \vdots & \vdots & & \vdots \\ \rho^{T-1} & \rho^{T-2} & \rho^{T-3} & \cdots & 1 \end{bmatrix} = \sigma_U^2 \Omega.$$

If we invert this matrix, we have

$$\Omega^{-1} = \frac{1}{1-\rho^2} \begin{bmatrix} 1 & -\rho & 0 & \cdots & 0 & 0 \\ -\rho & 1+\rho^2 & -\rho & \cdots & 0 & 0 \\ 0 & -\rho & 1+\rho^2 & \cdots & 0 & 0 \\ \vdots & \vdots & \vdots & & \vdots & \vdots \\ 0 & 0 & 0 & \cdots & -\rho & 1 \end{bmatrix}.$$

(This can be checked by multiplying Ω^{-1} times Ω.) For this autocorrelated model, it will be easier to give an intuitive description of the meaning and implication of Ω^{-1} after the next section.

6.6 Generalized Least Squares and Weighted Regression

Before trying to discuss the implications of generalized least squares and to provide an expanded description of its logic, we want to present an alternative method of computing the GLS estimates. This alternative method—weighted regression—is often simpler because it can be done within

[3]We must assume that the series extends prior to the beginning of our sample of observations, so that even our first observation corresponds to a high value for t. As t becomes large with $|\rho| < 1$, the series $(1 + \rho^2 + \rho^4 + \cdots) = 1/(1-\rho^2)$.

most standard computer programs. It gets its name because the generalized least squares estimates are computed by weighting or transforming each observation of X and Y and applying OLS to the transformed values rather than computing Ω^{-1} and $X'\Omega^{-1}X$. Once this alternative method is understood, it is much easier to give a description of what GLS is doing and why it is preferable to OLS.

The key to this alternative procedure is to define a new matrix F, such that $F'F = \Omega^{-1}$, where F is a $T \times T$ matrix. Once we have obtained this matrix, we create new variables $X^* = FX$ and $Y^* = FY$ and compute the ordinary least squares estimators based on X^* and Y^*,

$$b = (X^{*\prime}X^*)^{-1}X^{*\prime}Y^* = (X'F'FX)^{-1}X'F'FY = (X'\Omega^{-1}X)^{-1}X'\Omega^{-1}Y. \quad (6.9)$$

Thus by choosing the appropriate values for F, so that $F'F = \Omega^{-1}$, we can compute the generalized least squares estimator simply using a transformation of the original variables and an ordinary least squares regression program.

The key question is, How do we determine F? For the heteroskedastic case it is simple. Remember that

$$E(UU') = \sigma^2\Omega = \sigma^2 \begin{bmatrix} h_1 & 0 & 0 & \cdots & 0 \\ 0 & h_2 & 0 & \cdots & 0 \\ 0 & 0 & h_3 & \cdots & 0 \\ \vdots & \vdots & \vdots & & \vdots \\ 0 & 0 & 0 & \cdots & h_T \end{bmatrix} \quad \text{and}$$

$$\Omega^{-1} = \begin{bmatrix} 1/h_1 & 0 & 0 & \cdots & 0 \\ 0 & 1/h_2 & 0 & \cdots & 0 \\ 0 & 0 & 1/h_3 & \cdots & 0 \\ \vdots & \vdots & \vdots & & \vdots \\ 0 & 0 & 0 & \cdots & 1/h_T \end{bmatrix}.$$

It should be easy to see that F is simply a diagonal matrix with the reciprocal of the square root of each h_t along this diagonal:

$$F = \begin{bmatrix} 1/\sqrt{h_1} & 0 & 0 & \cdots & 0 \\ 0 & 1/\sqrt{h_2} & 0 & \cdots & 0 \\ 0 & 0 & 1/\sqrt{h_3} & \cdots & 0 \\ \vdots & \vdots & \vdots & & \vdots \\ 0 & 0 & 0 & \cdots & 1/\sqrt{h_T} \end{bmatrix}.$$

Since h_t is the ratio of σ_t^2 to σ^2 ($\sigma_t^2 = h_t\sigma^2$), when we use F to create X^* and Y^*, we are effectively weighting each observation by one over the square root of this ratio. The larger the variance of U_t, the smaller is $1/h_t$ and the smaller the relative weight given to that observation in computing the estimated coefficients. Conversely, the smaller the variance of U_t, the larger is $1/h_t$ and the more weight that observation receives.

In general, one must first decide the appropriate values for each h_t, from which the values of F can be computed. Estimating h_t often is a difficult process based both on prior knowledge of the model, data, and observations and on ex post analysis of these data. In the example using aggregate data mentioned above, h_t is simply $1/N_t$, the number of individuals composing the aggregate unit. Consequently, the diagonal elements of F equal $\sqrt{N_t}$, and we want to weight each observation by this value. Additional patterns for h_t and diagonostic procedures are discussed subsequently.

We want to digress here to point out that the motivation for the weighted analysis just mentioned is quite different from the motivation in many social science studies. In studies employing sample specific types of statistics, such as cross-tabulation tables, correlations, and standardized coefficients, observations drawn from different groups are assigned weights to make their proportion of the weighted sample equal to their proportion in the total population. Presumably the variables' means and variances then estimate the population values. For example, a sample survey may have oversampled blacks to provide information for separate studies of their political attitudes and behavior. A researcher wanting to use the entire sample to make T as large as possible will then assign white respondents a greater weight than blacks to make the black proportion of the weighted sample (given by the number of blacks divided by the sum of the weights) similar to the proportion black in the country's population. This weighting is required when sample specific statistical procedures are being used. In the case of our estimation model, we get unbiased estimates of the population parameters regardless of which members of the population are included. (The core assumption of our analysis of course is that all units can be represented by the same population model. If this is not the case, we need to estimate separate coefficients for each population, and no weighting scheme is appropriate for combining observations from different populations.) The only times we talk about estimating weighted regressions is when the error terms associated with the observations have unequal variances. This weighting is done then to improve the efficiency of our estimates. It has no effect on the expected value of the coefficients, which still have the population values as their means.

The autocorrelated case is more difficult to describe. The conventional practice for the first order autocorrelated model described in Eq. (6.8) is to define F as

$$F = \begin{bmatrix} -\rho & 1 & 0 & 0 & \cdots & 0 & 0 \\ 0 & -\rho & 1 & 0 & \cdots & 0 & 0 \\ 0 & 0 & -\rho & 1 & \cdots & 0 & 0 \\ \cdot & \cdot & \cdot & \cdot & & \cdot & \cdot \\ \cdot & \cdot & \cdot & \cdot & & \cdot & \cdot \\ \cdot & \cdot & \cdot & \cdot & & \cdot & \cdot \\ 0 & 0 & 0 & 0 & & -\rho & 1 \end{bmatrix},$$

where F is $(T-1) \times T$. This is equivalent to creating the new variables

$$X_{t,k}^* = X_{t,k} - \rho X_{t-1,k} \quad \text{and} \quad Y_t^* = Y_t - \rho Y_{t-1}.$$

However, this definition of F gives a product $F'F$ that does not exactly equal Ω^{-1}, but differs only in the upper left-hand corner element.[4] The consequences of using $F'F$ rather than Ω^{-1} are only slight, particularly if there are a large number of observations.

The intuitive explanation for this transformation is quite easy to see at this point. We can write $U_{t-1} = Y_{t-1} - X_{t-1}\beta$. Given that our model is $U_t = \rho U_{t-1} + \varepsilon_t$, we can write the expression for Y_t as

$$Y_t = X_t\beta + U_t = X_t\beta + \rho U_{t-1} + \varepsilon_t = X_t\beta + \rho(Y_{t-1} - X_{t-1}\beta) + \varepsilon_t \quad (6.9a)$$

or

$$Y_t - \rho Y_{t-1} = (X_t - \rho X_{t-1})\beta + \varepsilon_t. \quad (6.9b)$$

This gives the simple expression $Y_t^* = X_t^*\beta + \varepsilon_t$. Since ε_t is assumed to be independently distributed across observations with mean zero and variance σ^2, we can appropriately use OLS to estimate the transformed equation. In the autocorrelated case the "weighted" regression model is actually using partially first differenced variables, the amount of differencing being dependent upon the covariance among successive error terms.

6.7 Monte Carlo Simulation
of Generalized Least Squares

We have simulated both the above problems to give an alternative view of the results associated with using OLS and to illustrate how GLS remedies these difficulties. The first simulation is the heteroskedastic case and illustrates the problems encountered using aggregate data from different sized groups. The second simulation is the simple first order autocorrelated model illustrated above.

[4] The proper element of Ω^{-1} is $w^{11} = 1$ while $(F'F)_{11} = \rho^2$. These terms are equal only if $\rho = 1$, or if U_t and U_{t-1} are perfectly correlated.

Heteroskedasticity with Aggregate Data

The heteroskedastic simulation is done using the same 50 observations used in the earlier simulation in Section 3.3 with $T = 50$, the two variables X_2 and X_3, and the model $Y = 15 + X_2 + 2X_3 + U$. To simulate the situation with aggregate data, we drew 50 random numbers from a uniform distribution with the range 50 to 10,050 and treated each number as the population N_t of one of the 50 units from which we are drawing observations. We then treated the values for X_2 and X_3 as the mean values for those variables in each observational unit. Thus the systematic population model is $E(\bar{Y}) = 15 + \bar{X}_2 + 2\bar{X}_3$ and the implicit error term is the mean error term for all individuals in the unit $\bar{U}$. As pointed out above, the variance of $\bar{U}$ about the true value of zero is inversely related to the number of people in the unit. To complete one replication of the simulation, we effectively drew error terms for each person in the aggregate from a normal distribution with mean zero and standard deviation equal 1000, averaged them, and added this average to the systematic component given by $E(\bar{Y})$. This was done for all 50 observations. These 50 observations, which now simulate the usual situation one faces with aggregate data, are used to estimate the specified equation. The expectation with heteroskedasticity is that the coefficients estimated for each of the 400 OLS replications are unbiased, but that each will have a greater variance than coefficients estimated with GLS and that we shall not estimate each variance correctly. The estimates are also made with the appropriate GLS procedure, multiplying each value of X_{t2}, X_{t3}, and Y_t by the corresponding value of $\sqrt{N_t}$.

TABLE 6.1

Heteroskedastic Simulation [a]

	$\bar{b}_1$	$\hat{\sigma}_{b_1}$	$\bar{s}_{b_1}$	$\bar{b}_2$	$\hat{\sigma}_{b_2}$	$\bar{s}_{b_2}$	$\bar{b}_3$	$\hat{\sigma}_{b_3}$	$\bar{s}_{b_3}$	$\bar{s}$
OLS	15.31	8.12	11.81	1.00	0.35	0.27	1.94	1.13	1.33	24.84
GLS	15.21	7.02	6.65	0.99	0.17	0.16	1.98	0.73	0.73	1002.23

[a] $\bar{b}_k$ is the mean of the estimated coefficients; $\hat{\sigma}_{b_k}$ is the observed standard deviations of the b_k; $\bar{s}_{b_k}$ is the mean estimated standard error of each coefficient, Eq. (6.7); and $\bar{s}$ is the mean standard error of estimate.

The results of both the OLS and GLS estimates are shown in Table 6.1. The parameters estimated by OLS and GLS are unbiased as shown by the closeness of the mean estimated coefficients to the true values. What these simulations show quite clearly is the inefficiency of OLS when faced with the heteroskedastic problem: the standard deviations of the coefficients estimated by OLS are considerably higher than those of the GLS estimates, by a factor of two in the case of b_2. Finally, we can see that OLS does not provide accurate estimates of the variances of the estimated coefficients while GLS does. Furthermore, the OLS estimates of these coefficient variances do not deviate from the true values in any systematic fashion. In the case of b_1 and b_3 the variance is overestimated, while it is underestimated for b_2. We can see from Eq. (6.3) that both the extent of this bias and its direction are related to $(X'X)^{-1}$, $X'\Omega X$, and their products.

Autocorrelation Simulation

We have simulated the simple first order autocorrelated model for varying degrees of correlation among the error terms. This is done for our model with two explanatory variables and $T = 50$ and is accomplished by specifying each error term to be a function of the preceding error term plus a random component:

$$U_t = \rho U_{t-1} + \varepsilon_t.$$

We have let ρ take on values ranging from 0.00 to 0.90 and have assumed the variance of ε_t to be 36. The generalized least squares estimates were computed using the transformations $X_t^* = X_t - \rho X_{t-1}$ and $Y_t^* = Y_t - \rho Y_{t-1}$. The results of these simulations, with 200 replications, are shown in Table 6.2.

TABLE 6.2

Autocorrelated Simulation[a]

	OLS			GLS				Relative variances (GLS/OLS)
ρ	$\bar{b}_1$	$\hat{\sigma}_{b_1}$	$\bar{s}_{b_1}$	$\bar{b}_1$	$\hat{\sigma}_{b_1}$	$\bar{s}_{b_1}$	σ_{b_1}	
0.00	14.99	3.07	3.08	14.91	3.72	3.61	3.69	1.21
0.30	14.76	3.40	3.19	14.60	3.51	3.47	3.53	1.03
0.60	15.43	3.95	3.69	15.31	3.47	3.56	3.61	0.88
0.90	14.47	10.43	6.00	14.63	9.17	9.01	9.03	0.88
	$\bar{b}_2$	$\hat{\sigma}_{b_2}$	$\bar{s}_{b_2}$	$\bar{b}_2$	$\hat{\sigma}_{b_2}$	$\bar{s}_{b_2}$	σ_{b_2}	
0.00	0.997	0.058	0.060	0.998	0.067	0.066	0.067	1.16
0.30	1.005	0.069	0.062	1.007	0.066	0.064	0.065	0.96
0.60	0.995	0.083	0.072	0.995	0.056	0.057	0.058	0.67
0.90	1.000	0.158	0.117	0.997	0.049	0.050	0.050	0.31
	$\bar{b}_3$	$\hat{\sigma}_{b_3}$	$\bar{s}_{b_3}$	$\bar{b}_3$	$\hat{\sigma}_{b_3}$	$\bar{s}_{b_3}$	σ_{b_3}	
0.00	2.001	0.034	0.032	2.002	0.038	0.036	0.037	1.12
0.30	2.002	0.039	0.034	2.003	0.038	0.034	0.035	0.97
0.60	1.998	0.041	0.039	2.001	0.034	0.031	0.031	0.83
0.90	2.002	0.067	0.063	2.001	0.029	0.027	0.027	0.43

[a] $\bar{b}_k$ is the mean of the coefficients estimated for each replication; $\hat{\sigma}_{b_k}$ is the computed standard deviation of the estimated coefficients; $\bar{s}_{b_k}$ is the mean estimated standard error of the coefficient in each replication; and σ_{b_k} is the true coefficient standard error from the differenced regression.

Again, we can see the predictions of our theoretical development. Both the OLS and the GLS estimators are unbiased, as expected. Furthermore, except for the similation where $\rho = 0.00$ (no serial correlation) the OLS estimates have a greater variance than the GLS estimates.[5] The last column in Table 6.2 shows the relative variances of the coefficients estimated by OLS and the GLS procedure. For no autocorrelation or small levels of autocorrelation ($\rho = 0.3$), the GLS procedure has little advantage over OLS. However, as the strength of the autocorrelation increases, GLS offers a considerable reduction in the variance of the estimated parameters. The ratio of the variance of the coefficients estimated under different procedures is called the relative efficiency of the different estimation methods. With high levels of autocorrelation, the GLS estimator becomes much more efficient than the OLS estimator. It is also true that the OLS estimates of these coefficient variances are biased, while GLS provides accurate estimates of these terms. Finally, Table 6.3 shows the estimates for σ^2 with each procedure. The OLS estimates are badly biased in cases of more serious autocorrelation. They do not estimate either σ or σ_U, but some value between them. The GLS estimates on the other hand are unbiased estimates of σ_ε. The most serious consequence of misestimating σ_U is that it further biases the OLS estimates of the coefficient variances, making hypothesis testing meaningless.

TABLE 6.3
OLS *and* GLS *Estimates of* σ *for Autocorrelated Model*

	0.00	0.30	0.60	0.90
OLS	5.89	6.10	7.04	11.45
True σ_U	6.00	6.29	7.50	13.76
GLS	5.88	5.90	5.92	5.99
True σ_ε	6.00	6.00	6.00	6.00

The bias in the estimated variance of the coefficient depends both upon the autocorrelation in the error terms and the autocorrelation in the explanatory variables. When there is positive autocorrelation in both the error terms and the explanatory variables, OLS estimates of the coefficient variances will be underestimated. Thus, we tend to reject null hypotheses about particular parameters when we actually should not (for a given level of confidence). In other words, we accept the alternative hypothesis more often than we should, and our statistical tests are weakened. Just the opposite is true if the explanatory variables and the error terms are both negatively autocorrelated. When the autocorrelation of the error terms and the explanatory variables are of opposite signs, there is no simple rule about the direction of bias. However,

[5]The GLS variance is greater than the OLS variance because the procedure entails the loss of one observation. With the exact GLS estimator, which did not entail the loss of the first observation, OLS and GLS would be identical for $\rho = 0$.

by far the most common situation is the case where both error terms and explanatory variables have positive autocorrelation—a simple reflection of the fact that many social phenomena tend to be smooth over time and do not change radically from period to period.

6.8   Generalized Least Squares in Practice

The immediate question is, How do I know if the data and model violate the assumption that $E(UU') = \sigma^2 I$? And if they do, How do I determine Ω? The first, and best, way is on the basis of prior knowledge about the data and the model. The example of heteroskedasticity in cases of aggregate data based on different sized units illustrates this approach. The knowledge that the data consist of means calculated over different numbers of individuals provides the basis for assuming that the error terms are not of equal variance. The rules of statistical sampling then indicate what the appropriate values of Ω should be. However, there are many cases where prior knowledge and statistical rules do not provide all the answers. This section addresses ways of testing the hypothesis that the error terms are uncorrelated and of equal variance and of estimating the appropriate corrections.

All of the tests and corrective procedures rely on the fact that the OLS estimates are not biased and inconsistent, but only inefficient. In this regard the assumptions about $E(UU')$ are much easier to deal with than the assumption that $E(U) = 0$. This latter assumption cannot be tested with the observed data because $\bar{e} = 0$ and $X'e = 0$ by construction. In the case of heteroskedasticity and autocorrelation we can use the results of an OLS estimate to test the assumption that $E(UU') = \sigma^2 I$. Because the estimated coefficients are consistent, as the sample size increases the estimated coefficients come closer and closer to the true values. Consequently, the residuals from the OLS estimates come closer and closer to being the true error terms. This fact enables us to test assumptions about their variances and covariances.

Heteroskedasticity

The test for heteroskedastic disturbances must begin with some hypotheses about which observations are expected to have error terms with different variances. For example, in some cases we may want to consider whether the variance of the error term varies with the values of one of the explanatory variables,[6] $\sigma_t^2 = f(X_{t,k})$. With a very large sized sample, containing many observations for each discrete value of X_k, all we need do is obtain the residuals from the OLS estimation, group these according to the values of X_k,

---

[6]This specification has been used in estimating consumption functions, where a family's consumption C is a function of either current or lagged income Y_t or Y_{t-1}. The hypothesis is that the variance about the true model $C = \beta_1 + \beta_2 Y$ will increase as income increases because of the wider range of options available to higher income individuals. Thus $\sigma_t^2 = \sigma^2 Y_t$.

and calculate the estimated variance for the residuals corresponding to these values for the explanatory variables. These estimates then become the diagonal entries in Ω. We can also use these estimated variances and the appropriate form of the F-test to test statistically the null hypothesis that the variances for each value of X_k are equal.[7]

In most situations, however, we cannot hope to have a sample large enough to contain many values for each explanatory variable. The best we can do at this point is to group the observations for given ranges of values of the explanatory variable. We can then compute the variance of the residuals within each group. Although this does not permit us to estimate the value of σ^2 for each observation, it will give us an idea about how the error variance varies over the range of the values of the explanatory variable. If one assumes that the error variances for all the observations in a given grouping are equal, then the F-test is appropriate for statistically testing the null hypothesis that the variances are equal across groups, and this residual analysis estimates σ_t^2 for each group. If the variance of U_t changes systematically with the values of the explanatory variable, this computation and the F-test are inappropriate. It is possible to construct both parametric and nonparametric tests of heteroskedasticity, although the power of these tests is unknown. Of course, if the explanatory variable is already grouped, such as with dichotomous and other ordinally grouped variables, this grouping has already been accomplished by the construction of the variable, and we can proceed as if we have a large number of observations for each value of X_k.

If one accepts the idea that the error variances increase systematically with the values of X_k, $\sigma_t^2 = aX_{t,k}$, then the appropriate weighting scheme is to deflate each variable, including the dependent variable and the constant term, by the square root of X_k. To see this, note that since $\sigma_t^2 = aX_{t,k}$,

$$E(UU') = \Sigma_U = a \begin{bmatrix} X_{1,k} & 0 & 0 & \cdots & 0 \\ 0 & X_{2,k} & 0 & \cdots & 0 \\ 0 & 0 & X_{3,k} & \cdots & 0 \\ \cdot & \cdot & \cdot & & \cdot \\ \cdot & \cdot & \cdot & & \cdot \\ \cdot & \cdot & \cdot & & \cdot \\ 0 & 0 & 0 & \cdots & X_{T,k} \end{bmatrix} = a\Omega,$$

$$\Omega^{-1} = \begin{bmatrix} 1/X_{1,k} & 0 & 0 & \cdots & 0 \\ 0 & 1/X_{2,k} & 0 & \cdots & 0 \\ 0 & 0 & 1/X_{3,k} & \cdots & 0 \\ \cdot & \cdot & \cdot & & \cdot \\ \cdot & \cdot & \cdot & & \cdot \\ \cdot & \cdot & \cdot & & \cdot \\ 0 & 0 & 0 & \cdots & 1/X_{T,k} \end{bmatrix}.$$

[7]Different tests are proposed in Goldfeld and Quandt (1965, pp. 539–547). Their parameter test involves ranking the observations by X_k, dividing them into three groups, and estimating separate OLS regressions for the two extreme groups. An F-test for equality of residual variances is then used. See, also, Theil (1971, pp. 196–198).

Remember the most computationally efficient way to calculate the GLS estimates for this model is to create the new variables $Y^* = FY$ and $X^* = FX$, where $F'F = \Omega^{-1}$, and use OLS to compute $b = (X^{*\prime}X^*)^{-1}X^{*\prime}Y^*$. In this case the appropriate transformation is

$$
F = \begin{bmatrix}
1/\sqrt{X_{1,k}} & 0 & \cdots & 0 \\
0 & 1/\sqrt{X_{2,k}} & \cdots & 0 \\
\vdots & \vdots & & \vdots \\
0 & 0 & \cdots & 1/\sqrt{X_{T,k}}
\end{bmatrix},
$$

so that $Y_t^* = Y_t/\sqrt{X_{t,k}}$, $X_{t,1}^* = 1/\sqrt{X_{t,k}}$, $X_{t,2}^* = X_{t,2}/\sqrt{X_{t,k}}$, $X_{t,k}^* = \sqrt{X_{t,k}}$, and so on. It should be easy at this point to see how this procedure can be expanded to accommodate other hypotheses about the relationship between σ_t^2 and $X_{t,k}$. For example, if $\sigma_t^2 = aX_{t,k}^2$, $F_t = 1/X_{t,k}$ and $Y_t^* = Y_t/X_{t,k}$, $X_{t1}^* = 1/X_{t,k}$, $X_{t,K}^* = 1$, and $X_{t,K}^* = X_{t,K}/X_{t,k}$.

Clearly, the transformations change the variance of Y. Therefore, comparisons of the R^2 values in the transformed cases are inappropriate.

Heteroskedasticity—Two Examples

We want to present two examples where the error terms in a simple model are hypothesized to have unequal variances and where generalized least squares, or weighted regressions, can be used to overcome the problem.

The first example is drawn from a study by Bruce Russett attempting to answer the question, Who pays for defense? by estimating the effect of defense expenditures on other expenditures in the economy, Russett (1970). Russett's basic model analyzes the effect of defense expenditures on aggregate private consumption by estimating the model

$$C_t/GNP_t = \beta_1 + \beta_2 D_t/GNP_t + U_t,$$

where C_t is the aggregate private consumption in year t, D_t the national defense expenditures in year t, and GNP_t the gross national product in year t. The implication of the model is that, for each additional one percent of GNP spent on defense, consumption as a percent of GNP changes by β_2, or for each additional dollar of defense expenditures, consumption changes by β_2 dollars. The model of aggregate consumption implied by this model is

$$C_t = \beta_1 GNP_t + \beta_2 D_t + V_t,$$

so that β_1 is the aggregate marginal propensity to consume from current income. Recognizing the similarity of this model to aggregate consumption functions estimated in macroeconomics, a more appropriate consumption function should at least include a constant term:

$$C_t = \beta_0 + \beta_1 GNP_t + \beta_2 D_t + V_t.$$

This model[8] implies that consumption equals a constant term β_0 plus some proportion of national income β_1 minus some fraction of each dollar spent in the defense budget β_2.

If V_t satisfies our previous assumptions about having an expected value of zero and equal variances for all t, it would be a simple matter to estimate the equation for the post World War II period with the data in Table 6.4. However this second assumption may not be appropriate. It is quite conceivable that the standard deviations of each V_t increase with the magnitude of GNP_t. For example, in 1946 with a gross national product of only three hundred billion dollars and consumption of two hundred billion, the likelihood of an error term over a given magnitude, say ten billion, is probably less than the probability of V_t exceeding this level in 1974 when GNP was almost three times as great and consumption was over five hundred billion dollars. This expectation leads to the following hypothesis about the relationship between the variance of V_t and the level of GNP:

$$E\left(V_t^2\right) = \sigma_{V_t}^2 = \sigma^2 GNP_t^2,$$

or the variance of the error term V_t is proportional to the square of the gross national product in year t, which clearly violates the homoskedastic assumption.

As discussed above, the way to correct for this heteroskedasticity is to divide the observations for each variable, including the constant term, by GNP and estimate the following model:

$$C_t / GNP_t = \beta_0 / GNP_t + \beta_1 + \beta_3 D_t / GNP_t + U_t$$

where $U_t = V_t / GNP_t$. If our hypothesis about the variance of each V_t is correct, then the variance of U_t should be constant for all observations:

$$E\left(U_t^2\right) = E\left(V_t^2 / GNP_t^2\right) = \sigma_{V_t}^2 / GNP_t^2 = \sigma^2.$$

The data in the remainder of Table 6.4 are used to estimate this equation. The term $1/GNP$ must be included, unless one a priori assumes that $\beta_0 = 0$, meaning that aggregate consumption is simply proportional to GNP and to defense expenditures. The two estimated equations and the standard errors are

$$C_t = 26.19 + 0.6248\,GNP_t - 0.4398\,D_t, \qquad R^2 = 0.999, \quad SEE = 0.406,$$
$$(2.73) \quad (0.0060) \qquad\quad (0.0736)$$

$$C_t / GNP_t = 25.92(1/GNP_t) + 0.6246 - 0.4315\,D_t / GNP_t,$$
$$(2.22) \qquad\qquad\quad (0.0068) \quad (0.0597)$$

$$R^2 = 0.875, \quad SEE = 0.008.$$

[8]We shall pursue this model for pedagogical purposes, although there are clear deficiencies in the equation. The most obvious problem is that the definition of gross national product includes consumption, so that the model uses a variable that includes consumption expenditures to explain consumption. Consequently, it is hard to say whether β_1 measures the average proportion of GNP accounted for by consumption, the marginal propensity to consume, or some combination of both. Thus, there is a problem of simultaneous determination. Estimating models with similar problems is discussed in Chapter 9.

TABLE 6.4
Data for Consumption–Defense Expenditure Example (1958 dollars)[a]

Year	C_t	GNP_t	D_t	C_t/GNP_t	$1/GNP_t$	D_t/GNP_t
1946	215.0	312.6	22.0	0.688	0.0032	0.071
1947	215.3	309.9	12.2	0.695	0.0032	0.039
1948	218.2	323.7	13.4	0.674	0.0031	0.042
1949	223.5	324.2	16.8	0.689	0.0031	0.052
1950	238.3	355.3	17.6	0.671	0.0028	0.050
1951	240.9	383.5	39.2	0.628	0.0026	0.102
1952	247.8	395.1	52.5	0.627	0.0025	0.133
1953	260.4	412.8	55.1	0.631	0.0024	0.134
1954	263.9	407.0	46.0	0.648	0.0025	0.113
1955	280.0	438.0	42.5	0.639	0.0023	0.097
1956	283.8	446.0	42.9	0.636	0.0022	0.096
1957	288.6	452.5	45.3	0.638	0.0022	0.100
1958	290.2	447.4	45.9	0.649	0.0022	0.103
1959	306.1	475.8	45.2	0.643	0.0021	0.095
1960	314.8	487.7	43.5	0.646	0.0021	0.089
1961	320.4	497.1	45.7	0.644	0.0020	0.092
1962	335.7	529.7	48.8	0.634	0.0019	0.092
1963	349.9	551.0	47.4	0.635	0.0018	0.086
1964	368.6	581.0	45.9	0.634	0.0017	0.079
1965	390.4	617.9	45.2	0.632	0.0016	0.073
1966	409.3	658.2	53.3	0.622	0.0015	0.081
1967	418.5	675.1	61.6	0.620	0.0015	0.091
1968	438.4	706.6	64.0	0.620	0.0014	0.091
1969	452.0	725.6	61.1	0.623	0.0014	0.084
1970	456.7	722.5	55.2	0.632	0.0014	0.076
1971	471.9	746.3	50.4	0.632	0.0013	0.067
1972	498.9	792.5	51.2	0.630	0.0013	0.065
1973	521.8	839.2	48.2	0.622	0.0012	0.057
1974	515.2	821.1	46.2	0.627	0.0012	0.056
1975	521.1	804.6	46.0	0.648	0.0012	0.057

[a]Data from Council of Economic Advisors, *Economic Report of the President, 1976*. All data were converted to real 1958 dollars.

We can see that both give nearly identical coefficient estimates, but quite different estimated coefficient standard errors. (Note that the transformation of the variables changes the variance of the dependent variable. Thus, the R^2 values and the standard errors of estimate are not comparable.) With only 30 observations, it is virtually impossible to apply any statistical tests for whether the error variance is proportional to GNP squared, so one must choose between these two estimated equations on the basis of prior belief about the error structures.

Our second example is a further exploration of verbal achievements, this time with a new data source. Given the limited and nonexperimental nature of social science data, the only possible way to further test and refine the results from a study is to reestimate the model with data from another situation that measures the same behavior among a similar population.

Unfortunately, it is likely that the data and circumstances of this new "experiment" differ from the previous study, and the model must be altered to reflect these new circumstances, even though both studies are trying to estimate the same causal relationships among several variables. In some cases, the alterations may necessitate different estimation techniques to deal properly with these differences.

The example we want to discuss illustrates these adaptions as well as an application of generalized least squares to the heteroskedastic case. Chapter 5 contained an equation relating students' verbal achievement to teacher characteristics, several school variables, and previous achievement levels. This equation was estimated with longitudinal data on individual students in one California school system. Those results indicated that teachers' verbal abilities were important determinants of verbal achievement. This is however only one study, based on only one sample of third graders in one school district. If the result could be shown to hold for other students under different educational and environmental conditions, it would substantially add to our confidence in the conclusion.

The question is, What data are available to permit this further analysis? Perhaps the most extensive data come from Coleman *et al.* (1966).[9] These data were collected for a stratified probability sample and contain approximately 600,000 observations of students and their achievement. The fact that it is a stratified probability sample implies that certain groups are overrepresented in the sample; they were sampled more heavily than would be expected given their representation in the population. Data were also collected about students' family backgrounds and the schools that they attended. These data have several advantages for the purposes of replicating the earlier estimations. It is a much more extensive data set, covering many more children and school environments. Consequently, if the estimates are consistent with those from the California study, it substantially reinforces the previous estimates.

There are several difficulties with these data, however, necessitating alterations to the model in Chapter 5. Because achievement scores at an earlier point are missing, other variables are included in the model to measure the background and environmental characteristics important in the educational process. Also, several variables are included to measure the extreme heterogeneity among the schools. These factors affect children's performance and are correlated with teacher ability. Finally, the measure of teacher ability is not obtained with the same test in both studies, so that this important variable, and its estimated coefficients, are not comparable in the two models.

One further important disadvantage of the Coleman survey is that it does

---

[9]This is commonly referred to as the Coleman Report. Data pertained to students in grades 1, 3, 6, 9, and 12 for the year 1965.

not allow matching of the scores and backgrounds of the individual students with the characteristics of their individual teachers and programs. Thus the individual analysis of the California study cannot be replicated exactly. What is done instead is to relate the average achievement of students in a particular school to the average teacher, school, and background variables for that school.[10] This, then, is a conventional aggregate data sample rather than a sample of individuals.

The aggregation over individuals within schools is just the situation that we have considered in this chapter. For a sample of schools with five or more white sixth graders, the number of sixth grade students ranged from 5 to 202. (The sample is stratified by race, as is the California study.) If we believe that the error terms in the true model for individuals are homoskedastic (and we have no reason to believe otherwise), then we should use weighted regression (GLS) to calculate the estimates with aggregate data. On the other hand, the overrepresentation of certain groups in the probability sample has no similar effect on the estimation of the model. The implication of overrepresenting some groups or schools is that we can expect larger samples for such students and thus may be able to obtain more precise parameter estimates for them. However, we would not expect this overrepresentation to affect the variances of the error terms in any way. In fact, using the observational weights supplied by the survey (which are required to estimate population means, etc.) would *cause* heteroskedasticity in the weighted regression.

When weighted regression, with weights equal to the square root of the number of students in each school, is applied to the Coleman data, the estimated coefficients on teacher characteristics improved in precision. The standard error on teacher verbal test score goes from 0.105 to 0.086 when OLS is compared to GLS. The GLS should provide the better, more precise, estimate of the effect of teacher ability on the performance of students.

In terms of the comparability of the two estimates of teacher effects, which is the basic purpose of this analysis, we must take into account the fact that two different teacher ability variables are used. Consequently, we use the estimated elasticity of achievement with respect to teacher ability from each equation rather than the estimated coefficients. Thus we compare what a 1% increase in either measure of teacher ability means in terms of the percentage increase in student achievement. In this case, using the sample means as the point of reference, the estimated elasticities are 0.108 in the California study and 0.125 in the reanalysis of the Coleman data. Both studies then predict essentially the same increase in student performance for a 1% increase in teacher ability. This implies that the estimates are quite consistent between both data sets.

---

[10] A more detailed discussion of the reasons for aggregation, the specific models, and the results is found in Hanushek (1972).

Autocorrelation

The procedures for diagnosing and correcting problems associated with the serial correlation of error terms have received considerable attention in econometric texts and articles. We shall give only a brief summary of these procedures here and refer the reader to some of the more complete discussions elsewhere.[11]

A straightforward diagnostic procedure for autocorrelation is to make use of the residuals from an OLS estimate of the model. The residuals from the OLS estimation are consistent estimates of the underlying error terms. They may also provide some indication of the structure of the error process in small samples. The most common test statistic for autocorrelation is the Durbin–Watson (or d) statistic. This statistic is used to test for first order autocorrelation and is computed as

$$d = \sum_{t=2}^{T} (e_t - e_{t-1})^2 \bigg/ \sum_{t=1}^{T} e_t^2, \tag{6.10}$$

where the e_t are residuals from OLS estimators.

By rearranging Eq. (6.10) we see that the Durbin–Watson statistic is closely related to the correlation between successive error terms. Since $\sum_{t=2}^{T} e_t^2$ is approximately equal to $\sum_{t=2}^{T} e_{t-1}^2$ and since each of these is approximately equal to $\sum_{t=1}^{T} e_t^2$, we have

$$d = \frac{\sum_{t=2}^{T} (e_t - e_{t-1})^2}{\sum_{t=1}^{T} e_t^2}$$

$$\cong \frac{\sum_{t=2}^{T} (e_t - e_{t-1})^2}{\sqrt{\sum_{t=2}^{T} e_t^2} \sqrt{\sum_{t=2}^{T} e_{t-1}^2}} = \frac{\sum_{t=2}^{T} e_t^2 - 2\sum_{t=2}^{T} e_t e_{t-1} + \sum_{t=2}^{T} e_{t-1}^2}{\sqrt{\sum_{t=2}^{T} e_t^2} \sqrt{\sum_{t=2}^{T} e_{t-1}^2}}$$

$$\cong \frac{2\sum_{t=2}^{T} e_t^2}{\sum_{t=2}^{T} e_t^2} - \frac{2\sum_{t=2}^{T} e_t e_{t-1}}{\sqrt{\sum_{t=2}^{T} e_t^2} \sqrt{\sum_{t=2}^{T} e_{t-1}^2}} \cong 2 - 2r_{t,t-1},$$

where $r_{t,t-1}$ is the simple correlation of the successive residuals for the second

[11]One of the best and most complete discussions is Hibbs (1974).

through *T*th observations. As the sample becomes large (asymptotically), $d = 2(1 - \rho)$, where ρ is the serial correlation parameter from Eq. (6.8).

In the absence of serial correlation of the residuals (i.e., $r_{t,t-1} = 0$), $d = 2$. If successive error terms are perfectly positively correlated ($r_{t,t-1} = 1$), $d = 0$. Values of d between zero and two represent decreasing amounts of positive correlation between successive error terms. Values of d greater than two imply a negative correlation between successive pairs, and reaches a maximum value of 4.0 if $r_{t,t-1} = -1.0$.

The Durbin–Watson stastistic has two functions in this context. The first is as a statistical test of the null hypothesis that successive error terms are uncorrelated, i.e., that $\rho = 0$. The second is as a way to estimate this correlation, which can then be used to compute the generalized least squares estimate. Unfortunately, the distribution of the Durbin–Watson statistic depends not only on the sample size and number of coefficients being estimated (as does the *t*-statistic), but also upon the sample values of the explanatory variables.[12] Nevertheless, the distribution does have upper and lower bounds that are functions only of the sample size, the number of exogenous variables, and ρ. These upper and lower bounds, displayed in Appendix III, can be used in a test of the hypotheses of no serial correlation. However, some ambiguity will result if the test statistic falls between the two bounds. Since the distribution is symmetric about the value 2, we need consider only half of the distribution. This discussion looks at the range 0–2, which corresponds to positive serial correlation. Thus, for a given data set and model to be estimated and assuming the more usual case of $\rho > 0$, if the value of the Durbin–Watson statistic is greater than this upper bound for a specified confidence level, we shall not reject the null hypothesis of no serial correlation. Likewise, if the value of the *d*-statistic is less than the lower bound, we shall reject this hypothesis and proceed to use generalized least squares. If the value falls between the lower and upper bounds, we are uncertain whether to accept or reject the null hypothesis.

We have computed the Durbin–Watson statistic for the autocorrelated simulations reported in Table 6.2. In Table 6.5 we show the mean of these estimates for each value of ρ, the degree of serial correlation. The lower and upper bounds at the 5% level for an estimation with 50 observations and two coefficients plus a constant are 1.38 and 1.54, respectively. Given this information, we can see that we would on average accept the null hypothesis that $\rho = 0$ in the case with no serial correlation using a 95% significance level. In the case where $\rho = 0.30$, the average Durbin–Watson statistic falls into this indeterminant area. The two more highly autocorrelated simulations have mean Durbin–Watson's well below the lower bound, indicating that we would generally reject the null hypothesis of no serial correlation.

---

[12]It should be noted that the Durbin–Watson test is not appropriate in a model that contains lagged values of the dependent variable, e.g., when Y_{t-1} is an explanatory variable. This situation is discussed in Section 6.10.

TABLE 6.5

Mean Estimated Durbin–Watson Statistics

	ρ			
	0.00	0.30	0.60	0.90
$\bar{d}$	2.01	1.49	0.98	0.43
$\hat{\rho}$	-0.005	0.245	0.510	0.785

In the simulations used here, the d-statistic tends to underestimate the extent of the serial correlation. The second row in Table 6.5 shows the correlation $\hat{\rho}$ implied by the Durbin–Watson. We can see that we consistently underestimate this correlation. This is undoubtedly due to the fact that with only 50 observations the approximations used in computing the Durbin–Watson do not hold. To show this effect, the simulations with $\rho = 0.30$ and $\rho = 0.90$ were redone for a sample size of 200 and the d-statistics computed. The mean values were 1.43 and 0.26, respectively, implying correlations of 0.28 and 0.87, respectively. These values are much closer to the set level of serial correlation. It is also true that the value of 1.43 in the simulation with $\rho = 0.30$ lies below the 5% lower bound for a sample size of 200, so that we would reject, on average, the null hypothesis for this simulation as well.

We have already mentioned that the second use of the residuals from the OLS estimates, as reflected in the Durbin–Watson statistic, is to estimate the value of ρ, the amount of first order serial correlation. This estimate for ρ can then be used to obtain the generalized least squares estimates of the model's coefficients. This can be done by using the transformations shown in Eqs. (6.9a) and (6.9b). To illustrate the results of this procedure, we took the OLS residuals from each replication in our simulations, estimated the amount of correlation and used this as an estimate of ρ. GLS coefficient estimates were then computed according to the procedure outlined above. Table 6.6 shows the results from this two-step procedure. The table also shows the true GLS estimates from Table 6.2 for comparison. These comparisons indicate that even though we have underestimated the amount of serial correlation (see the previous discussion), our two-step procedure has virtually identical statistical properties to the true generalized least squares estimates using the specified values for ρ.

In situations where one suspects a different pattern of serial correlation than the first order scheme used above, the same procedures are applicable. For example, a quarterly economic model may want to hypothesize that U_t and U_{t-4} are correlated in addition to the previous notion that successive error terms are correlated. This additional correlation could be estimated by setting $s = 4$ in Eq. (6.10) and estimating $\rho_{t,t-4}$. If the conclusion from this calculation is that U_t and U_{t-4} are correlated, this estimate can be used in forming Ω and performing the GLS estimation. (In such situations, a mixture of first and fourth order serial correlation is probably appropriate. Thus,

TABLE 6.6

Results from Simulated Two-Step GLS Procedure [a]

ρ	GLS with true ρ			GLS with estimated ρ		
	$\bar{b}_1$	$\hat{\sigma}_{b_1}$	$\bar{s}_{b_1}$	$\bar{b}_1$	$\hat{\sigma}_{b_1}$	$\bar{s}_{b_1}$
0.00	14.91	3.72	3.61	14.81	3.72	3.54
0.30	14.60	3.51	3.47	14.60	3.59	3.48
0.60	15.30	3.47	3.56	15.33	3.47	3.50
0.90	14.63	9.17	9.01	14.60	8.93	8.57
	$\bar{b}_2$	$\hat{\sigma}_{b_2}$	$\bar{s}_{b_2}$	$\bar{b}_2$	$\hat{\sigma}_{b_2}$	$\bar{s}_{b_2}$
0.00	0.998	0.067	0.066	1.000	0.066	0.064
0.30	1.007	0.066	0.064	1.008	0.067	0.063
0.60	0.995	0.056	0.057	0.994	0.057	0.059
0.90	0.997	0.049	0.050	0.996	0.049	0.053
	$\bar{b}_3$	$\hat{\sigma}_{b_3}$	$\bar{s}_{b_3}$	$\bar{b}_3$	$\hat{\sigma}_{b_3}$	$\bar{s}_{b_3}$
0.00	2.002	0.038	0.036	2.002	0.038	0.035
0.30	2.003	0.038	0.034	2.003	0.039	0.034
0.60	2.001	0.034	0.031	2.001	0.034	0.032
0.90	2.001	0.029	0.027	2.002	0.029	0.028
	$\bar{s}$			$\bar{s}$		
0.00	5.88			5.82		
0.30	5.90			5.84		
0.60	5.92			5.86		
0.90	5.99			5.96		

[a] $\bar{b}_k$ is the mean of coefficients estimated for each replication; $\hat{\sigma}_{b_k}$ is the computed standard deviation of the estimated coefficients; and $\bar{s}_{b_k}$ is the mean estimated standard error of the coefficient in each replication.

simple transformations such as the generalized difference approach become very difficult, and it is necessary to resort to GLS using Ω^{-1} directly.)

The previous equation estimating the relationship between aggregate consumption and defense expenditures offers a simple example of how to test and correct for autocorrelations. Once we have created the appropriate variables to give a homoskedastic error term, we are still left with the problem that successive error terms may be correlated. For example, consumer attitudes may go through fluctuating periods of optimism and pessimism that are uncorrelated with either *GNP* or defense spending but which lead to higher or lower levels of consumer spending. If these fluctuating moods remain for more than a single year, then successive error terms in the estimated model will be correlated. If we hypothesize a first order autocorrelated pattern among the U_t (after dividing V_t by GNP_t), then we can estimate the magnitude of this serial correlation with the Durbin–Watson statistic from the previous estimates. This value is 1.47, which implies some correlation with a value for $\hat{\rho}$ of 0.265. As we did with the Monte Carlo simulation, we can correct for the presence of this autocorrelation by estimating the equation[13]

$$\left(\frac{C_t}{GNP_t} - \hat{\rho}\frac{C_{t-1}}{GNP_{t-1}} \right) = \beta_0 \left(\frac{1}{GNP_t} - \hat{\rho}\frac{1}{GNP_{t-1}} \right) + \beta_1 (1 - \hat{\rho})$$

$$+ \beta_2 \left(\frac{D_t}{GNP_t} - \hat{\rho}\frac{D_{t-1}}{GNP_{t-1}} \right) + \varepsilon_t.$$

This estimation yielded the following parameter estimates (and standard errors): $b_0 = 24.371$ (3.13); $b_1 = 0.6265$ (0.0086); and, $b_2 = -0.4210$ (0.0720). Again we can see that the coefficient estimates are virtually the same as those in the earlier equations, but that the estimated standard errors have changed. In this case, they have increased with the corrections for autocorrelation. If the error terms are correlated, as the previous Durbin–Watson suggested, then these standard errors are more accurate estimates of the true variances of the estimated coefficients.

6.9   Visual Diagnostics

The foregoing has described some formal statistical tests that might be applied to test for heteroskedasticity or autocorrelation. On the formal statistical side there are also other tests (e.g., the von Neumann ratio for autocorrelation). However, there are also alternatives in the form of visual inspection for heteroskedasticity or autocorrelation.

For situations in which there is a manageable number of observations (i.e., excluding extremely large cross-sectional analyses), most computerized programs will allow the calculation and display of the residuals from an ordinary least squares regression. Such displays can be inspected for patterns in the residuals.

In the case of suspected autocorrelation, the residuals can be listed or actually plotted against time. If the residuals tend to run in streaks (i.e., consecutive residuals tend to have the same sign, either positive or negative), further analysis of the possibility of autocorrelation would be called for.[14]

In the case of heteroskedasticity, the residuals should be plotted against the dimension that is suspected of being correlated with the variance. For

---

[13] This model is developed from the first order autocorrelated scheme for U_t:

$$U_t = \rho(GNP_t/GNP_{t-1})U_{t-1} + \varepsilon_t \qquad \text{and} \qquad \sigma^2_{\varepsilon_t} = \sigma^2(GNP_t^2).$$

(See Problem 6.2.)

[14] An actual test for autocorrelation is the calculation of the number of "runs" in the residuals.

example, if X_k is firm or jurisdiction size, and the error variance is suspected to increase with size, the residuals should be plotted against X_k. As was shown previously, the residuals will be uncorrelated with X_k (by the arithmetic properties of OLS). This implies that we shall observe no clear linear relationship between X_k and e. However, if the errors are heteroskedastic, the variance, or width of the plot of residuals, should change with X_k. Figure 6.2 shows a hypothetical plot where the variance of the residuals is related to X_k.

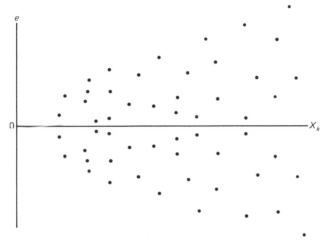

FIGURE 6.2   *Hypothetical plot of heteroskedastic residuals, where* $\sigma^2 = f(X_k)$.

6.10   Dynamic Models

Within the context of single-equation models, the presence of autocorrelation by itself implies only inefficiency in the parameter estimates. However, this is not the situation with one class of models commonly used in time series analysis. A natural formulation of many time series problems involves explicit recognition that adjustment to variations in exogenous factors occurs over some time period. In other words, individuals do not adjust their behavior immediately or completely to changed conditions, either because adjustment is costly and difficult or because their expectations of future conditions are not completely revised on the basis of the current values of the exogenous variables. As will be demonstrated, this type of behavior—with certain assumptions about the form of the adjustment process—often results in estimating models with lagged values of the dependent variable (i.e., Y_t is a function of Y_{t-1} and a set of exogenous variables). In such models, called dynamic models, serial correlation of the errors is no longer a simple problem of efficiency but instead is a problem of bias and inconsistency in the parameter estimates and of misleading test statistics. Therefore, corrective action in dynamic models becomes much more important than in models with fixed regressors.

The formulation of the partial adjustment model and the adaptive expectations model are somewhat different, but they both lead to the same basic form of the model to be estimated. In the partial adjustment model, assume that the desired or optimal level of Y, denoted by Y^*, is

$$Y_t^* = \beta_1 + \beta_2 X_t. \tag{6.11}$$

However, because of adjustment costs, ignorance, or inertia, we might assume that Y_t does not move instantaneously to the desired level but instead that a constant proportion of the gap (γ) between the previous value of Y and the desired value is closed each period as in

$$Y_t - Y_{t-1} = \gamma(Y_t^* - Y_{t-1}) + U_t, \quad \text{where} \quad 0 < \gamma \leqslant 1. \tag{6.12}$$

The desired level of Y is not observed; however, substituting Eq. (6.11) into (6.12) yields a model that can be estimated on the basis of observed variables. That is,

$$Y_t = \beta_1\gamma + \beta_2\gamma X_t + (1 - \gamma)Y_{t-1} + U_t. \tag{6.13}$$

The adaptive expectations model is somewhat different in formulation. It may be unreasonable to think that the desired level of Y is a function simply of the current exogenous variables. An alternative formulation is that the dependent variable depends upon the expected level of the exogenous variable. This expected level, denoted X^*, is unobserved, but a reasonable hypothesis may be that the expected level is updated on the basis of the difference between the observed X and the previous value of the expected level of X. Thus,

$$X_t^* - X_{t-1}^* = \gamma(X_t - X_{t-1}^*) \quad \text{with} \quad 0 < \gamma \leqslant 1. \tag{6.14}$$

If the true model determining Y is

$$Y_t = \beta_1 + \beta_2 X_t^* + U_t, \tag{6.15}$$

we can substitute for the formulation of expectations from Eq. (6.14) to yield

$$Y_t = \beta_1\gamma + \beta_2\gamma X_t + (1 - \gamma)Y_{t-1} + U_t - (1 - \gamma)U_{t-1} \tag{6.16}$$

(by substituting $Y_{t-1} - \beta_1 - U_{t-1}$ for $\beta_2 X_{t-1}^*$). Comparing this with Eq. (6.13), we see that both models yield the same form of the estimating equation. The only difference comes in a different form of the model's error term.

These dynamic models imply that the present value for Y_t depends upon the past values of the exogenous variables. This model is essentially the same as a model first proposed by Koyck (1954) that explicitly models Y_t as a linear function of the past X values and imposes a particular form on the coefficients for past values of X. [This model is often called the Koyck lag model.] If we substitute into Eq. (6.13) or (6.16) for the lagged values of the endogenous variable, we can solve for the relationship between Y_t and past values of X_t. Such substitution indicates that the coefficient on X_{t-s} (for $s > 0$) is simply $\beta_2\gamma(1-\gamma)^s$. In other words, the coefficients on the exogenous variables are assumed to follow a declining pattern that is a geometric series in terms of $1-\gamma$. This implies that the effect on Y of any permanent change in the exogenous variable increases over time. The immediate effect of a unit change in X is $\beta_2\gamma$; this increases[15] over time to β_2. (The long-run effect in either the partial adjustment model or the expectations model is simply β_2 since the long-run complete adjustment would be accomplished or expectations would match observed values.)

The similarity of the model with a lagged dependent variable and a model that includes the history of the X's suggests that an alternative estimation scheme would be to estimate directly a model with lagged X's. In other words, one could estimate

$$Y_t = \beta_1 + \beta_2 X_t + \beta_3 X_{t-1} + \cdots + U_t. \tag{6.17}$$

This would allow coefficient patterns that were different from the Koyck pattern of a declining geometric series. However, direct estimation of Eq. (6.17) is often very difficult since problems of multicollinearity may seriously affect the coefficient estimates. (The values of X are likely to be highly correlated with each other over time.) Estimation of the model with lagged endogenous variables improves our chances of obtaining precise coefficient estimates at the cost of imposing a rigid structure on the pattern of coefficients.

It is possible to modify some of the more stringent assumptions imposed by the Koyck lag scheme. Equation (6.13) implies that the current value of X has the largest impact on Y and that values farther from the current period all have less impact. Such an assumption may be inappropriate for certain adjustment processes. Reactions to changed exogenous factors may not be instantaneous, implying that the impact of the current X may be less than that of a previous X. Such possibilities can be incorporated by modifying the model such that the Koyck pattern begins at some past period. For example, we could allow the first coefficient to be free from the geometric pattern by estimating a model such as

$$Y_t = \beta_1 + \beta_2 X_t + \beta_3 X_{t-1} + \gamma Y_{t-1} + U_t. \tag{6.18}$$

The effect of this specification is to allow the coefficient for the present value

---

[15]The general method for calculating the long-run effect in models with lagged endogenous variables is to divide the coefficient on the exogenous variable by 1 minus the coefficient on the lagged endogenous variable. This is found from the general formula for the sum of an infinite geometric series where the common ratio between successive values in the series falls between 0 and 1. Long-run effects in this situation mean the total change in the endogenous variable resulting from a permanent shift in the exogenous variable, while everything else remains the same.

of the exogenous variable (X_t) to diverge from the geometric scheme and, in fact, to be less than the impact of earlier X's.

This model has been described in terms of a single exogenous variable. It is possible to include other exogenous variables. However, simply adding another exogenous variable to Eq. (6.13) or (6.16) implies that both exogenous variables have the same pattern of coefficients over time. Allowing for different patterns of coefficients is difficult within this framework.

There is an important difference between models with lagged endogenous variables and models without them in terms of the effect of serial correlation in the error terms. With no lagged endogenous variables, serial correlation implied inefficient parameter estimates and some difficulty in performing statistical tests. With lagged endogenous variables, serial correlation implies biased (and inconsistent) parameter estimates. This is easily seen by observing that serial correlation in the error terms implies that the assumption about the independence of the error term and the explanatory variables is untenable. Since Y_{t-1} contains U_{t-1} and since U_t and U_{t-1} are correlated, Y_{t-1} and U_t will be correlated. Furthermore, this problem is not corrected by having large samples; thus, the parameter estimates will be inconsistent.[16] With lagged endogenous variables, attention to serial correlation becomes more important.

A further difficulty is introduced by lagged endogenous variables. The derivation of the distribution for the Durbin–Watson statistic is based upon fixed regressors and is inappropriate when lagged dependent variables are included in the model. Thus, the standard test may give quite misleading impressions and the standard statistical tests for a null hypothesis of no serial correlation are inappropriate in this situation. Tests with high significance levels, say the 95% level, guard well against type I errors (incorrectly rejecting the true null hypothesis) but do relatively poorly at type II errors (accepting a false null hypothesis). In this situation, type II errors become more important. Correcting for serial correlation when none exists (a type I error situation) leads to inefficiency, but the coefficients remain consistent. However, with dynamic models not correcting for serial correlation when it does exist (a type II error situation) leads to inconsistent estimates. Under a wide range of circumstances (i.e., differing values of the serial correlation coefficient and of the coefficient on the lagged endogenous variable), the bias and inconsistency of OLS estimates are more serious problems than any efficiency losses.

Monte Carlo studies indicate that OLS is superior to alternative estimation

---

[16]Consistency of parameter estimates is a large sample property. This is discussed in Appendix I. For our purposes, it implies that as the sample size increases the estimated coefficient approaches the true parameter with probability one. Large sample properties are often resorted to in situations where results for small samples are not generally available. This will be the case in Chapters 7–10. While the small sample behavior of estimators for which only large sample properties are available is not generally known, many estimation experiences show reasonably good small sample behavior of consistent estimators. This is the case with the estimator described below for serial correlation in dynamic models.

schemes for very low values of serial correlation.[17] Assuming that the error process is

$$U_t = \rho U_{t-1} + \varepsilon_t, \qquad (6.19)$$

OLS is generally superior to other methods when $\rho \leqslant 0.2$ (comparing only positive values of ρ, which are the most likely to occur in practice). At larger values of ρ, several alternative methods are superior to OLS.

In the list of alternative estimators, one estimator that performs well across a wide range of circumstances is a variant of the estimator discussed earlier in this chapter. If we consider estimating the model in a generalized difference form where the dependent variable is $Y_t - \hat{\rho} Y_{t-1}$ and each of the explanatory variables including the lagged endogenous variable are similarly differenced, the only problem is finding a suitable estimate of the serial correlation coefficient $\hat{\rho}$. The Cochrane–Orcutt method of estimating ρ and the parameters of the model is to iterate on different values of $\hat{\rho}$; in other words,[18] to estimate the model for a series of different values $\hat{\rho}$. This procedure provides a scheme for choosing the successive values of $\hat{\rho}$. Alternatively, we could search across all possible values of $\hat{\rho}$ and choose the estimated equation for which the sum of squared residuals is smallest. The resulting estimator will be the maximum likelihood estimator of the model in the presence of serial correlation.[19] For large samples, this estimator is consistent and is most efficient. Further, Monte Carlo studies have shown that this estimator performs quite well in small samples.

One other issue is important in dynamic models. With serial correlations, the error structure becomes intertwined with the dynamics of the model, and it is very difficult to disentangle the two. In situations where the coefficient on the lagged endogenous variable is really zero, serial correlation can make it appear that there is really a significant coefficient. For example, the cited Monte Carlo studies of estimating a model with serially correlated errors when there is no dynamic structure (i.e., a zero coefficient on Y_{t-1}) indicate

[17]The Monte Carlo results described in this chapter are all due to Peck (1975). The experiments involve altering the size of the coefficient on the lagged endogenous variable, altering the serial correlation of the errors and of the exogenous variable, and altering the variance in the systematic component relative to the variance of the error component of the model. All experiments involve a single exogenous variable and first order serial correlation.

[18]Cochrane and Orcutt (1949). The Cochrane–Orcutt procedure is an iterative solution technique that relies upon successively estimating new values of the serial correlation coefficient and parameter values associated with different $\hat{\rho}$'s.

[19]In order to be truly maximum likelihood, some allowance must be made for the first observation (which would be lost by a differencing technique). If the first observation is transformed by dividing by $\sqrt{1 - \hat{\rho}^2}$, the technique would be maximum likelihood. There is also a problem of what starting values in the search should be used if the Cochrane–Orcutt iterative technique is applied. An alternative is to search across possible values of ρ within the 0 to 1 interval. This could be done by starting with fairly large increments (say 0.1) and reducing the increment in the appropriate range uncovered by the coarser search.

that the OLS coefficient estimates for Y_{t-1} have a biased term approaching 0.8 for larger values of ρ. Thus it becomes very difficult to distinguish true dynamics from simple serial correlation.

This supports the notion that, if there is any reason to suspect serial correlation, one should apply corrective actions. In fact, with dynamic models it may be superior *not* to test for serial correlation but instead to proceed as if it were present. At the very least, any tests for serial correlation should be applied at significance levels much below those usually used, say at the 25% level.

Finally, as with the previous section on serial correlation, we have only specified the models in terms of first order serial correlation. It may well be that the current error terms are related to more than just the immediately preceding error terms. The corrective action suggested here becomes much more complicated if there is higher order serial correlation. However, correction for first order serial correlation is a logical starting place for many situations and can be considered as an approximation for more complex error structures.

6.11   Conclusion

We have continued the investigation of the assumptions made in Chapters 3 and 5 justifying the properties of the ordinary least squares estimator by considering cases where the individual error terms do not have equal variances and are not drawn independently of each other. The first case is referred to as the problem of heteroskedasticity, while the second is called autocorrelation. In considering the effects of violating these assumptions and the ways to diagnose and correct the problems, you should remember that we are maintaining the assumption that all error terms have expected values of zero, $E(U) = 0$. If this assumption is also violated, then none of the discussion of this chapter is applicable.

The consequences of not satisfying the assumption $E(UU') = \sigma^2 I$ are unbiased and consistent but inefficient estimators. The results of the Monte Carlo simulations in Tables 6.1 and 6.2 illustrate this quite clearly. The mean estimate of each coefficient equaled the true value in both the heteroskedastic and autocorrelated models. These simulations also pointed out that the OLS estimator is not the best, or most efficient, estimator. In some ways, this may seem like a less important problem than that posed by violations of the assumption about the mean of the error term. However, given the fact that in real situations there is only one replication with which to estimate the coefficients, one certainly wants to use the information in that sample as efficiently as possible. The major topics of this chapter are how to diagnose the presence of these problems and then how to compensate for their presence to regain the efficiency of our estimator.

The appropriate procedure when faced with situations where the error terms are correlated or have unequal variances is referred to as generalized least squares (GLS). This method uses the information about the variances and covariances of the error terms to increase the efficiency of the estimators. In the case of observations with error terms of unequal variance, the procedure effectively gives greater weight to those observations whose error terms have smaller variances. In the autocorrelated case, this procedure transforms the variables in such a way that the error terms implicit in the transformed variables are uncorrelated.

When applying the GLS method, it is certainly best to have clear prior assumptions about the structure of the error terms. This is the situation in our example of the heteroskedastic case. There we are willing to state that the error terms are independent of each other and have variances that are inversely proportional to the number of individuals composing each aggregate unit. This assertation led directly to the application of GLS. In other cases, we may not have sufficient prior information to go directly to using GLS. We may have strong suspicions that the equal variance and the uncorrelatedness assumptions are violated, but not have enough prior information to state the magnitude of the problem. This is clearly the case with autocorrelation where it is hard to state a priori the magnitude of the correlation among successive error terms.

It is here that the consistency of the OLS estimators [provided that $E(U) = 0$] is of great help. Because of this consistency, so long as the sample size is sufficiently large, we can use the residuals from the OLS estimation to estimate the information required by the GLS procedure. Analysis of these residuals can be used to estimate the magnitude of any correlations and in some cases even the variance of the error terms for different observations. These estimates of Ω can then be used as part of the GLS estimation to improve the efficiency of the estimates.

The proof that the GLS estimates are BLUE holds strictly only when Ω is known. However, when we have a consistent estimate of Ω, GLS using $\hat{\Omega}$ is asymptotically efficient. That is, as the sample size becomes large, the estimates will be BLUE. The small sample properties of using an estimated error matrix depend upon the accuracy of the estimates. Using the OLS residuals to estimate the error matrix is also possible only if we can impose some structure on the errors. We cannot use the full set of residuals from an OLS procedure in a second stage since that error matrix will be singular.

The value of the two-stage procedure when we had some idea of the error structure was most clearly demonstrated in the autocorrelated simulations where we used the residuals from the OLS estimates to calculate the amount of serial correlation and then used this estimate to make the appropriate transformation of the original variables. The coefficient estimates obtained in this fashion are as efficient as those obtained from the GLS estimators which used the true correlation.

It is important to keep in mind the various situations that lead to violations of the basic assumption that $E(UU') = \sigma^2 I$ and how the method of generalized least squares can be used to overcome the consequences of these violations. In the next chapter we shall give several examples of models and types of data that produce these violations and show how different applications of GLS are used.

APPENDIX 6.1

Derivation of Generalized Least Squares Estimator

This appendix both shows that OLS is not the most efficient estimator when $E(UU') \neq \sigma^2 I$ and derives the GLS estimator. To do this, we rely on the proof of best in Chapter 5, to show that GLS and not OLS is the minimum variance, linear unbiased estimator. Any alternative estimator is

$$b^\# = [(X'X)^{-1}X' + D]Y = \beta + (X'X)^{-1}X'U + DU = \beta + [(X'X)^{-1}X' + D]U \quad (6.1.1)$$

with the condition that $DX = 0$ (see Appendix 5.1). The variance of $b^\#$ is

$$E(b^\# - \beta)(b^\# - \beta)' = [(X'X)^{-1}X' + D]E(UU')[X(X'X)^{-1} + D']$$

$$= \sigma^2[(X'X)^{-1}X' + D]\Omega[X(X'X)^{-1} + D']$$

$$= \sigma^2\big[(X'X)^{-1}X'\Omega X(X'X)^{-1} + (X'X)^{-1}X'\Omega D'$$

$$+ D\Omega X(X'X)^{-1} + D\Omega D'\big]. \quad (6.1.2)$$

We want to minimize this expression with respect to D, subject to the constraint that $DX = 0$ (to ensure unbiasedness).

By using the more advanced mathematics of constrained optimization, we can search over all values for the elements of D that satisfy the constraint $X'D' = 0$ to find those values that minimize (6.1.2). To accomplish this, we minimize the function

$$L = \sigma^2[(X'X)^{-1}X'\Omega X(X'X)^{-1} + (X'X)^{-1}X'\Omega D' + D\Omega X(X'X)^{-1} + D\Omega D']$$

$$- 2\Lambda X'D' \quad (6.1.3)$$

with respect to D and Λ. (The additional term $-2\Lambda X'D'$ ensures that the constraint $DX = 0$ is met.) The partial first derivatives are

$$\partial L/\partial D' = 2(X'X)^{-1}X'\Omega + 2D\Omega - 2\Lambda X' = 0, \quad (6.1.4)$$

$$\partial L/\partial \Lambda = -2X'D' = 0. \quad (6.1.5)$$

From (6.1.4) we get

$$D = \Lambda X'\Omega^{-1} - (X'X)^{-1}X'. \quad (6.1.6)$$

However, we know from the constraint and Eq. (6.1.5) that $DX = 0$. Combining this condition with (6.1.6) gives

$$DX = \Lambda X'\Omega^{-1}X - I = 0 \qquad \text{or} \qquad \Lambda = (X'\Omega^{-1}X)^{-1}.$$

Substituting this in Eq. (6.1.6) yields

$$D=(X'\Omega^{-1}X)^{-1}X'\Omega^{-1}-(X'X)^{-1}X' \tag{6.1.7}$$

and the estimator

$$b^\# =[(X'X)^{-1}X'+D]Y=[(X'X)^{-1}X'+(X'\Omega^{-1}X)^{-1}X'\Omega^{-1}-(X'X)^{-1}X']Y$$

$$=(X'\Omega^{-1}X)^{-1}X'\Omega^{-1}Y, \tag{6.1.8}$$

which is the GLS estimator.

This demonstration shows that there is a linear unbiased estimator with a lower variance than OLS (shown by the fact that $D\neq 0$), thus OLS is not BLUE. Furthermore, the alternative linear unbiased estimator that minimizes the variance of the estimated coefficients is the generalized least squares estimator discussed throughout this chapter.

APPENDIX 6.2

Unbiased Estimator of σ^2

The demonstration that $[1/(T-K)](e'\Omega^{-1}e)$ is an unbiased estimator of σ^2 follows the similar demonstration for the OLS case in Appendix 5.2. The point is to show that

$$E\left(\frac{e'\Omega^{-1}e}{T-K}\right)=\sigma^2 \qquad \text{given that} \qquad E(UU')=\sigma^2\Omega.$$

To do this, we first get an expression relating the residuals e to the unobserved error terms U:

$$e=Y-\hat{Y}=X\beta+U-Xb=X(\beta-b)+U=-X(X'\Omega^{-1}X)^{-1}X'\Omega^{-1}U+U$$

$$=[I-X(X'\Omega^{-1}X)^{-1}X'\Omega^{-1}]U.$$

From this,

$$e'\Omega^{-1}e=U'[I-\Omega^{-1}X(X'\Omega^{-1}X)^{-1}X'][\Omega^{-1}-\Omega^{-1}X(X'\Omega^{-1}X)^{-1}X'\Omega^{-1}]U$$

$$=U'[\Omega^{-1}-\Omega^{-1}X(X'\Omega^{-1}X)^{-1}X'\Omega^{-1}]U$$

$$=U'\Omega^{-1}U-U'\Omega^{-1}X(X'\Omega^{-1}X)^{-1}X'\Omega^{-1}U.$$

Because $e'\Omega^{-1}e$ is a scalar,

$$e'\Omega^{-1}e=\text{tr}[U'\Omega^{-1}U]-\text{tr}[U'\Omega^{-1}X(X'\Omega^{-1}X)^{-1}X'\Omega^{-1}U]$$

$$=\text{tr}[\Omega^{-1}(UU')]-\text{tr}[\Omega^{-1}X(X'\Omega^{-1}X)^{-1}X'\Omega^{-1}(UU')].$$

Thus with the assumptions of fixed X and Ω and the properties of the trace operator,

$$E(e'\Omega^{-1}e)=\text{tr}[\Omega^{-1}E(UU')]-\text{tr}[\Omega^{-1}X(X'\Omega^{-1}X)^{-1}X'\Omega^{-1}E(UU')]$$

$$=\sigma^2\left\{\text{tr}(\Omega^{-1}\Omega)-\text{tr}[\Omega^{-1}X(X'\Omega^{-1}X)^{-1}X'\Omega^{-1}\Omega]\right\}$$

$$=\sigma^2\left\{\text{tr}(I_T)-\text{tr}[(X'\Omega^{-1}X)^{-1}X'\Omega^{-1}X]\right\}=\sigma^2[T-\text{tr}(I_K)]=\sigma^2(T-K).$$

This result gives

$$E\left(\frac{e'\Omega^{-1}e}{T-K}\right) = \frac{1}{T-K}\sigma^2(T-K) = \sigma^2,$$

and the estimator of σ^2 is unbiased.

REVIEW QUESTIONS

1. The accompanying table gives federal nondefense and state and local expenditures in 1958 dollars. Using the Russett model, estimate the impact of defense outlays on these expenditures by both OLS and GLS.

Year	Nondefense	State–Local	Year	Nondefense	State–Local
1946	3.75	14.69	1961	9.18	47.98
1947	4.69	16.88	1962	11.16	50.77
1948	7.29	18.85	1963	12.60	54.31
1949	8.59	23.37	1964	13.96	58.37
1950	5.36	24.33	1965	15.15	63.23
1951	4.79	25.10	1966	15.01	69.33
1952	6.75	26.19	1967	15.65	76.03
1953	9.51	27.85	1968	16.76	82.42
1954	6.92	30.57	1969	16.15	86.26
1955	6.05	33.13	1970	16.19	90.32
1956	5.64	35.11	1971	17.89	94.60
1957	5.44	37.54	1972	18.89	99.75
1958	7.70	40.61	1973	17.84	104.78
1959	7.48	42.59	1974	19.50	107.67
1960	8.33	44.63	1975	20.50	108.64

2. For the Russett consumption function, we hypothesized the error structure

$$U_t = \rho(GNP_t/GNP_{t-1})U_{t-1} + \varepsilon_t,$$

where $\sigma^2_{\varepsilon_t} = \sigma^2(GNP_t)^2$ for all t, $\sigma_{\varepsilon_t\varepsilon_s} = 0$ for all $t \neq s$, and $\sigma^2_v = \sigma^2/(1-\rho^2)$. Show:
(a) $U_t = \sum_{i=0}^{t}\rho^i(GNP_t/GNP_{t-i})\varepsilon_{t-i}$.
(b) $E(U_t^2) = \sigma^2(GNP_t)^2\sum_{i=0}^{t}\rho^{2i} = \sigma^2(GNP_t)^2/(1-\rho^2) = (GNP_t)^2\sigma^2_v$.
(c) $E(U_tU_{t-s}) = \rho^{|s|}(GNP_t)(GNP_{t-s})[\sigma^2/(1-\rho^2)] = \rho^{|s|}(GNP_t)(GNP_{t-s})\sigma^2_v$.
(d) Write out $E(UU')$ and its inverse. (Hint: the Ω^{-1} for this model closely resembles that for the simple autocorrelated model.)
(e) What elements of F will give $F'F \cong \Omega^{-1}$? (The difference between them again is the first element.)
(f) What is the form for the transformed variables Y_t^* and X_t^*?

7 | Models with Discrete Dependent Variables

7.1 Introduction

The estimation techniques to this point are very flexible; they are applicable in a wide range of cases. We did not have to develop specialized techniques to handle different data or types of behavior. The methods developed are suitable for many types of single-equation models including models with a range of nonlinear relationships and models that include the influence of various qualitative factors. However, there is one increasingly important class of single equation models presenting special estimation problems. This is the class of models with discrete, or qualitative, dependent variables.

Social scientists frequently are interested in analyzing behavior observed as discrete outcomes or discrete events. For example, throughout this book we have spent considerable time attempting to understand the voting behavior of individuals. Voters make a choice between two discrete objects (Johnson or Goldwater); they can neither choose 0.9 Johnson and 0.1 Goldwater nor express anything but a binary choice, representable by ones or zeros. Similarly, if one is interested in college or occupational choices, in mobility, or in unemployment, one must observe all or none decisions or events. This focus on individual choice has become more important as microdata samples (samples pertaining to individuals) have become more plentiful.

Unfortunately, the single equation techniques presented in previous chapters are not always well suited for analysis of such data. Linear probability models (models that relate the probability of an event to a series of exogenous factors in a linear fashion) are often unrealistic, and attempts to estimate such models by OLS based upon individual observations quite generally lead to biased and inconsistent estimates.

The difficulties with estimating probability models are lessened with aggregate data, where the dependent variable is the proportion of the individuals making a given choice or falling into a particular category, rather than the binary variable observed at the individual level. However, the problems discussed in this chapter are not absent even with aggregate data, although they may be easier to solve. The most important reason for this chapter is the fact that individual level data are usually richer than aggregate data, permit estimation of more elaborate models, and thus are to be preferred when available.[1] Consequently, it is important to discuss the problems presented by these data and how to think about overcoming them.

Unhappily, we now understand more about the difficulties than we do about their solution. The basic problem is to specify plausible models to describe the probabilities of discrete events. Estimation of these models using microdata often requires techniques that are not well developed yet. However, the area is just beginning to receive significant amounts of attention. This chapter is meant more to display the nature of problems and the direction of approaches than to present complete solutions.

The choice on the presentation of the material reflects its basic difficulty. The alternative methods formulate the estimation problem differently than we have up to this point. They also require rather sophisticated estimation procedures—often relying upon nonlinear estimation techniques. The details of these procedures are often beyond the scope of this book, leading to a presentation of the general formulations and solutions without detailed developments. More complete presentations are referenced frequently.

7.2   The Problem of Estimating Models with Discrete Dependent Variables

Consider the analysis of voting behavior in the 1964 election. The focus of the analysis is understanding the determinants of individual choice behavior. As such, we are interested in the distribution of expected choices, i.e., probabilities, and how these probabilities change with different exogenous factors. Until now we have estimated a linear model that uses a dichotomous dependent variable ($V = 1$ if the individual votes for Johnson and $V = 0$ if the individual votes for Goldwater). A natural interpretation is then that these values represent outcomes of probabilistic events, and the coefficients for the linear model represent the marginal changes in the probability of voting for Johnson associated with each of the independent variables. However, the form of the model and the nature of the *observed* behavior make it difficult to estimate accurately the true probability for all values of the explanatory variables. This makes interpretation of the estimated coefficients as marginal changes in the probabilities tenuous in many cases.

---

[1]In simplest terms the hypotheses investigated with the aggregate data often relate to the behavior of individuals. Aggregation across individuals may obscure the relationship of interest by suppressing variations in the X's across individuals.

A Historic Approach

The fact that discrete dependent variables cause estimation problems in simple linear models has been recognized for some time. Goldberger (1964, pp. 248–250) addresses the problem in the following manner. Suppose Y is the observed outcome of a discrete event and that the process is characterized by a linear regression model such as

$$Y = X\beta + U, \tag{7.1}$$

where Y is $T \times 1$, X is $T \times K$, β is $K \times 1$, U is $T \times 1$, and $X\beta$ represents the true probability of the particular outcome for each individual observation. Since Y takes on values of only 0 (the event does not occur) or 1 (the event does occur), there are only two possible values for each U_t. Given the values of X_t for any observation, U_t must be either $-X_t\beta$ or $1 - X_t\beta$—depending upon whether the observed Y_t is zero or one. (X_t here implies a given row of the X matrix and is simply the particular values of the exogenous variables for a given observation.)

In order for the OLS estimator of β to be unbiased, $E(U_t)$ must equal zero. If $E(U_t) = 0$, the implied distribution of U_t must be given by

$$U_t: \quad -X_t\beta \quad 1 - X_t\beta$$
$$f(U_t): \quad 1 - X_t\beta \quad X_t\beta$$
$$\left[E(U_t) = \sum U_t f(U_t) = (-X_t\beta)(1 - X_t\beta) + (1 - X_t\beta)X_t\beta = 0 \right].$$

From this distribution, we can calculate the variance of each U_t as

$$E(U_t^2) = \sum U_t^2 f(U_t) = (-X_t\beta)^2(1 - X_t\beta) + (1 - X_t\beta)^2 X_t\beta = (X_t\beta)(1 - X_t\beta). \tag{7.2}$$

Clearly, in this case the variance depends upon the values of X_t, and any assumption of homoskedasticity is untenable. This suggests that generalized least squares should be applied to deal with the heteroskedasticity.[2]

Generalized least squares could not be applied directly, however, since β (and hence $X_t\beta$) is unknown. In this situation, Goldberger suggests a two-step procedure. In the first step, OLS is used to estimate β. This yields a consistent estimate of $X_t\beta$, namely the predicted value of $\hat{Y}_t = X_t b$. The second-step estimation incorporates the estimated variance for each U_t, i.e., $\hat{Y}_t(1 - \hat{Y}_t)$. This second step is a generalized least squares estimation[3] of the parameters

---

[2] As described in Chapter 6, the OLS estimates in this case with $E(U_t) = 0$ would be unbiased, but the heteroskedasticity would imply that the estimator was no longer BLUE. The variance of b would be larger than necessary.

[3] As outlined in Chapter 6, for this case the simplest way of obtaining GLS estimates is to do weighted regression analysis where the weights are simply the inverses of the variances for each observation. This is also equivalent to multiplying each of the Y_t and the X_t by $1/(\hat{Y}_t(1 - \hat{Y}_t))^{1/2}$ and performing OLS regression on the newly obtained observations.

β. This two-step estimation is performed in order to improve the efficiency (reduce the variance) of the estimated[4] b's.

The example of voting behavior used throughout this book falls in the category of models with discrete dependent variables. The interest is in estimation of the voting behavior of a population where the dependent variable equals one for a Johnson vote and zero for a Goldwater vote. The OLS estimates[5] are

$$V = 0.087 + 0.636P + 0.360E. \tag{7.3}$$
$$ (0.021) \quad (0.028) \quad (0.031)$$

Reestimating this model using the two-round Goldberger procedure yields

$$V = 0.074 + 0.647P + 0.362E. \tag{7.4}$$
$$ (0.020) \quad (0.028) \quad (0.031)$$

For this particular example, there is very little change in the coefficient estimates. Further, the estimated standard errors of the coefficients are very similar.[6] (Such small differences are not necessarily the rule however.)

Two issues must be considered in evaluating this procedure. First, there is no requirement that $\hat{Y}_t$ falls between 0 and 1. This is somewhat embarrassing when one attempts to apply a probability interpretation to the model since predicted probabilities outside of the [0, 1] interval have no meaning. Further, it introduces technical difficulties since the estimated error variance [$\hat{Y}_t(1 - \hat{Y}_t)$] is then negative. As a practical manner, whenever the estimated value of Y_t falls outside of the allowable range, a value of $\hat{Y}_t$ that is very close to the limit (say 0.005 or 0.995) can be substituted for the second round estimation. (This in fact was necessary in the voting example.)

Nevertheless, a more serious issue cannot be sidestepped as easily. The thrust of this technique treats the central problem of estimation as the efficiency of the parameter estimates. The starting point of the analysis is an assumption that the model specification is correct, including the implications about the error distribution. However, there are strong reasons to believe that this model specification might generally be incorrect. If such is the case, this procedure leads in the wrong direction.

---

[4] Since the weights are also based upon sample estimates from the first round, these estimates are actually asymptotically efficient; i.e., they are minimum variance (unbiased linear) estimators for large samples.

[5] P is again the party identification variable and is coded in the form of 0 = strong Republican, 0.25 = weak Republican, 0.75 = weak Democrat, and 1.0 = strong Democrat. E is the evaluation variable ranging from 0 for individuals who think the Republicans can best handle all issues to 1 for people who think Democrats can best handle all issues.

[6] The estimated standard errors in the OLS case are the usual estimates that assume homoskedasticity. They are not the true standard errors that would result from a correct calculation based on heteroskedastic error variances. Thus, they should be interpreted as indicating how much one would be misled by using the (biased) OLS estimates. In this case, the answer is not much.

Functional Form

Suppose that we are indeed interested in estimating a probability model that relates the probability of a given event to a series of exogenous factors. With simple regression analysis and a continuous dependent variable, it is common to begin with a linear, additive model, i.e., a model that is linear in the exogenous variables. For the current problem, there is strong reason to believe that such a functional form is unlikely to be a reasonable approximation of the true model.

One way to write the model of interest is to consider $P_t = \text{Prob}(Y_t = 1) = F(X_t)$. In this case, the interest centers upon how the exogenous variables affect the probabilities of the event in question, and $F(X_t)$ is simply the cumulative density function expressed in terms of X_t. There are several reasons to suspect that $F(X_t)$ is a nonlinear function. First, $F(X_t)$ must fall between 0 and 1. This implies the relationship must at least be nonlinear at the boundaries. Certainly, it is possible to constrain the relationship to always fall within the 0, 1 limits; however, that implies a discontinuous function at those limits. In many cases such discontinuities appear implausible. Secondly, while such a linear probability function is acceptable from a technical standpoint, it does not seem to be very realistic in application. Empirical observation suggests that nonlinear functions of a general S-shape are more reasonable. Such functions imply that a given marginal change in probability is more difficult to obtain when the probability is closer to one of the limits. For example, changing P from 0.97 to 0.98 requires a larger change in X than changing P from 0.49 to 0.50. Finally, in the case of several exogenous variables, an additive form seems particularly inappropriate. Instead, one would expect some interaction among the variables; the marginal change in probability associated with a given variable almost surely depends upon the values of the other exogenous factors.

In the simplest case, presume that the value of one variable is sufficiently large so that the event is almost certain to occur; then increases in the level of other variables would have no effect on the probabilities of the event occurring. An analysis of the effect of income and price on the probability of automobile ownership illustrates this point. Almost all individuals with annual income above $15,000 own an automobile, and a $300 rebate on the purchase price of new autos would have virtually no effect on the probability of ownership, while such a rebate might have substantial effects on ownership by people with $7000 incomes. Such behavior indicates an interaction effect between the exogenous variables that cannot be captured by a linear probability model.

Figure 7.1 illustrates the problem of estimating a linear probability model when the true probability function assumes a general S-shape. For the case of one exogenous variable and a nonlinear $F(X)$, estimating a linear probability model introduces a systematic error into the model. Assume $P_t = \text{Prob}(Y_t =$

1)$= F(X_t)$ is nonstochastic, i.e., that the underlying probability is completely determined by the single exogenous variable, X_t. Then the error in the linear probability model is $U_t = F(X_t) - (\beta_1 + \beta_2 X_t)$. The deviations of the linear model from the cumulative distribution function represent the errors in the model being estimated. However, as Fig. 7.1 vividly illustrates, the errors in the linear model are *not* independent of the exogenous variable (although with a particular sample they may be uncorrelated). For large values of X, $(X > X^u)$, U is negative and its magnitude is an increasing function of X; these values are indicated by U^- in the figure. Similarly, U tends to be positive and also negatively correlated with X for small values of X $(X < X^l)$. Thus, even if we observe P_t, the true probability of a given choice for each different value of X, the deviations implied by the linear model vary systematically with X and preclude obtaining good estimates of the parameters of the distribution.

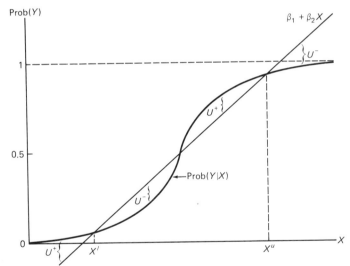

FIGURE 7.1   *Linear approximation of probability model.*

Observational Errors

The nature of the observed dependent variable adds to the estimation problems. The basic model being considered is a probability model where the probability of a given choice or given event is a function of some exogenous variables. But we do not observe these probabilities. Instead we simply observe the outcome of the choice of the event, which can be represented only by a two-valued variable. A series of ones and zeros is observed rather than a range of values which can approximate the true model.[7] Having binary observations creates two problems, one related to the nature of the error

---

[7] It is possible to represent the two outcomes by any two numerical values. We always use zeros and ones because of the convenience of interpretation in a probability sense.

terms implicit in each observation and the other related to the functional form of the model.

Because the observed values are dichotomous, there are only two possible values for error terms for any value of X. In the case of the true probability model they are $1 - F(X)$ and $-F(X)$; for the linear model they are $1 - X\beta$ and $-X\beta$. Unless the occurrence of each outcome, and thus the resulting error terms, follows a very specific distribution such as the one assumed by Goldberger, the expected value of this error term will not be independent of the values of X. As expected, the OLS estimates of the linear probability model give inaccurate estimates of the true probability function in these circumstances.

Both types of error terms, those introduced by using the wrong functional form and those created by the dichotomous nature of the observations, are likely to be correlated with the values of the explanatory variables and thus to lead to violations of the basic OLS assumptions. The consequences of these violations are estimated coefficients whose expected values depend upon the particular sample used in the estimation. Figure 7.2a–c illustrates this result. As Fig. 7.2a shows, if most of the observations are well distributed in the center of the probability distribution, the linear estimation model may be quite successful at estimating the central part of the distribution function, but not the extremes. However, as Figs. 7.2b and 7.2c show, for other samples, it is possible to get quite misleading estimates that do not approximate any part of the true probability model.

This is disconcerting. All previous discussions emphasized the importance of using estimators whose expected values are *not* affected by the sample of X's. With the model just discussed, the observed sample of X's can have a powerful effect on the coefficient estimates in the sense of producing misleading estimates of the distribution function. The linear least squares model simply cannot take into account the distributional properties of the implicit error terms. Particularly in small samples, this can have serious implications for the estimates of the distribution function.

The dichotomous nature of the dependent variable seriously impedes efforts to overcome these estimation problems. The least squares estimation model is not restricted to models that are linear in the variables. Chapter 4 shows several common transformations of the observed variables; these can be used to fit nonlinear equations (which are linear in the parameters). However, the dichotomous nature of the dependent variable renders most of these ineffective. Since only two values for Y are observed, as in Fig. 7.3, transformations of the dependent variable for the individual observations will be ineffective because transformations of dichotomous variables are still dichotomous. The points in Fig. 7.3 do not, and cannot, be made to suggest any particular functional form. However, in order to introduce the 0, 1 bounds into the model, transformations involving the dependent variable are generally required.

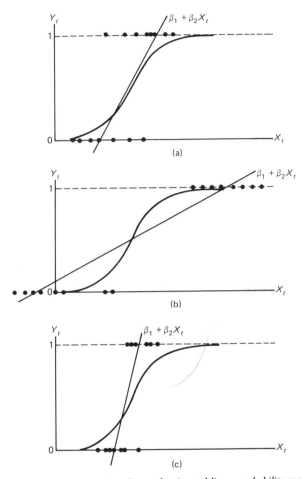

FIGURE 7.2   *Sample dependency of estimated linear probability models.*

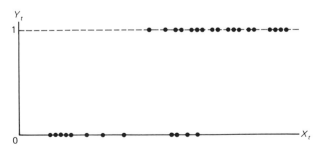

FIGURE 7.3   *Observed data with dichotomous dependent variable.*

Polytomous Dependent Variables

The estimation difficulties created by linear models with discrete dependent variables become unmanageable in situations where outcomes are measured by categorical variables with multiple responses or are the joint outcomes of several separate events. In modeling voting in a multicandidate election such as the 1968 presidential election or occupational choices or in explaining a series of joint outcomes such as party affiliation and vote or residential locations and type of housing unit, the dependent variables cannot be ordered with each value indicating more or less of something than the previous value. These cases, called polytomies, can be treated as dichotomies by comparing one outcome to all others. The probability of voting for Nixon rather than Humphrey or Wallace or a probability of living in Boston in a three-bedroom apartment rather than all other possibilities are examples of this procedure. However, these methods lose all information about the joint nature of the events and do not accurately model the desired range of behavior. We return to polytomous situations and discuss ways to try to estimate such models after developing the dichotomous model. The treatments of polytomous dependent variables are extensions of the simpler, dichotomous models.

7.3 Alternative Models—Dichotomous Dependent Variables

Consideration of modeling behavior with discrete dependent variables can be divided into consideration of the functional form for the underlying probabilities and consideration of the appropriate estimation technique for alternative models and data sets. This section sketches alternative functional forms that have been suggested for the dichotomous case, while subsequent sections develop estimators of these models and extend them to the polytomous case.

We write the model under consideration as

$$P_t = \text{Prob}(Y_t = 1) = F(X_t\beta) \quad \text{and} \quad 1 - P_t = \text{Prob}(Y_t = 0) = 1 - F(X_t\beta),$$

$$(7.5)$$

where $F(X_t\beta)$ is simply the cumulative distribution function that describes how the probabilities are related to the exogenous variables. There are clearly many different distribution functions that could be used; here we shall concentrate upon two historically important functions.

Perhaps the most frequently assumed form for the underlying distribution functions (in terms of current usage and where development is taking place) is the logistic distribution. This is defined as

$$P = 1/(1 + e^{-X\beta}). \quad (7.6)$$

This distribution ranges from 0 to 1 as $X\beta$ goes from $-\infty$ to $+\infty$. Figure 7.4 graphs a logistic distribution function. The popularity of this distribution

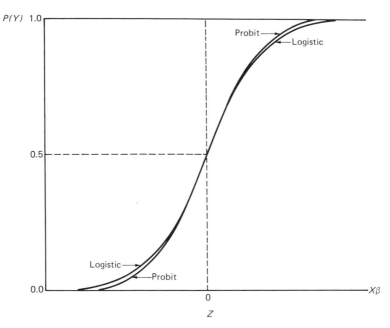

FIGURE 7.4   *Comparison of logistic and probit forms.*

arises from its convenient mathematical properties along with its desirable shape. Note that if $P = 1/(1 + e^{-X\beta})$ as in Eq. (7.6), then

$$1 - P = e^{-X\beta}/(1 + e^{-X\beta}) = 1/(1 + e^{X\beta}). \qquad (7.7)$$

Rearrangement of these expressions gives

$$L = \log \frac{P}{1 - P} = \log P - \log(1 - P)$$

$$= -\log(1 + e^{-X\beta}) - \left[\log(e^{-X\beta}) - \log(1 + e^{-X\beta})\right] = X\beta. \qquad (7.8)$$

L is called the logit or the log of the odds ratio, and analysis based upon the logistic distribution is often called logit analysis. As P goes from 0 to 1 ($X\beta$ goes from $-\infty$ to $+\infty$), L goes from $-\infty$ to $+\infty$; thus, while the probabilities are bounded, the logits are unbounded with respect to the values of X.

Several properties of the logistic function might be pointed out at this time. First, while the logits are a linear function of the exogenous variables, the probabilities themselves are not.[8] The relationship between the change in probabilities and a change in one of the exogenous variables can be seen by taking the partial derivative of P with respect to one of the exogenous

---

[8] In reality, transformations of the exogenous variables are permitted so that we can regard the logits as linear in the parameters and not necessarily the original variables. We can replace $X\beta$ by $g(X)$, where g is one of the functions discussed in Chapter 4. Thus, in the same way as the linear model discussed previously is quite flexible, so is the logit model.

variables, say X_k:

$$\frac{\partial P}{\partial X_k} = \frac{\partial\left[1/(1+e^{-X\beta})\right]}{\partial X_k}$$

$$= \frac{1}{(1+e^{-X\beta})^2} \beta_k e^{-X\beta} \left[\text{since } \frac{\partial(1/y)}{\partial x} = -\frac{1}{y^2}\frac{\partial y}{\partial x} \text{ and } \frac{\partial e^y}{\partial x} = e^y\frac{\partial y}{\partial x}\right]$$

$$= \beta_k P(1-P) \quad \left[\text{from (7.7) and (7.8)}\right]. \tag{7.9}$$

Secondly, we can also see that this form automatically allows for interactions among the variables since the value of the derivative depends upon where it is evaluated (in terms of P) and this level of P will depend upon the values both of X_k and the other exogenous variables.[9]

An alternative model is based upon the assumption that $F(X\beta)$ is cumulative normal. This model is called the probit model[10] and is also displayed in Fig. 7.4. The interpretation of this model is straightforward. The definition of the cumulative normal distribution gives

$$P = \text{Prob}(Y_i = 1) = F(X\beta) = \frac{1}{(2\pi)^{1/2}} \int_{-\infty}^{X\beta} e^{-U^2/2} dU, \tag{7.10}$$

where U is a random variable distributed as a standard normal deviate, i.e., U is $N(0,1)$. Thus, the probability of the event in question occurring, $\text{Prob}(Y_t = 1)$, is the area under the standard normal curve between $-\infty$ and $X\beta$. The larger the value of $X\beta$, the more likely the event is to occur.

In the same manner as for the logistic distribution, it is possible to show how the probabilities change with the exogenous variables. This can be shown to equal

$$\frac{\partial P}{\partial X_k} = \beta_k f(X\beta), \tag{7.11}$$

where $f(X\beta)$ is the value of the normal density function at the point $X\beta$. In other words, since this ordinate is larger near the center of the distribution, the largest response to a change in a given exogenous variable is obtained there. The change in probabilities becomes progressively smaller as one approaches either $P=0$ or $P=1$.

The normal and logistic distributions are very similar. The differences in the functions show up in the tails of the distributions with the normal curve approaching the axes more quickly than the logistic curve.[11]

---

[9] The interaction can be seen directly by finding $\partial^2 P/\partial X_k X_j$. This is a function of β_k, β_j, and P; $\partial^2 P/\partial X_k \partial X_j = \beta_j\beta_k P(1-P)(1-2P)$.

[10] This model is also referred to as a normit model. Historically, probit and normit models differed by a normalization of $X\beta$.

[11] In fact, the logistic distribution is very similar to the t-distribution with seven degrees of freedom, while the normal distribution is a t-distribution with infinite degrees of freedom.

7.4   Logit Analysis—Grouped Data

Let us consider the simple case of a single dichotomy. The discussion of the problems with linear models indicated that two issues are important. The first, and most critical, is measuring the actual probability of a given outcome for values of the explanatory variables. The other is approximating the functional form of the relationship between the probabilities and the values of X, as with the functional forms of the previous section, once the probability estimates are determined.

In addressing the measurement question, if one has many observations of the actual outcomes for given values of X, it is possible to calculate the relative frequency of the outcome and to use this as an estimate of the probability of the occurrence for each X. This is in fact a standard approach to probabilities—the probability of an event is the limit (as the sample size gets infinitely large) of the relative frequency of occurrence. This method is the basis for many experimental designs, say to test the effectiveness of varying amounts of rubella serum in controlling measles. One large group of individuals would be given no serum, other groups 5 cc, 10 cc, and so on of the serum. The relative frequency of measles for each of these groups is then interpreted as an estimate of the probability of measles associated with given amounts of serum.

In a nonexperimental setting, if many observations for each value of X are available, it is possible to use a similar approach to estimating the underlying probabilities. For example, imagine that the exogenous variables are themselves categorical, such as race, occupation, or religion. It is then possible to observe the frequency of voting for a given candidate within each categorical group. As the sample size of each group increases, the observed frequency becomes a better estimate of the true probability. When the exogenous variables are not categorical, they can be grouped into categories in order to obtain large numbers of observations. For example, if the exogenous variable is income, it is possible to group income into classes (e.g., less than $3000, $3000–10,000, and greater than $10,000), and proceed in the same manner.

Once the problems of measuring the probabilities are overcome, say through large numbers of observations for each category, different functional forms can be used to relate these probabilities to values of the exogenous variables. The dependent variable is no longer dichotomous, so transformations of the type discussed in Chapter 4 are possible, and one can even use least squares methods to estimate choice models such as the logit model.

This in fact is just what is done in one class of logit models. In particular, if for each category X_j $(j = 1, \ldots, J)$ we have N_j observations of a dichotomous (0 or 1) variable Y_{ij} $(i = 1, \ldots, N_j)$, the estimated probability for the jth group is

$$\hat{P}_j = \frac{1}{N_j} \sum_{i=1}^{N_j} Y_{ij}.$$

Assuming a logistic distribution of the probabilities, we have from Eq. (7.8),

$$L_j = \log \frac{P_j}{1 - P_j} = X_j \beta.$$

In other words, $L_j = \log(P_j/(1 - P_j))$ is a linear function of the X's and can be estimated by least squares using $\hat{P}_j$.

This method, which represents a natural extension of the previous regression models, has been developed by Berkson (1953) and extended by Theil (1970). It begins by grouping the observations on the basis of the explanatory variables and estimating the probability of Y for each group. The log of the odds transformation converts the probability estimates to a continuous unbounded variable which becomes the dependent variable in a linear model with the categorical definitions as explanatory variables. Because of the need to group observations, this method is best applied to situations where there are many observations and where the explanatory variables themselves fall into natural groupings. This makes it very appropriate for many models being estimated with survey data.

TABLE 7.1

Table for Logit Analysis

	Variable I			
Variable J	1	2	3	
1	F_{11}	F_{12}	F_{13}	$F_{1.}$
2	F_{21}	F_{22}	F_{23}	$F_{2.}$
3	F_{31}	F_{32}	F_{33}	$F_{3.}$
	$F_{.1}$	$F_{.2}$	$F_{.3}$	$F_{..}$

The key to logit analysis is the calculation of a contingency table or cross tab for the explanatory and dependent variables. Table 7.1 lays out the schematic for such a problem for the case where there are two explanatory variables, each with three categories, and a dichotomous dependent variable. The cell definitions refer to different combinations of the (categorical) explanatory variables. The cell entries show the relative frequency of a given outcome among the observations satisfying the cell definition. Such a table is shown in Chapter 1 for the voting example. Tables can be easily expanded to more variables and categories. In Table 7.1, F_{ij} refers to the proportion of the observations in the joint category (i,j) exhibiting the particular behavior being studied. If we continue to let Y be defined as a dichotomous $(0, 1)$ dependent variable, then $F_{ij} = (\sum_{t=1}^{N_{ij}} Y_{t,ij})/N_{ij}$, where N_{ij} is the number of observations in cell (i,j). In the voting example, weak Democrats with a Democratic issue preference are in cell 33, $N_{33} = 189$, and $F_{33} = 0.975$. (The contingency table for the voting example is reproduced as Table 7.2.)

TABLE 7.2

Cross Tabulation of 1964 Voting by Issue Evaluation
and Party Identification (Proportion Voting for Johnson)

Party Identification	Evaluation			
	Republican	Indifferent	Democrat	All
Strong Republican	0.102 (109)	0.115 (35)	0.214 (33)	0.126 (177)
Weak Republican	0.258 (75)	0.544 (50)	0.667 (52)	0.461 (177)
Weak Democrat	0.606 (70)	0.890 (56)	0.975 (189)	0.879 (335)
Strong Democrat	0.727 (31)	0.893 (56)	0.990 (344)	0.958 (431)
All	0.334 (285)	0.687 (217)	0.917 (618)	0.724 (1120)

The estimated model considers the logit to be a linear function of the categorical variables used to make the table, as in Eq. (7.8). In the most straightforward case, the equation is written as a series of dummy variables, one for each category of each variable. The estimated model then omits one of these dummies for each explanatory variable to prevent singularity. For the example in Table 7.1, the simple logit model is

$$L_{ij} = \beta_1 X_1 + \beta_2 X_2 + \beta_3 X_3 + \beta_4 X_4 + \beta_5 X_5 = X\beta, \qquad (7.12)$$

where $X_1 = 1$ for all cells; $X_2 = 1$ for cells where variable $j = 2$, $= 0$ otherwise; $X_3 = 1$ for cells where variable $j = 3$, $= 0$ otherwise; $X_4 = 1$ for cells where variable $i = 2$, $= 0$ otherwise; and $X_5 = 1$ for cells where variable $i = 3$, $= 0$ otherwise.

The interpretation of this model is quite straightforward and follows that of the dummy variable regression model discussed in Chapter 4. Each cell is denoted by a unique combination of values for the dummy variables, and the logit for each cell is obtained from the coefficients on each variable in the appropriate combination. For example, the cell $(1, 1)$ is denoted by 0's for all X's except X_1, implying $L_{11} = \beta_1$, and the cell $(2, 3)$ is denoted by values of 1 for X_1, X_3, and X_4, and 0's for the other variables, thus $L_{23} = \beta_1 + \beta_3 + \beta_4$. A similar process holds for each cell in Table 7.1 with the following logit values:

$$L_{11} = \beta_1, \qquad L_{12} = \beta_1 + \beta_2, \qquad L_{13} = \beta_1 + \beta_3,$$
$$L_{21} = \beta_1 + \beta_4, \qquad L_{22} = \beta_1 + \beta_2 + \beta_4, \qquad L_{23} = \beta_1 + \beta_3 + \beta_4,$$
$$L_{31} = \beta_1 + \beta_5, \qquad L_{32} = \beta_1 + \beta_2 + \beta_5, \qquad L_{33} = \beta_1 + \beta_3 + \beta_5.$$

We refer to this as an additive model because the effect on the logit value of any one variable does not depend upon the values for other explanatory variables. For example, $\beta_5 - \beta_4$ is the expected difference in the logit between observations in group 3 rather than group 2 on variable i, regardless of how the observation is classed on variable j. We can see this specification better by examining the logit values for each of the cell entries. Thus

$$L_{31} - L_{21} = L_{32} - L_{22} = L_{33} - L_{23} = \beta_5 - \beta_4.$$

The same interpretation holds for all coefficients in the model. (We shall discuss interaction terms subsequently.)

To estimate Eq. (7.12), we must estimate P_{ij}, which of course is not observed. However, if all observations in each cell are drawn independently of each other, the observed frequencies in each cell, the F_{ij}, are binomially distributed about the true probability. Thus $E(F_{ij}) = P_{ij}$ and $\text{var}(F_{ij}) = P_{ij}(1 - P_{ij})/N_{ij}$. If we let $L_{ij}^* = \log(F_{ij}/1 - F_{ij})$, the general model can be written as

$$L_{ij}^* = X_{ij}\beta + (L_{ij}^* - L_{ij}) = X_{ij}\beta + U_{ij}, \qquad (7.13)$$

where L_{ij} is defined in terms of the true probabilities as in Eq. (7.8). When all the observations are drawn independently, the asymptotic properties of the U's are[12]:

$$E(U_{ij}) = 0, \qquad \text{var}(U_{ij}) = 1/(N_{ij}P_{ij}(1 - P_{ij})), \qquad \text{and} \qquad \text{cov}(U_{ij}U_{kl}) = 0.$$

With these assumptions, the model in Eq. (7.13) is heteroskedastic since the variance of U for each cell depends upon the probability of occurrence in each cell, P_{ij}, and on the number of observations in each cell, N_{ij}. In this form, Eq. (7.13) satisfies all the requirements of generalized least squares with $E(UU') = \Omega$. The diagonal elements of Ω are $1/N_{ij}P_{ij}(1 - P_{ij})$, and the off-diagonal elements are zero. However, we do not know P_{ij}. Instead, we use the F_{ij} to estimate the P_{ij} in Ω. This is justified for large N_{ij}, i.e., this is asymptotically appropriate since the F_{ij} are consistent estimates of the P_{ij}. The GLS estimates can be computed either with a weighted regression where the weights are $N_{ij}F_{ij}(1 - F_{ij})$ or with OLS on transformed variables where the transformation is accomplished by multiplying each variable, including the constant term, by the square root of the above weight. The estimates of the coefficients in Eq. (7.13) are

$$b = (X'\hat{\Omega}^{-1}X)^{-1}X'\hat{\Omega}^{-1}L^*. \qquad (7.14)$$

The asymptotic variances and covariances of b in Eq. (7.4) are estimated by[13]

$$(X'\hat{\Omega}^{-1}X)^{-1}. \qquad (7.15)$$

---

[12] The mean and variance for both the P_{ij} and the U_{ij} follow directly from the binomial distribution. See Theil (1970).

[13] In using most OLS or weighted regression programs to compute the logit coefficients and their variances, one correction needs to be made to obtain the correct variances and covariances. In our discussion of GLS in the previous chapter, the conventional statement of the generalized error term variance-covariance matrix includes the scalar σ^2 and $E(UU') = \sigma^2\Omega$, where σ^2 is estimated by the square of the standard error of the estimate, $s^2 = (1/(T - K))(e'\Omega^{-1}e)$. In computing the variance of each coefficient, the diagonal elements of $(X'\Omega^{-1}X)^{-1}$ were multiplied by s^2. This is not correct for the logit model however. In this model there is no scalar in $E(UU')$ because no additive randomness is allowed; the only randomness comes from observing the $(0, 1)$ outcomes instead of the underlying probabilities. The correct estimates of the coefficient variances are simply the diagonal elements of $(X'\Omega^{-1}X)^{-1}$. Consequently, with most programs, the correct estimates of the coefficient standard errors are the printed standard errors divided by $\hat{s}$.

Inspection of the variance expressions and thus the weights shows that cells where F_{ij} equals 0 or 1 or where $N_{ij}=0$ cannot be used. With $N_{ij}=0$, this is not bothersome; however, with $N_{ij}>0$, this is a loss of information.

We can easily summarize the basic structure of the logit model. Each cell in the original contingency table, Table 7.1, is treated as a separate observation, with the observation's characteristics given by the various classifications of the explanatory variables. The dependent variable for each "observation" is the observed proportion or relative frequency within the cell. By the log of the odds transformation, we create an unbounded continuous dependent variable that is modeled by the standard linear model. The error terms implicit in each of these observations asymptotically have a zero mean, but a variance that is inversely related to $P_{ij}(1-P_{ij})$ and the number of individual units contained in the cell. Because of these different variances, generalized least squares estimation that weights each cell by the reciprocal of these variances is appropriate. Thus the logit model is a straightforward application of the generalized least squares model.

We can illustrate this entire procedure with the table in the voting example (Table 7.2). The layout of the appropriate vectors and matrices is shown in Table 7.3. This table shows the sample size, the proportion voting for Johnson, the estimated logit, and the appropriate weight for each cell in that table (the observations). The table also shows the set up for the dummy explanatory variables corresponding to the logit model:

$$L^* = \beta_1 + \beta_2 Indiff + \beta_3 Dem + \beta_4 WR + \beta_5 WD + \beta_6 SD + U. \qquad (7.16)$$

where $Indiff=1$ for voters with issue evaluations of 0.5 and 0 otherwise, $Dem=1$ for voters with a Democratic evaluation and 0 otherwise, $WR=1$ for weak Republicans and 0 otherwise, $WD=1$ for weak Democrats and 0 otherwise, and $SD=1$ for strong Democrats and 0 otherwise.

TABLE 7.3

1964 Voting Logit Model

N_{ij}	F_{ij}	$L_{ij}^{*\,a}$	Weight	X_1 Const	X_2 Indiff	X_3 Dem	X_4 WR	X_5 WD	X_6 SD
109	0.102	−2.175	9.984	1	0	0	0	0	0
35	0.115	−2.041	3.562	1	1	0	0	0	0
33	0.214	−1.301	5.551	1	0	1	0	0	0
75	0.258	−1.056	14.358	1	0	0	1	0	0
50	0.544	0.176	12.403	1	1	0	1	0	0
52	0.677	0.740	11.731	1	0	1	1	0	0
70	0.606	0.431	16.713	1	0	0	0	1	0
56	0.890	2.091	5.482	1	1	0	0	1	0
189	0.975	3.664	4.607	1	0	1	0	1	0
31	0.727	0.979	6.153	1	0	0	0	0	1
56	0.893	2.122	5.351	1	1	0	0	0	1
344	0.990	4.595	3.406	1	0	1	0	0	1

$^a L_{ij}^* = \log(F_{ij}/1 - F_{ij})$.

The estimated equation is

$$L^* = -2.75 + 1.16\,Indiff + 2.17\,Dem + 1.60\,WR + 3.47\,WD + 4.05\,SD, \quad (7.17)$$
$$\;\;\;(0.25)\;\;(0.24)\qquad(0.25)\qquad\;(0.28)\qquad\;(0.30)\qquad\;(0.35)$$

where the estimated coefficient standard errors are in parentheses.

This estimated equation predicts the logit for a strong Republican with a Republican evaluation to be −2.75. By undoing the arithmetic of the logit transformation, this corresponds to a predicted probability of voting for Johnson of 0.06. The logit increases by 1.16 for those who are indifferent in their evaluations and by 2.17 for those who prefer the Democrats, regardless of party affiliations. Likewise, the logit increases by 1.60 for weak Republicans, 3.47 for weak Democrats, and 4.05 for strong Democrats, independent of evaluations. Note, however, that these logit differences do not directly relate to probability differences because the probabilities themselves depend upon the values for the other variables; i.e., while the model is additive in the logits, it is nonlinear in terms of the probabilities. For example, the expected difference in the probability of voting for Johnson between a strong and a weak Republican is 0.18 if both have Republican evaluations, 0.33 if both are indifferent, and 0.38 if both prefer Democrats.

The estimates in Eq. (7.17) rely upon the asymptotic weights derived from the assumption that the observed frequencies are represented by a binomial distribution. The weighting, as described in Chapter 6, is done to improve the efficiency of the estimates. Ordinary least squares estimates should, however, be consistent estimates, albeit inefficient. The OLS estimates of the model in Eq. (7.16) are

$$L^* = -2.98 \;\; + 1.04\,Indiff + 2.38\,Dem + 1.79\,WR + 3.90\,WD + 4.40\,SD \quad (7.18)$$
$$\;\;\;(0.484)\;\;(0.484)\qquad(0.484)\qquad\;(0.558)\qquad(0.558)\qquad(0.558)$$

The estimated coefficients in Eq. (7.18) are reasonably close to those in Eq. (7.17). However, the estimated standard errors with the OLS estimation differ by a factor of two. The standard errors presented for the OLS estimates are those routinely produced and are thus based upon an assumption of homoskedastic error terms. As such, they are biased estimates of the true standard errors of the coefficients. In this context, the ratio of the standard errors of the generalized least squares and ordinary least squares coefficients does not represent the efficiency gains from using GLS. The OLS standard error estimates do indicate the possibility of erroneous statistical tests and inferences. In this case we would be less likely to reject a null hypothesis of no relationship with the OLS information than with the GLS information.

The logit model can also be specified to include interaction terms among the explanatory variables. Interaction terms require additional dummy variable terms. In the voting model, if we want to continue the hypothesis that voters who see no difference in their evaluations of the parties rely more on their party affiliations in deciding how to vote than those with evaluations, we must simply specify an additional set of party affiliation variables for voters with indifferent evaluations. This model is

$$L^* = \beta_1 + \beta_2 Indiff + \beta_3 Dem + \beta_4 WR + \beta_5 WD + \beta_6 SD$$
$$+ \beta_7 IWR + \beta_8 IWD + \beta_9 ISD.$$

The explanatory variables IWR, IWD, and ISD are one for the cells corresponding to weak Republicans, weak Democrats, and strong Democrats with indifferent evaluations, respectively, and zero for all other cells. With this specification, the predicted logits for voters without indifferent evaluations are the same as before. The logits for indifferent voters are $\beta_1 + \beta_2$ for a strong Republican, $\beta_1 + \beta_2 + \beta_4 + \beta_7$ for a weak Republican, and so on. Thus the expected difference between a weak Republican and a weak Democrat is $\beta_5 - \beta_4$ if both have an evaluation and $(\beta_5 - \beta_4) + (\beta_8 - \beta_7)$ if both are indifferent. The estimated interaction model is

$$L^* = -2.64 + 0.60 Indiff + 2.17 Dem + 1.42 WR + 3.30 WD + 4.13 SD$$
$$+ 0.80 IWR + 0.83 IWD + 0.03 ISD.$$

The hypothesis that voters with no issue preferences give more weight to party affiliation is not generally supported with this analysis. The hypothesis is supported among Republicans, with there being a greater difference in the expected logits of strong and weak Republicans if they have no issue preferences than if they both prefer one of the parties (2.22 versus 1.42). However, the difference between weak Republicans and weak Democrats is the same (1.91 versus 1.88) regardless of whether they have an issue preference or not, and there is a much smaller difference between weak and strong Democrats if there is no issue preference than if one exists (0.03 versus 0.83). Thus, based upon the estimated coefficients, the hypothesis is only partially supported. (Subsequent statistical tests will confirm this conclusion.)

This example also permits us to illustrate the hypothesis testing procedures used in logit analysis. Contrary to the regression model, the usual hypotheses are not about single coefficients but about alternative specifications of the model. In this logit specification the individual coefficients refer to specific values of a given variable. Our interest is generally in testing the effect of a variable, and this generally involves testing a set of coefficients. (In this regard they correspond to the types of hypotheses tested with the F-statistic described at the end of Chapter 5.) The simplest hypothesis is that the differences between the observed and the predicted frequencies could occur by chance if the estimated model is the correct one. The tests of different specifications are that one variable, or a set of variables, or some interaction terms should be excluded from the model. The null hypothesis is that the effect of these variables or terms is zero. These statistical tests are based on the fit of the observed frequencies for each cell to the probabilities predicted by the logit model. In the simple case, the better the fit, the more likely we are to conclude that the model is the correct one. In the other application, the greater the differences in the fits with and without the terms in question, the more likely we are to decide they are important.

We shall only outline the development of the appropriate procedures. The statistical justification for the tests relies even more heavily on N_{ij} being large than does the derivation of the estimators. The essence of the method is that if each N_{ij} is sufficiently large, the observed frequencies F_{ij} are normally distributed about the true probabilities P_{ij}, and an expression involving their squared difference has a chi-squared distribution.[14] The actual calculations substitute the estimated P_{ij} from the logit model for the true values; i.e., if $\hat{L}_{ij}^{*}$ is the predicted logit, the corresponding predicted probability is $\hat{P}_{ij} = e^{\hat{L}_{ij}^{*}}/(1 + e^{\hat{L}_{ij}^{*}})$. Theil (1970) shows that with these assumptions, the expression

$$\sum_{i} \sum_{j} \frac{N_{ij}\left(F_{ij} - \hat{P}_{ij}\right)^{2}}{\hat{P}_{ij}\left(1 - \hat{P}_{ij}\right)} \tag{7.19}$$

is distributed as a chi-squared with degrees of freedom equal to the number of cells in the original table minus the number of coefficients estimated.

The test of the overall specification of the model is essentially asking, What is the probability that the observed frequencies could occur by chance if the estimated structure is the correct one? The better this fit, the smaller the deviations $F_{ij} - \hat{P}_{ij}$, the smaller the χ^{2} value, and the more likely we are not to reject the null hypothesis that the specified structure is the correct one. Conversely, the worse the fit, the larger the deviations, the higher the χ^{2} value, and the more likely we are to reject the null hypothesis. In the case of the first voting model estimated by GLS, the χ^{2} value is 19.19, with six degrees of freedom. The critical value for the χ^{2} statistic with that many degrees of freedom is 12.59 for $\alpha = 0.05$, and 16.81 for $\alpha = 0.01$. Thus we are likely to reject the null hypothesis that the simple model adequately accounts for the observed voting behavior.

In testing two alternative specifications, the differences in the value of Eq. (7.19) for each specification is distributed as a χ^{2} statistic with degrees of freedom equal to the number of coefficients being tested. In this way, we can test the specification of the interaction terms between party affiliation and indifferent evaluations.

The value of Eq. (7.19) for the model with interaction terms is 16.26, which gives a difference from the simpler model of 2.93. The critical value of the χ^{2} statistic at the 5% level for three degrees of freedom is 7.82. Our value is considerably below this value. Consequently, we would not reject the null hypothesis that there is no difference in the behavior of voters with no issue preferences, i.e., we would conclude that indifferent voters give the same weight to party affiliation as voters with an issue preference.

The logit model described is an analysis of variance model in that all the exogenous variables in the estimation are dummy variables related to categories of the explanatory factors. This format is convenient when there is no natural scale for the X's, for example, when one of the explanatory variables

---

[14] As N gets large, the binomial distribution approaches the normal distribution.

is the race of an individual. At times, however, the X's have a natural scale, such as years of schooling or age. When there is a scale relating different cells of a variable, this information can be used in the estimation to increase the efficiency, and often the interpretability, of the coefficient estimates. Instead of defining a separate dummy variable for each value of a particular variable, the specific values of the variable would be introduced. For example, imagine that one is interested in the probability of an individual going to college and that one explanatory variable is father's education, which comes in three categories: 10–12, 13–16, and 17 + yr. The analysis of variance format would introduce a separate variable for each of the three categories. The alternative approach is to define one variable that takes on the values of 11, 14.5, and 18, depending upon into which specific category the individual falls. This latter approach has the advantage of estimating fewer coefficients, which in turn can increase the precision of the estimates. Further, it provides a method of predicting probabilities for sample points outside of the given categories (such as the probability of college attendance for someone whose father has a Ph.D.). Such predictions are not easily obtained in the analysis of variance format.[15]

There are two potential difficulties that may limit the uses of the logit model just discussed. The development of the statistical procedures and the justification for the tests of significance depend upon having a large number of individual observations within each cell of the contingency table. The estimation of P_{ij} by F_{ij} is the basis for the derivation of the variances of the error terms in the logit model [Eq. (7.13)] and thus the key to the generalized least squares estimation procedure. The estimation of the variance of the logits by $N_{ij}F_{ij}(1 - F_{ij})$ for subsequent use in the GLS procedure is precise only for large samples. The large sample assumption applies to *each* table entry, the N_{ij}, not to the total sample size N. In many applications, even though there may be a large number of total observations, by the time several

---

[15] Efficiency gains may be offset by biases if in fact the underlying probability function is not linearly related to the values of the exogenous variables. Here, as in the standard regression model, however, it is required only that the relationship between the log of the odds and the exogenous variables be linear in the parameters. Thus, transformation of the exogenous variables may be applied.

Extrapolation in this case is also subject to the same concerns as extrapolation in the standard linear case. When one is predicting outside of the observed range of the exogenous variables, there is always the danger that the estimated relationship will not hold up, i.e., that the functional relationship is different.

When a scaled variable is introduced, the estimated standard errors for the coefficient seem appropriate for use in testing hypotheses about the exogenous variable. The estimated coefficients are asymptotically normal in this case, and the standard errors are asymptotically true. There is thus a question as to whether one should use normal probabilities for hypothesis testing or whether one should use a t-distribution. There is no absolutely correct rule in this case, but it seems sensible to use the t-distribution which takes into account the number of cells and parameters used in estimation. This can however provide only rough guidance since it is not a technically correct test.

explanatory variables are included in the contingency table, many of the individual cells are too small to support this assumption. The problem with small cells is one of inefficiency (or poor weights), however, and not one of bias or inappropriateness of the model.

The second, and more serious, difficulty is the assumption underlying the use of the binomial distribution in the first place. In order to assume that F_{ij} has the binomial distribution with mean P_{ij}, and thus that $\log[F_{ij}/(1-F_{ij})]$ asymptotically has mean $\log[P_{ij}/(1-P_{ij})]$, it is necessary to assume that each individual observation in cell (i,j) has the same probability P_{ij} of exhibiting the behavior being modeled. In the logit model, $F_{ij} \neq P_{ij}$ simply because only N_{ij} individuals are sampled, not because some of the individuals might have a different P_{ij} during the course of the experiment. If the P_{ij} must also be subscripted for each individual in N_{ij}, then

$$E\left[\log\left(\frac{F_{ij}}{1-F_{ij}}\right)\right] \neq \log\left[\frac{E(F_{ij})}{1-E(F_{ij})}\right] = \log\left(\frac{P_{ij}}{1-P_{ij}}\right)$$

even though $E(F_{ij}) = P_{ij}$ and even with large samples.

Differences in the P_{ij} for each individual in a cell clearly introduces error into the weights that are used in the GLS procedure. Thus, the efficiency of the estimates would be affected in much the same way as having small cells. However, it introduces a second problem: biased estimates, because the error term in the estimating equation will generally not be independent of the exogenous variables. Thus, this concern is a more important limitation in the estimation.

There are several reasons why the P_{ij} may not be equal for each individual in N_{ij}. We have already discussed how social behavior and the models used to describe it must be probabilistic because of the partial nature of the theories and the inability to measure and include all possible influences on individual behavior. In all modeling to this point, we simply assumed that the expected values of the errors introduced by this stochastic behavior is zero, not that each error term is zero. The estimated logit model, however, is based on the assumption that each individual has the same probability, not merely the same expected probability.

There is a second fundamental reason why the assumption of constant P_{ij} for each individual is inappropriate besides the presence of stochastic errors. In many cases, the variables used to construct the contingency table underlying the logit estimates are obtained by grouping continuous variables. This grouping may be done explicitly, such as when income, age, or education is stratified to make a table. It may also be done implicitly when people are asked to group themselves in the course of responding to survey questions. The voting example illustrates both processes. Take the case of the party identification variable. We previously discussed how true party identification is a continuous, but unobserved, variable and that the 1964 survey simply asked people to classify themselves according to the categories of strong and weak affiliations with either party. With issue evaluations, trichotomous categories are formed by grouping all Republican and all Democratic evaluations, regardless of their magnitude. If behavior is really determined by the values of the true underlying variables, and not the assigned classifications, then each cell of the contingency table contains individuals with systematically different P_{ij}, depending upon their true values of the continuous variables. In other words, in the voting model, the probability of someone voting for Johnson depends on the *extent* to which they evaluate the Democrats as being better able to handle issues (not just whether the Democrats are better) and on the *strength* of their affiliation with one of the parties. This situation leads to quite serious deviations from the required assumptions of the logit model. One way of reducing this problem is by creating more categories where needed. However, this reduces N_{ij} and makes the asymptotic analysis of the logit model more difficult to accept. Use of a continuous exogenous variable can be viewed as a limiting case. When all real numbers are acceptable values of an exogenous variable, a typical "cell" will have one observation. Since this observation must be either 0 or 1, it is not possible to calculate $\log[F_{ij}/(1 - F_{ij})]$, and the procedure breaks down.

The consequences of this problem may not be trivial from the standpoint of accurately estimating the influence of the explanatory factors. Take the case of our voting model. The categorical evaluation variable does not admit the fact that among voters with Republican evaluations, Republican identifiers may be more Republican in their evaluations than Democrats, and strong Republicans even more so than weak ones. Consequently, we would expect strong Republicans to have a lower probability of voting for Johnson, not just because they are strong Republicans, but because they also are more Republican in their evaluations. (We leave aside until Chapter 9 the question of estimating whether party identification causes the different evaluations or whether the evaluations cause the party affiliations, or both.) A symmetric argument holds among people with Democratic evaluations. The net effects of the measurement problems induced by assuming that all Republican evaluations are identical are biased coefficients of the direct effect of party and evaluation on voting behavior. Similar problems will hold anytime underlying continuous variables are arbitrarily grouped into categorical variables, regardless of whether the explanatory variables are causally related or simply correlated by the nature of the data sample.

7.5   Logit Analysis—Microdata

There has been recent work aimed in alleviating some of the above difficulties (McFadden, 1974; Nerlove and Press, 1973). While postulating the same probability model, estimation techniques that preclude the need to categorize, or group, the explanatory variables are developed. One immediate gain of this approach is that continuous exogenous variables are permissible.

The alternative estimator uses the same logistic form for the underlying probability. Only now, instead of estimating the log of the odds function, the procedure deals directly with the probability function.[16] For observation t with values of the explanatory variables denoted by the vector X_t, from Eqs. (7.6) and (7.7) the probabilities that $Y_t = 1$ and that $Y_t = 0$ are

$$P_t = \frac{1}{1 + e^{-X_t\beta}} \quad \text{and} \quad 1 - P_t = \frac{e^{-X_t\beta}}{1 + e^{-X_t\beta}} = \frac{1}{1 + e^{X_t\beta}}.$$

The difference in the estimator to be considered here—a maximum likelihood estimator—and the GLS logit estimator of the previous section arises from the criterion used to select the b's.

The maximum likelihood procedure differs from the grouped logit model by treating each unit as a separate observation, rather than grouping them to get estimates of P_{ij}. The logic behind this procedure is to derive an expression for the likelihood of observing the pattern of successes ($Y_t = 1$) and non-successes ($Y_t = 0$) in a given data set. The value of this expression (the likelihood function) depends upon the unknown parameters of the probability function (β). If all observations are obtained independently, as is reasonable in a cross-sectional analysis, the likelihood of obtaining the given sample is found from the product of the probabilities of the individual observations having the observed outcomes.[17] If we arbitrarily order the observations so that the T_1 cases where $Y_t = 1$ (which each occur with probability P_t) come prior to the $T - T_1 = T_2$ cases where $Y_t = 0$ (which each occur with probability $1 - P_t$), the likelihood function $\mathcal{L}$ is

$$\mathcal{L} = \text{Prob}(Y_1, Y_2, \ldots, Y_{T_1}, \ldots, Y_T) = \prod_{t=1}^{T_1} P_t \prod_{t=T_1+1}^{T} (1 - P_t)$$

$$= \prod_{t=1}^{T} P_t^{Y_t} (1 - P_t)^{1 - Y_t}. \tag{7.20}$$

Equation (7.20) indicates the likelihood of obtaining the given sample based upon the unknown probability elements P_t. [The last term in Eq. (7.20) is a

---

[16] A variety of functions could be substituted for the logistic function. However, the logistic function is popular because it satisfies the boundary constraints associated with the probability function, it is mathematically tractable, and it can be related to theoretical choice models (cf. McFadden, 1974). It looks very similar to the normal probability function. Other alternatives include trigonometric transformations such as

$$F(X\beta) = \tfrac{1}{2} + (1/\pi)\tan^{-1}(X\beta) \quad \text{or} \quad F(X\beta) = \tfrac{1}{2}(1 + \sin(X\beta)).$$

The major differences between these probability functions are found in the tails of the distributions.

[17] A general discussion of maximum likelihood techniques is contained in Appendix I on statistical concepts.

simplification which relies upon the fact that Y_t equals either zero or one; therefore, the Y_t and $1 - Y_t$ used as exponents merely indicate which probability term is relevant for a given observation.]

The maintained hypothesis here is that P_t and $1 - P_t$ are specific functions of X_t and β, namely the logistic function.[18] This implies that the likelihood function can also be written in terms of the X_t and β. The criterion we shall use in choosing among potential values of b is the maximization of the likelihood function in Eq. (7.20); in other words, we shall choose that b for which $\mathcal{L}$ is a maximum. The maximum likelihood (ML) criterion is frequently used in statistics because it is known usually to be the asymptotically efficient estimator, but it is also an intuitively appealing criterion. The criterion is simply answering the question, What underlying parameters would be "most likely" to have produced the observed data?[19]

The actual process of maximizing the likelihood function can be quite difficult. The general approach most frequently used (and followed here) is to maximize the logarithm of the likelihood function instead of the likelihood function itself. Since the logarithmic transformation is a monotonic transformation (i.e., it preserves the ordering of the variables after transformation), the maximum of $\log \mathcal{L}$ will occur at the same value of b as that at which the maximum of $\mathcal{L}$ occurs. By analyzing $\log \mathcal{L}$ (denoted $\mathcal{L}^*$), the complicated product in Eq. (7.20) becomes a simple sum of logs of the arguments. This is easier to handle mathematically. The log likelihood function for the logistic model is

$$\mathcal{L}^* = \log \mathcal{L} = \sum_{t=1}^{T} Y_t \log P_t + \sum_{t=1}^{T} (1 - Y_t) \log(1 - P_t)$$

$$\left[\text{since } \log(Z^a) = a \log Z \text{ and } \log(Z_1 Z_2) = \log Z_1 + \log Z_2 \right]$$

$$= \sum Y_t \log P_t - \sum Y_t \log(1 - P_t) + \sum \log(1 - P_t)$$

$$\left[\text{since } \sum (1 - W_t) Z_t = \sum Z_t - \sum W_t Z_t \right]$$

$$= \sum Y_t \log \frac{P_t}{1 - P_t} + \sum \log(1 - P_t)$$

$$\left[\text{since } \log W - \log Z = \log(W/Z) \right]$$

$$= \sum Y_t X_t \beta - \sum \log(1 + e^{X_t \beta})$$

$$\left[\text{by Eqs. (7.6)–(7.8)} \right]. \tag{7.21}$$

To find the estimated values for β that maximize $\mathcal{L}^*$, we take the first partial

---

[18] In general, we cannot estimate both the form of the underlying probability model and the parameters of this distribution. Thus, the maintained hypothesis is generally that the probability distribution takes on some specific form such as the logistic model in Eq. (7.6). Subsequently, we shall discuss alternative forms for the maintained hypothesis.

[19] The previous least squares (LS) development applied the criteria of minimizing the sum of squared errors. When the error term is normally distributed, the LS estimator is identical to the maximum likelihood estimator. See Appendix I.

derivatives of $\mathcal{L}^*$ with respect to b, and set these equal to zero:

$$\frac{\partial \mathcal{L}^*}{\partial b_k} = \sum Y_t X_{tk} - \sum \frac{X_{tk} e^{X_t b}}{1 + e^{X_t b}} = \sum Y_t X_{tk} - \sum \frac{X_{tk}}{1 + e^{-X_t b}} = 0$$

$$\text{for } k = 1, \dots, K. \quad (7.22)$$

If there are K exogenous variables in the model, Eq. (7.22) provides K equations that can be solved for the estimated values of β. However, inspection shows that these simultaneous equations are nonlinear and may be exceedingly complex to solve. In fact they are virtually impossible to solve without a modern computer which can be programmed to solve iteratively complex nonlinear systems.

The maximum likelihood estimation technique can be applied where the explanatory variables are truly categorical, such as sex, region, or the four seasons; or where the only data available have been previously categorized, as with the party identification variable; or where continuous and categorical explanatory variables are mixed together.[20]

The full maximum likelihood estimator just developed alleviates several of the problems associated with the grouped logit model and leaves one unresolved. This method does not require prior grouping and aggregation of explanatory variables and the resulting assumption that people with different characteristics (but falling in the same cell) have the same probabilities. It is also not dependent upon the requirement of large cell sizes, which necessitated the grouping in the first place.[21] However, this method does assume that each individual's probability of $Y_t = 1$ is exactly given by the expression in Eq. (7.6). There is no allowance for the fact that actual probabilities may fluctuate randomly about some expected value. Unfortunately, adding a stochastic term for individuals to the model is not a simple task, and it is not currently known how to solve the problem.[22]

[20] When all exogenous variables are categorical, this analysis is consistent with the recent work by Goodman (1970, 1972) on log-linear analysis of contingency tables. The Goodman contingency table analysis concentrates upon maximum likelihood estimation of relationships between several discrete variables and delves into testing for models that introduce more explicit interaction of the X's. In that sense, this analysis is a special case of his analysis.

[21] To be precise, the maximum likelihood estimation method permits small cells and uses the information in small cells. However, the justification for the method is asymptotic and, therefore, relies upon large sample properties. Under many circumstances, maximum likelihood estimators have been found to have desirable properties even when applied to small samples.

[22] Conceptually, it would be possible to modify the underlying logistic distribution by inserting an individual stochastic term in Eq. (7.6) such as $P_i = 1/(1 + e^{-(X\beta + \varepsilon_i)})$ where i denotes an individual. This implies that $L_i = \log[P_i/(1 - P_i)] = X\beta + \varepsilon_i$. However, when one attempts to estimate this model based upon 0, 1 observations, the problem becomes intractable. Essentially, even with standard assumptions of $E(\varepsilon_i) = 0$ and $E(\varepsilon\varepsilon') = \sigma^2 I$, the individual error terms neither drop out of the estimation equations nor simplify to be described by a few parameters, leaving the impossible task of estimating T error terms and K parameters with T observations. Amemiya and Nold (1975) developed the model for grouped data. Their method introduced an additional error term with variance σ^2 to Eq. (7.13) to give a composite error term with variance $\sigma^2 + [N_{ij} P_{ij} (1 - P_{ij})]^{-1}$. The model is then estimated with a two-step generalized least squares procedure.

7.6   Probit Analysis

Historically, probit analysis has been commonly used whenever one has individual, or micro, data and is considering a model with a discrete dependent variable. Recent developments, as described in Section 7.5, provide an alternative, logit analysis. Since the ML logistic estimator is very similar to the probit estimator, the choice between logit and probit is largely one of convenience and program availability.

The probit model assumes that the underlying probability function is normal rather than logistic. (As shown in Fig. 7.4, these two functions are very similar except for small differences in the tails of the distributions.) The probit model with microdata is also estimated by maximum likelihood techniques. The derivation of this estimator can be obtained in a fashion that exactly parallels the ML logistic estimator. Starting with the likelihood function in Eq. (7.20), the standard normal distribution for the P_t can be substituted instead of the logistic distribution that was used in the previous section.

Instead of following that line of development, we shall develop probit in a slightly different manner. Consider the exogenous variables as forming an index related to the choice such as

$$Z_t = X_t \beta, \tag{7.23}$$

where the parameters of the index (β) are the same for all individuals.[23] The index value itself is unobserved. If we assume that the observed choice ($Y_t = 1$ or 0) depends upon a threshold that is specific to the individual (U_t), we have the following model of behavior:

$$\begin{aligned} Y_t &= 1 && \text{if} \quad Z_t \geqslant U_t, \\ &= 0 && \text{if} \quad Z_t < U_t. \end{aligned} \tag{7.24}$$

The thresholds U_t are also unobserved; however, if the U_t are normally distributed with the same mean and variance, it is possible to estimate the parameters of the underlying index, or choice function.

In this problem Y_t and the X_t are observed; the β's, the Z_t, and the U_t are not. On the basis of this information, one modification to Eq. (7.24) is needed. It is clearly not possible to distinguish between one value of β and any scalar multiple of this (say $c\beta$ where c is any real number) since a parameter vector β and a variance of U equal to σ^2 would yield the same observed data as a parameter vector of $c\beta$ and a variance of U equal to $c^2\sigma^2$. Therefore, we shall standardize[24] Eq. (7.24) by dividing both β and U_t by σ.

---

[23] As in similar situations before, the only real requirement is that the index be linear in the parameters, not linear in the variables.

[24] Historically, probit estimates have often added a constant to the index to aid computation, i.e., estimates have often been based upon $X_t\beta* + 5$ so that the calculations involve only positive numbers. With modern computers this transformation is not necessary. Estimates that do not include a constant but which are directly estimated as in the text are also referred to as normit analyses.

This yields

$$Y_t = 1 \qquad \text{if} \quad X_t\beta/\sigma \geq U_t/\sigma \quad \text{or} \quad X_t\beta^* \geq U_t^*,$$
$$= 0 \qquad \text{if} \quad X_t\beta/\sigma < U_t/\sigma \quad \text{or} \quad X_t\beta^* < U_t^* \qquad (7.24')$$

where asterisks indicate standardized values. For most problems, this standardization of the coefficients offers no problems. If we also assume that U_t has a mean of zero, U_t^* is a standard normal variable [i.e., $U_t^* \sim N(0, 1)$].[25]

Substituting the normal distribution, we have

$$P_t = \text{Prob}(Y_t = 1) = \text{Prob}(X_t\beta^* > U_t^*) = \frac{1}{\sqrt{2\pi}} \int_{-\infty}^{X_t\beta^*} e^{-U^{*2}/2} dU^* = F(X_t\beta^*),$$

$$(7.25)$$

$$1 - P_t = \text{Prob}(Y_t = 0) = \text{Prob}(X_t\beta^* < U_t^*) = 1 - \frac{1}{\sqrt{2\pi}} \int_{-\infty}^{X_t\beta^*} e^{-U^{*2}/2} dU^*$$

or

$$= \frac{1}{\sqrt{2\pi}} \int_{X_t\beta^*}^{+\infty} e^{-U^{*2}/2} dU^* = 1 - F(X_t\beta^*). \qquad (7.26)$$

If Eqs. (7.25) and (7.26) are substituted into the likelihood function expressed in Eq. (7.20) and that function maximized with respect to β^*, the resulting coefficients will be the probit estimates.

7.7 An Example

The voting model can now be reestimated using the maximum likelihood techniques and both the probit and logit models. For these purposes we estimate the probability of voting for Johnson as a function of the continuous issue evaluation variable (E) and the scaled party preference variable ($P = 0$, 0.25, 0.75, and 1.0 for strong Republican, weak Republican, weak Democrat, and strong Democrat, respectively). These results can be compared with the linear probability model in Eq. (7.3) [or in Eq. (7.4) where corrected for heteroskedasticity].

These estimates are

$$V = -2.864 + 3.060E + 4.176P \qquad \text{(ML logit)}, \qquad (7.27)$$
$${(0.142) \quad (0.198) \qquad (0.044)}$$

$$V = -1.62 + 1.73E + 2.36P \qquad \text{(probit)}. \qquad (7.28)$$
$${(0.074) \quad (0.107) \quad (0.095)}$$

The numbers in parentheses are estimates of the asymptotic standard errors for each coefficient. As discussed before, there is uncertainty as to exactly

---

[25] As in the case of simple linear models, the assumption of a mean equal to zero for the threshold is not very costly as long as the index $(X_t B^*)$ contains an intercept. The intercept would then include the mean of the error term. Since we are usually interested in the slope coefficients, failure of this assumption has little practical significance.

what statistical tests should be applied; however, if we use either the normal or the t-distribution for testing the coefficients, we find that they are clearly significantly different than zero.

One important aspect of these estimated models is that the coefficients on the variables cannot be directly compared. Each of the coefficients is measuring something different. In the linear probability model, the coefficient indicates the marginal change in the probability associated with a unit change in the exogenous variable. In the logit model, the coefficient is the change in the log of the odds associated with a unit change in the exogenous variable. And, in the probit model the coefficient is the change in standard deviations of the normally distributed variable. For example, consider a weak Republican ($P = 0.25$) with a Democratic evaluation ($E = 1$). The OLS prediction is $V = 0.606$; this is the predicted probability of voting for Johnson. The logit prediction is $V = 1.24$; this is the predicted log of the odds and corresponds to a predicted probability of 0.776. The probit prediction is $V = 0.700$; this is the predicted value of a standard normal variable and corresponds to a predicted probability of 0.758.

It is possible to compare the predictions that are made from the different models at different points. Such comparison of estimated probabilities demonstrates the similarity of the logit and probit models and also demonstrates some of the problems with the linear probability model. Table 7.4 and Fig. 7.5 display the predicted probabilities of voting for Johnson for selected values of the exogenous variables. They show the problem of the linear probability model in that it gives predictions of probabilities that are greater than one. For values of the exogenous variables not at the extreme, the estimates of the logit and probit are very similar to each other but considerably different from the OLS model. Finally, as suggested earlier, the probit

TABLE 7.4

Comparisons of Predicted Probabilities of Voting for Johnson

Variable values		Probit[a]	Logit[b]	OLS[c]
P	E			
0(SR)	0	0.053	0.054	0.09
	0.2	0.102	0.096	0.16
	0.4	0.178	0.163	0.23
	0.6	0.282	0.264	0.31
	0.8	0.408	0.398	0.38
	1.0	0.544	0.550	0.45
1(SD)	0	0.770	0.789	0.73
	0.2	0.862	0.873	0.80
	0.4	0.924	0.927	0.87
	0.6	0.961	0.959	0.95
	0.8	0.983	0.977	1.02
	1.0	0.993	0.988	1.08

[a] $V = -1.62 + 1.73 E + 2.36 P$ [Eq. (7.28)].
[b] $V = -2.86 + 3.06 E + 4.18 P$ [Eq. (7.27)].
[c] $V = 0.09 + 0.36 E + 0.64 P$ [Eq. (7.3)].

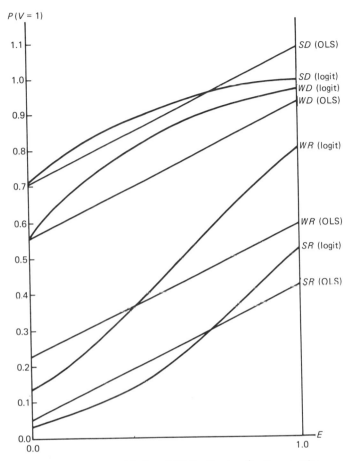

FIGURE 7.5   Graph of logit and OLS estimates of voting equation.

model tends to approach the axis somewhat quicker than the logit model; however, the differences are relatively small.

7.8   Monte Carlo Simulation
of Dichotomous Dependent Variables

There is one further way to illustrate the potential difficulties with using OLS to estimate equations with dichotomous dependent variables; this is with our Monte Carlo simulation. This will show quite clearly both the inability of the linear probability model to fit the true probabilities and the sensitivity of the OLS estimates to the particular sample at hand. In the simulation, the true probabilities are given by the following logistic model:

$$\log\frac{P}{1-P} = L = -4.595 + 6.127X_2 + 3.0635X_3,$$

where $P = \text{Prob}(Y = 1) = 1/(1 + e^{-L})$. The variables are specified so that X_2 is a continuous variable ranging from zero to one and X_3 is a trichotomy with values of 0.0, 0.5, and 1.0. Table 7.6 shows the true probabilities for various values of these explanatory variables. They range from 0.01 to 0.82 for observations with X_3 equal to zero and from 0.18 to 0.99 for cases where X_3 equals one.

The simulation generates observed dichotomous variables for Y by drawing a number at random from a uniform distribution with the range zero to one. If the value of the random number is less than or equal to the true probability for the particular observation P_t, then Y_t is assigned the value one. Conversely, if the random number exceeds P_t, Y_t is recorded as zero. For example, if $P_t = 0.7$, then any random number less than or equal to 0.7 will give $Y_t = 1$ for that simulation. (Given the density function for a uniform distribution bounded by zero and one, the probability of getting a draw less than or equal to 0.7 equals 0.7; thus the value 1 is assigned to Y_t.) This process is done for 200 observations and the observed values of Y, X_2, and X_3 are used to compute the OLS estimates for the linear probability model

$$Y = \beta_1 + \beta_2 X_2 + \beta_3 X_3 + U.$$

This process is repeated for 200 replications of the simulation, so that we have 200 OLS estimates of the coefficients.

The entire simulation is done twice, varying the sample distribution of values for X_3 to show the effects of sample differences. In the first simulation (sample A), half the observations have values of zero for X_3 with the remainder equally divided between values of 0.5 and 1.0. In the second simulation (sample B), half the observations have X_3 values equal to 0.5 and a quarter each have values of zero and one. Comparisons of the mean coefficients from each simulation illustrate how the distribution of OLS coefficients changes with different sample characteristics.

TABLE 7.5

Mean Coefficients for OLS Estimates
of Dichotomous Dependent Variable Models

Sample	$\bar{b}_1$	$\bar{b}_2$	$\bar{b}_3$
A	−0.22	0.98	0.46
B	−0.18	0.89	0.48

The mean OLS coefficients for each sample are shown in Table 7.5 and the true and predicted probabilities for various values of the explanatory variables in Table 7.6. Figure 7.6 plots these true and predicted probabilities to show further the potential difficulties with OLS.

The two important limitations to the OLS estimates are well illustrated by the table and figure. The two samples clearly give estimates with different

TABLE 7.6

Mean OLS Predictions for Selected Sample Values

Sample values		Probabilities		
		True	OLS mean predictions	
X_2	X_3		A	B
0.00	0.0	0.010	−0.217	−0.179
0.20	0.0	0.033	−0.022	−0.001
0.40	0.0	0.105	0.174	0.177
0.60	0.0	0.285	0.369	0.356
0.80	0.0	0.576	0.565	0.534
1.00	0.0	0.822	0.760	0.712
0.00	0.5	0.045	0.015	0.061
0.20	0.5	0.137	0.210	0.239
0.40	0.5	0.351	0.406	0.417
0.60	0.5	0.649	0.601	0.596
0.80	0.5	0.863	0.797	0.774
1.00	0.5	0.955	0.992	0.952
0.00	1.0	0.178	0.247	0.301
0.20	1.0	0.424	0.442	0.479
0.40	1.0	0.715	0.638	0.657
0.60	1.0	0.895	0.833	0.836
0.80	1.0	0.967	1.029	1.014
1.00	1.0	0.990	1.224	1.192

distributions, particularly for β_2, indicating how OLS estimates in this instance are sample specific. The second point is that OLS badly estimates the probabilities in the extreme ranges of the variables. For those observations where X_3 equals zero, the OLS model predicts negative probabilities for small values of X_2. Likewise, for X_3 equal to one, the predicted probabilities exceed one for large values of X_2. In both instances, the OLS procedure is vastly overestimating the true marginal probability for a change in the value of X_2 because of the inappropriateness of the linear probability model.

The conclusion to be drawn from these examples and simulations depends upon one's criteria for good estimates. On one hand, it can be argued that the OLS estimates do not differ greatly from those of the more sophisticated and expensive procedures or from the true value for most of the range of values of the explanatory variables. Neither do the estimates from the two simulated samples differ greatly. Consequently, when cost and program availability considerations are included in the decision on which estimation procedure to use, OLS can be considered and possibly used in a wide range of situations. However, there are potentially large errors in doing so, and the likelihood of these large errors depends upon the nature of the sample, the behavior being modeled, and the interpretations and applications desired from the estimated model. If the sample is skewed or if one intends to rely on the predicted

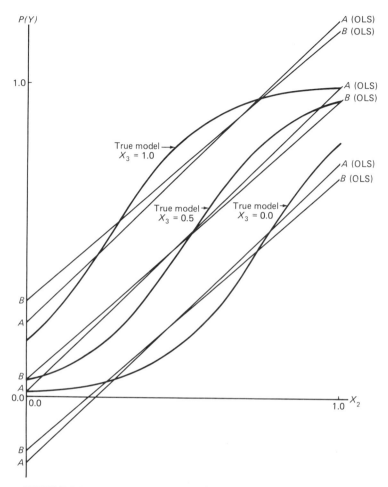

FIGURE 7.6   True and estimated probabilities for Monte Carlo simulation.

probabilities and their changes for extreme values of the explanatory variables, the errors inherent in the OLS estimates may outweigh the costs and inconveniences of using a logit or probit type estimator.

7.9   Polytomous Variables/Joint Distributions

The message of the preceding section has been that the application of least squares methodology to the dichotomous dependent variables problem potentially could be misleading. In general, the functional form will not be correct, and ignoring the distribution of the error terms may make the least squares line a poor approximation of an underlying probability function. The previous discussion should not, however, imply that one should never use least squares.

Nevertheless, there are other situations where use of least squares estimation simply cannot provide reasonable estimates even for large samples with well-behaved underlying distributions. Consider a situation where there are more than two possible choices or events. This is a frequent occurrence. For example, individuals may be employed in one of many occupations; choose between work, college, or vocational training; or vote for one of three candidates. This situation—polytomous choice—is not so easily analyzed in the least squares model as the dichotomous choice problem. The trick of dummy variables, whether as a dependent or an independent variable, is that no scale or units of measurement is imposed upon the variables. However, if we attempt to extend a single variable to measure polytomous events or choices, we must specify a metric. For example, with work/training choices, defining a variable $Y = 0$ if the choice is work, $Y = 1$ if college, and $Y = 2$ if vocational training implies that the difference between work and college is the same as the difference between college and vocational training and that vocational training is twice (as important? as likely?) as college. With dichotomies, it is possible to define a variable that simply indicates qualitative difference; with polytomies, we are forced to define a variable that indicates quantitative differences. We often are unable to do this, and in such cases least squares methodology simply cannot be adopted.[26]

Each of the above previously discussed techniques has been extended to handle models where the dependent variables are grouped into several categories, rather than simply dichotomies. It is at this point that the two logit models are often superior to the probit model. The n-chotomous probit model assumes an ordering to the various categories.[27] This is done by including additional threshold values. For a trichotomy, the model assumes the existence of two thresholds U_t^1 and U_t^2. If the observed dependent variable is coded simply as 1, 2, or 3, depending upon to which group the observation is assigned, then

$$Y_t = \begin{cases} 3, & U_t^{2*} \leqslant Z_t = X_t \beta^*, \\ 2, & U_t^{1*} \leqslant Z_t = X_t \beta^* < U_t^{2*}, \\ 1, & Z_t = X_t \beta^* < U_t^{1*}, \end{cases}$$

---

[26] There are of course situations where there is a natural metric. Consider for example the analysis of employment (full or part time) and unemployment. A natural metric would be the hours per week that an individual worked; similarly, if considering the purchase of alternative qualities of some good such as housing, the choice between low, medium, and high quality housing may be measured by housing value. Some problems may still exist such as when many individuals fall at a boundary such as working 0 hours or not purchasing a house. This situation is discussed by Tobin (1958).

[27] For an extended discussion of n-chotomous probit, see Zavoina and McKelvey (1969).

where asterisks again indicate standardization by σ, the variance of U. The model, then, considers all categories of the dependent variable to be aggregations of a range of the underlying variable, meaning they constitute *ordered* categories. The procedure is useful for models with ordinal dependent variables such as one gets with Guttman scales, the party identification variable, and some attitude survey variables. The ordering assumption is an important restriction because the model cannot be used for categorical variables such as the choice in multicandidate elections, the selection of an occupation, and so on.

The extensions of the two logistic models can handle any categorical dependent variable, e.g., the choice among occupations, or voting in multicandidate elections, but cannot readily incorporate information on ordering if such information is available. In logit models, the estimated functions express the log of the odds of one outcome versus another, called the conditional logit, as a linear function of the explanatory variables. With three possible outcomes, denoted as $Y=1$, 2, or 3 with respective probabilities of P_1, P_2, and P_3, the conditional logits are $L_1=\log(P_2/P_1)$ and $L_2=\log(P_3/P_1)$. The odds ratios P_2/P_1 and P_3/P_1 represent the odds that $Y=2$ rather than $Y=1$ and that $Y=3$ rather than $Y=1$, respectively. The conditional logits are simply the logarithms of these conditional odds. These definitions imply the third log of the odds ratio as

$$L_3=\log(P_3/P_2)=\log\left[(P_3/P_1)(P_1/P_2)\right]$$

$$=\log(P_3/P_1)-\log(P_2/P_1)=L_2-L_1.$$

The logistic model expresses the conditional log of the odds variables as linear functions of the explanatory variables. These log of the odds models are

$$L_1=\log(P_2/P_1)=X\beta_1 \tag{7.29}$$

$$L_2=\log(P_3/P_1)=X\beta_2. \tag{7.30}$$

By appropriate manipulation, we can express the individual probabilities as functions of X, β_1, and β_2. From (7.29) and (7.30) and the condition that $P_1+P_2+P_3=1$,

$$P_2=e^{X\beta_1}P_1, \qquad P_3=e^{X\beta_2}P_1, \qquad \text{and} \qquad 1=P_1+e^{X\beta_1}P_1+e^{X\beta_2}P_1.$$

These expressions give $P_1=1/(1+e^{X\beta_1}+e^{X\beta_2})$.

The polytomous logit model can be estimated either by generalized least squares or by maximum likelihood methods. The GLS method requires that the observations be in contingency table format, just as in the dichotomous version, limiting one to categorical explanatory variables. The two methods also again differ in estimating the values for β_1 and β_2. The GLS method starts with the expected error between the true logit, in Eqs. (7.29) and (7.30), and the logit calculated on the basis of the observed frequencies. However, in the multiple response situation, we in effect have two observations for each cell in the contingency table, one corresponding to L_1 and the other to L_2. Because both true probabilities and the observed frequencies (denoted f_i) sum to one ($P_1 + P_2 + P_3 = 1$ and $f_1 + f_2 + f_3 = 1$ for each cell), the error terms for the expressions for each logit from the same cell are not independent. If by some sampling error, f_2 exceeds the true P_2, then either or both f_1 and f_3 must understate the true P_1 and P_3. These interrelationships create covariances between the pairs of error terms for each cell. This means that the Ω matrix in the expression $E(UU') = \Omega$ is not diagonal as in the dichotomous case [see Eq. (7.13) and the accompanying discussion]. This necessitates a more complicated estimation procedure than the simple weighted least squares regression. Theil (1970) shows what the appropriate values for Ω are and how they can be estimated from the sample data.

The maximum likelihood model of Nerlove and Press (1973) proceeds in a fashion that parallels their treatment of the dicotomous case. If the first T_1 observations have $Y = 1$ and the next T_2 observations have $Y = 2$, and all observations are drawn independently, then the probability of observing the given sample is

$$\mathcal{L} = \prod_{t=1}^{T_1} P_{1_t} \prod_{t=T_1+1}^{T_1+T_2} P_{2_t} \prod_{t=T_1+T_2+1}^{T} P_{3_t}.$$

The log of this likelihood function is

$$\mathcal{L}^* = \log \mathcal{L} = \sum_{t=1}^{T_1} \log P_{1_t} + \sum_{t=T_1+1}^{T_1+T_2} \log P_{2_t} + \sum_{t=T_1+T_2+1}^{T} \log P_{3_t}$$

$$= -\sum_{t=1}^{T_1} \log(1 + e^{X_t\beta_1} + e^{X_t\beta_2}) - \sum_{t=T_1+1}^{T_1+T_2} \log(1 + e^{X_t\beta_1} + e^{X_t\beta_2})$$

$$+ \sum_{t=T_1+1}^{T_1+T_2} X_t\beta_1 - \sum_{t=T_1+T_2+1}^{T} \log(1 + e^{X_t\beta_1} + e^{X_t\beta_2}) + \sum_{t=T_1+T_2+1}^{T} X_t\beta_2$$

$$= -\sum_{t=1}^{T} \log(1 + e^{X_t\beta_1} + e^{X_t\beta_2}) + \sum_{t=1}^{T} D_1 X_t\beta_1 + \sum_{t=1}^{T} D_2 X_t\beta_2, \tag{7.31}$$

where $D_1 = 1$ if $Y_t = 2$ and 0 otherwise and $D_2 = 1$ if $Y_t = 3$ and 0 otherwise. The coefficient estimates are chosen so as to maximize the value of $\mathcal{L}^*$. This is done by setting the first partial derivations of $\mathcal{L}^*$ with respect to β_1 and β_2 to zero:

$$\frac{\partial \mathcal{L}^*}{\partial \beta_1} = -\sum_{t=1}^{T} \frac{X_t e^{X_t\beta_1}}{1 + e^{X\beta_1} + e^{X\beta_2}} + \sum_{t=1}^{T} D_1 X_t = 0, \tag{7.32}$$

$$\frac{\partial \mathcal{L}^*}{\partial \beta_2} = -\sum_{t=1}^{T} \frac{X_t e^{X_t\beta_2}}{1 + e^{X\beta_1} + e^{X\beta_2}} + \sum_{t=1}^{T} D_2 X_t = 0. \tag{7.33}$$

(Note: these are vector equations. If written out for the individual elements in β_1 and β_2, there would be as many equations as parameters.) As in the dichotomous case, these constitute a set of nonlinear simultaneous equations which are quite difficult to solve.

The discussions of the limitations and advantages of these different techniques presented with the dichotomous case are applicable here as well. The generalized least squares logit model requires large numbers of observations within each cell of the contingency table, categorical explanatory variables, and assumes that each individual in a cell has the same probability of a given response. The maximum likelihood in the probabilities model removes the necessity of large N_{ij} and thus the accompanying restriction of categorical explanatory variables. The probabilities however are still assumed to be nonstochastic.

Joint events described by more than one variable can be treated as an extension of the polytomous variable case. The simplest way to view this extension is to consider compressing the jointly determined variables into a single variable where different choices of the new variable represent the possible combinations of the two (or more) variables. For example, consider two dichotomous variables where the choice on each is represented as either 0 or 1. The joint occurrence of these can be represented by four points, i.e., $(0,0)$, $(0,1)$, $(1,0)$, and $(1,1)$. The 1964 voting and party identification variables can be replaced by a categorical variable with eight separate categories: strong Republican and Goldwater, strong Republican and Johnson, weak Republican and Goldwater, and so on. If we add the trichotomous evaluation variable, there would be 24 categories to this variable. The logit approach can then be applied to these new polytomies.

This discussion suggests why probit analysis has not been extended to handle several variables. Placing the joint variables in a single variable makes ordering them extremely difficult.

A full discussion of jointly determined dependent variables is premature at this point. The general topic of joint determination (with continuous dependent variables) is taken up in Chapters 8 and 9. Suffice it to note that joint determination with discrete dependent variables is still in a fairly rudimentary state and the types of models that can be estimated are quite restricted.[28]

[28] The central issue is the form of the underlying model for the jointly determined variables. We may believe that the value of one dependent variable influences the value of the others and vice versa. The system of equations that depicts these interactions (and the influence of the exogenous variables) is called the structural representation of the system. This set of equations can be rearranged through substitution so that each equation contains only one endogenous variable. This is the reduced form representation of the system. As discussed in Chapters 8 and 9, we are generally interested in estimating the structural representation of the system. However, the ML logit procedure is basically restricted to estimation of reduced form equations. The extension to determine dependent variables jointly can be found in Nerlove and Press (1973).

7.10 Conclusions

The message of this chapter is that estimation of models with discrete dependent variables demands special consideration. The general situations where models of discrete dependent variables are relevant are cases where we are interested in choice behavior or the occurrence/nonoccurrence of a particular event. When this is the case, we are not interested in estimating the value or numerical size of the dependent variable. Instead, we are interested in analyzing the underlying probability of a given event or choice; more specifically, how a series of exogenous variables influences the underlying probabilities.

We do not observe the probabilities themselves. Instead, we observe the event's occurrence or nonoccurrence. This implies a specific distribution for the observational errors (if we assume that the underlying probabilities are what we wish to measure).

The Monte Carlo simulation clearly illustrates the potential difficulties with the linear probability model and ordinary least squares estimation. The linear functional form can approximate only certain segments of the true probability function, and with only a dichotomous observed dependent variable none of the usual transformations can be made to approximate the true function. Unfortunately, the segments that the OLS estimates approximate will depend upon the characteristics of the particular sample used in the estimation, thus producing sample specific results.

The estimation problems become far worse when the behavior being explained is not measured by a dichotomy, but either by a categorical variable with several classes (the polytomous case) or by several jointly determined, grouped variables. In these situations it is not possible to order the potential outcomes and thereby create a dependent variable with any meaning. Obviously, the linear model and OLS completely fail in such circumstances.

Most estimation procedures follow one of two paths. In one, the observations are grouped to create estimates of the probabilities of each outcome for given values of the explanatory variables. Under certain assumptions about the behavior being modeled and about the nature of the explanatory variables the estimation can be accomplished with generalized least squares. The other method uses a maximum likelihood technique in which an expression for the likelihood of observing the given outcomes is derived as a function of a few underlying parameters of the true probability distribution. The procedure then chooses estimates for those parameters that would make the observed outcomes the most likely outcomes. This method does not require that the observations be grouped, consequently there is no need to categorize the true explanatory variables.

Both estimation procedures require assumptions about the true relationship between the probability of a given outcome and the values of the explanatory variables. The most common assumption is that this relationship follows either the cumulative normal distribution or a logistic function. The former assumption is associated with probit estimators, while the latter has led to the development of the various GLS and maximum likelihood logit estimators.

The choice of an estimation procedure depends upon the nature of the data, the model being estimated, and the cost and convenience of the alternatives. Some of these considerations were mentioned in the discussion of the Monte Carlo results. The choice is further complicated by the fact that the maximum likelihood estimators, regardless of which probability distribution is assumed, can become quite expensive and difficult to use because they require the solution of a series of nonlinear simultaneous equations. The GLS approach to estimating models assuming the logistic distribution are generally cheaper to use because it is still a linear estimator. However, this requires stronger assumptions about the nature of the behavior being modeled and the characteristics of the explanatory variables, namely that they can be legitimately categorized. Finally, if one is dealing with polytomous or jointly dependent variables, OLS is clearly inappropriate and the choice is between either the GLS or maximum likelihood estimators.

8 | Introduction to Multiequation Models

8.1 Introduction

All the models estimated so far involve only a single equation, with one endogenous or dependent variable, and several exogenous or explanatory variables. The interpretation given these models in Chapter 4 is that the coefficients estimate the direct effects of each exogenous variable with the values for the other variables held constant. This structure very nicely matches the experimental methodology being copied with the statistical analysis. However, both theory and reality often tend to be more complicated than such simple models. For example, there are many instances where it is likely that changes in one explanatory variable result in changes in other explanatory variables as well as in the dependent variable. In this case, the total expected change in the behavior being modeled from a change in an exogenous variable is not just the direct effect estimated by the model's coefficients, but must also include the changes resulting from the changes in the other explanatory variables brought about by the first variable. These latter changes are often referred to as the "indirect" effects. These indirect effects may either reinforce or negate the direct effects of the initial explanatory variable.

The voting model discussed throughout this book illustrates these points. Issue evaluations and party identifications are the variables influencing voting behavior, and we have estimated that a unit change in party affiliations (or evaluations) leads to a given change in the probability of voting for Lyndon Johnson, evaluations (or party affiliations) held constant. But is it not likely that changes in affiliations result in changes in evaluations, or vice versa, which result in further changes in the probability of voting for Johnson? To describe and predict voting accurately, both these direct and indirect effects must be estimated and considered. The relative magnitudes of the effects of evaluations on affiliations and of affiliations on evaluations are also important in evaluating different theoretical models of political behavior and are considered in the next chapter.

Substantially more complex models are required to represent and estimate both the direct and indirect effects and the reciprocal influences that may exist among a set of variables. All of the various relationships must be modeled as separate equations, with their own endogenous and exogenous variables and error terms, hence the name multiequation models. They are also called structural models because the equations represent the underlying behavioral structure. In addition to being more complex, these models often present estimation problems that cannot be overcome with the procedures of the preceding chapters. The purpose of the next three chapters is to describe these models and their interpretations, discuss the estimation difficulties, and present some of the alternatives developed to deal with these difficulties.

8.2 Two Examples of Structural Systems

We present two examples to illustrate the concepts and organization of structural models. The first is the simple three-variable voting model. The second is a model explaining the occupational and educational aspirations of high school males.

The simple three-variable voting model is quite easy to describe. One common explanation of voting behavior holds that party identifications are derived from people's social experiences and are not affected by behavior during any one election. These party affiliations then influence what positions people take on issues and how they perceive the competing candidates and parties. Individuals will adopt the positions advocated by the leaders of "their" party and when asked to evaluate which party or candidate will do a better job at handling some issue or do what the person wants on that issue, they will respond favorably toward their party. Evaluations then are influenced by party identifications. Both evaluations and party affiliations will determine voting behavior, as modeled previously. This simple set of hypotheses is diagramed in Fig. 8.1 and modeled by

$$E = q_1 P + U_1, \tag{8.1}$$

$$V = q_2 P + q_3 E + U_2. \tag{8.2}$$

We assume that all variables are measured as deviations about their means so that we can suppress the constant term to simplify the exposition. The arrows in Fig. 8.1 indicate the direction of hypothesized causation among the variables. The q terms labeling each arrow represent the magnitude of the expected influences.

To describe the total effect of party identification on voting behavior, we need to consider both the direct effect q_2 and the indirect effect arising from the influence of party identification on evaluations and the subsequent influence of evaluations on voting. This indirect effect is estimated as q_1 times

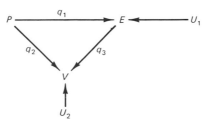

FIGURE 8.1   *Voting behavior path model.*

q_3, so that the total effect of party affiliation on voting is $q_2 + q_3q_1$. This is easily seen by substituting Eq. (8.1) into (8.2),

$$V = q_2P + q_3q_1P + q_3U_1 + U_2 = (q_2 + q_3q_1)P + q_3U_1 + U_2$$

It is also possible to argue that the causal structure runs from evaluations to party identifications and from both to voting. People may choose their party affiliation on the basis of which party is most likely to follow their preferred policies on different issues. These evaluations, and the resulting party affiliations, then determine voting decisions. This hypothesis would reverse the direction of the causal arrow labeled q_1. Unfortunately, the simple model outlined here and the subsequent techniques in this chapter do not permit us to test these two alternative structures. The estimation of such models is the subject of the next chapter.

The second example is taken from the work of Duncan, Haller, and Portes (1968) (DHP) on the aspirations of male adolescents. We modify their model slightly for reasons of pedagogical development. In the next two chapters we shall return to their full model, which includes the simultaneous influence of peers. The DHP study uses a sample of 17-year-old males in a central Michigan high school to estimate a model relating a boy's intelligence (I), socioeconomic background (S), perceived parental aspirations (P), occupational aspirations (O), and educational aspirations (E). The simplified model we start with is shown in Fig. 8.2 and in the equations

$$P = q_1I + q_2S + U_1 \tag{8.3}$$

$$O = q_3I + q_4S + q_5P + U_2 \tag{8.4}$$

$$E = q_6I + q_7S + q_8P + q_9O + U_3. \tag{8.5}$$

(Duncan, Haller, and Portes did not relate parental aspirations to intelligence and status, but considered it as an exogenous explanatory variable.) This model hypothesizes that parental aspirations are related to intelligence and socioeconomic status. The three family background variables I, S, and P, determine the boy's occupational aspirations, and all four variables then influence his educational aspirations.

In this five variable model, there are a large number of indirect effects to consider. For example, by substituting Eq. (8.3) into (8.4), we can see that the total effect of intelligence on occupational aspirations is $q_3 + q_5q_1$. Discussions

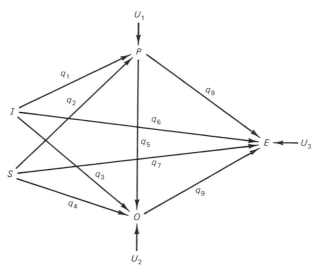

FIGURE 8.2   Simple aspiration model.

of the total effect of intelligence on educational aspirations must take into account the effects of all the interactions among the parental, occupational, and educational aspiration variables as well as the direct effects of intelligence on educational aspirations. After substitution of the appropriate equation, we estimate this total effect of I on E as $q_6 + q_8 q_1 + q_9 q_3 + q_9 q_5 q_1$. (As an exercise, derive the total effect of socioeconomic status on occupational and educational aspirations.)

The objective in estimating structural equations is not simply to obtain estimates of these total effects. Providing the conditions discussed in the preceding chapters are satisfied, the total effects of party identification on voting and of intelligence and socioeconomic status on aspirations can be estimated directly by the simple OLS regressions of V on P and O and E on I and S. In many cases however, knowledge of the causal structure is important. For example, does party identification influence voting directly, or primarily through people's evaluations of the parties' issue positions? Are educational aspirations a direct consequence of status and intelligence, or do parental and/or occupational aspirations intervene? In order to explore these, and similar questions, we must estimate the structural equations.

8.3   Path Analysis

There are a number of ways to proceed in estimating structural models. There are also a large number of problems encountered in obtaining these estimates. The simplest, and most straightforward, situation occurs if the error terms in each of the structural equations are independent of the explanatory variables in that equation. In this case, the basic assumption for ordinary least squares is met, and OLS can be used to estimate each equation in the system. A common method, referred to as path analysis, makes exactly this assumption. It differs from OLS only because it uses correlations rather than covariances, so that all the estimated coefficients are the standardized betas discussed in Chapter 4 rather than estimates of the population parameters.

In the simple path analysis procedure, both the hypothesized structural equations and the assumption that all error terms are independent of the explanatory variables in their respective equations are used to "predict" the correlations among the different variables. These predicted correlations, denoted by R's, are functions of the coefficients in the specified equations. The observed correlations among the variables in the given data set, denoted by r, are then used to estimate these path coefficients.

The three-variable structural equation voting model provides a simple illustration of path analysis. With three variables, we have three expected correlations: R_{PE}, R_{PV}, and R_{EV}. If we assume that all variables have been standardized to have zero means and unit variances, we can write the three expected correlations[1] as

$$R_{PE} = E(r_{PE}) = E\left(\frac{1}{T}\sum P_t E_t\right) = E\left(q_1 \frac{1}{T}\sum P_t^2 + \frac{1}{T}\sum P_t U_{t_1}\right)$$

$$= q_1 + E(r_{PU_1}) = q_1$$

[from Eq. (8.1) and $E(PU_1) = 0$],

$$R_{PV} = E(r_{PV}) = E\left(\frac{1}{T}\sum P_t V_t\right) = E\left(q_2 \frac{1}{T}\sum P_t^2 + q_3 \frac{1}{T}\sum P_t E_t + \frac{1}{T}\sum P_t U_{t2}\right)$$

[from Eq. (8.2)],

$$= q_2 + q_3 E(r_{PE}) + E(r_{PU_2}) = q_2 + q_3 R_{PE}$$

[from $E(PU_2) = 0$],

$$R_{EV} = E(r_{EV}) = E\left(\frac{1}{T}\sum E_t V_t\right) = E\left(q_2 \frac{1}{T}\sum E_t P_t + q_3 \frac{1}{T}\sum E_t^2 + \frac{1}{T}\sum E_t U_{t2}\right)$$

[from Eq. (8.2)],

$$= q_2 E(r_{PE}) + q_3 + E(r_{EU_2}) = q_2 R_{PE} + q_3$$

[from $E(EU_2) = 0$].

These equations make use of the previous assumptions that the variables are in standardized forms and are uncorrelated with the error terms meaning that $(1/T)\sum P_t^2 = 1$, $E(r_{PU_1}) = 0$, and so on.

We do not know the true value for these expected correlations because all we observe are the correlations among the three variables in one sample of

---

[1] For variables with zero mean and unit variance, the simple correlation coefficient is merely the sum of cross products of the variables.

observations. However, if we are willing to use these sample correlations as our "best guess" of the true values, we can substitute these observed values for $E(r_{PE})$, $E(r_{PV})$, and $E(r_{EV})$ and solve for the estimated values of q_1, q_2, and q_3,

$$0.504 = r_{PE} = \hat{q}_1, \qquad 0.678 = r_{PV} = \hat{q}_2 + \hat{q}_3 r_{PE}, \qquad 0.368 = r_{EV} = \hat{q}_2 r_{PE} + \hat{q}_3;$$

$$\hat{q}_1 = r_{PE} = 0.504, \tag{8.6}$$

$$\hat{q}_2 = (r_{PV} - r_{PE} r_{EV})/(1 - r_{PE}^2) = 0.537, \tag{8.7}$$

$$\hat{q}_3 = (r_{EV} - r_{PE} r_{PV})/(1 - r_{PE}^2) = 0.280. \tag{8.8}$$

These estimates imply that party identification has a direct effect of 0.504 on voting and an indirect effect of $0.14 = (0.504)(0.280)$ for a total effect of 0.68. Because all the variables have been standardized by their sample standard deviations, the units of the coefficients correspond to these sample standard deviations.

If we hypothesize the alternative model with evaluations determining party identifications, thus reversing the direction of the influence estimated by q_1, the estimated path coefficients are equal to those just estimated. This indicates that there are two alternative models of voting behavior, with distinctly different substantive implications, which produce the same patterns or correlations among the variables. Consequently, on the basis of the behavior observed so far and without additional information we have no way to choose empirically between the two models. This question of what conditions are required to choose empirically between alternative multiequation models is an important one and will occupy a considerable portion of Chapters 8–10.

The aspiration model in Eqs. (8.3)–(8.5) leads to a similar, but larger set of predicted correlations. In this case, with five variables there are ten correlations. However, we ignore the correlation between intelligence and socioeconomic status r_{IS} because it is not hypothesized to be the result of any causal influence included in the model. (Such a causal connection may exist, but outside the model we are considering.) The predicted relationships among observed correlations and estimated coefficients are

$$r_{IP} = q_1 + r_{IS} q_2 \tag{8.9a}$$
$$r_{SP} = r_{IS} q_1 + q_2 \tag{8.9b}$$
$$r_{IO} = q_3 + r_{IS} q_4 + r_{IP} q_5 \tag{8.10a}$$
$$r_{SO} = r_{IS} q_3 + q_4 + r_{SP} q_5 \tag{8.10b}$$
$$r_{PO} = r_{IP} q_3 + r_{SP} q_4 + q_5 \tag{8.10c}$$
$$r_{IE} = q_6 + r_{IS} q_7 + r_{IP} q_8 + r_{IO} q_9 \tag{8.11a}$$
$$r_{SE} = r_{IS} q_6 + q_7 + r_{SP} q_8 + r_{SO} q_9 \tag{8.11b}$$
$$r_{PE} = r_{IP} q_6 + r_{SP} q_7 + q_8 + r_{PO} q_9 \tag{8.11c}$$
$$r_{OE} = r_{IO} q_6 + r_{SO} q_7 + r_{PO} q_8 + q_9. \tag{8.11d}$$

These expected relationships are again derived from the hypothesized model [Eqs. (8.3–8.5)] plus the assumptions that the variables are standardized and that the exogenous variables are independent of the errors in each equation.

By successively solving these three sets of simultaneous linear equations using the actual correlations given as an appendix to Chapter 9, we get the following estimates:

$$\hat{q}_1 = 0.182, \quad \hat{q}_3 = 0.331, \quad \hat{q}_6 = 0.139,$$
$$\hat{q}_2 = 0.009, \quad \hat{q}_4 = 0.244, \quad \hat{q}_7 = 0.215,$$
$$\hat{q}_5 = 0.141, \quad \hat{q}_8 = 0.138,$$
$$\hat{q}_9 = 0.468.$$

The estimates of this model provide an interesting example of what can be learned from structural models that would be missed by simply estimating the total effects of some variables on others. The total effects of intelligence and socioeconomic status on educational aspirations are both estimated to be 0.33 (these values are also the standardized coefficients from the regression of E on I and S). However, the structural estimates indicate that most of the effect of socioeconomic status comes from its direct influence on educational aspirations q_7, while with intelligence, the direct effect q_6, is only a small part of this total effect. The largest contribution to the total effect of intelligence on educational aspirations comes through the effect of intelligence on occupational aspirations q_3, and the subsequent effect of occupational aspirations on educational aspirations q_9, and not the direct effect itself.

The correspondence between the estimated path coefficients and standardized regression coefficients is easy to see from the estimate for q_1 in the voting model in Eq. (8.6). This estimate for q_1 is the simple correlation coefficient between party identification and evaluations r_{EP} which is also the standardized regression coefficient from the regression of evaluations on identifications:

$$\hat{q}_1 = r_{PE} = \frac{\text{cov}(P,E)}{s_P s_E} = \frac{\text{cov}(P,E)}{\text{var}(P)} \frac{s_P}{s_E} = b_1 \frac{s_P}{s_E} = b_1^*, \tag{8.12}$$

where s_i is the standard deviation of variable i. Thus the estimate for q_1 is equivalent to the standardized regression coefficient for Eq. (8.1). Similarly, we can see from Eq. (8.7) and (8.8) that the estimates for q_2 and q_3 in the voting model are the standardized coefficients for the model in Eq. (8.8). We can do similar manipulations on Eq. (8.9)–(8.11) to show that the estimated path coefficients in the aspiration model are the standardized coefficients for the OLS regressions for Eqs. (8.3)–(8.5).

It is readily seen in these examples that what are conventionally referred to as simple path coefficients are merely the standardized regression coefficients obtained by separately applying OLS to each of the structural equations. This immediately points out the need to consider all of the potential problems discussed in Chapters 4–6 with respect to the specification of each equation.

The limitations of standardized coefficients discussed in Section 4.2 are easily overcome by using the unstandardized regression coefficients. This is only the beginning however. Can one assume that the error terms are uncorrelated with the explanatory variables? Have variables been omitted from each equation which may exert a systematic influence on the variable being explained? If so, can they legitimately be assumed to be uncorrelated with the explanatory variables? And so on. All of the questions and problems discussed in these earlier chapters must be dealt with before trying to estimate these structural equations with OLS, path analysis, or even more advanced methods.

Dealing with these previously mentioned problems is only the beginning of the difficulties in estimating structural equation models. The crux of the additional problems is that once variables that are endogenous to the model are introduced as explanatory variables in other equations, some of the explanatory variables are unlikely to be independent of the error term in the equation being estimated (except in one special case). For example, if U_2 and U_3 in the models for occupation and education aspiration are correlated, then U_3 and O in Eq. (8.5) will be correlated because U_2 constitutes part of the occupation aspiration variable. It should be clear from our development of the basic multivariate model that if the error terms and explanatory variables are correlated, serious estimation problems arise. These last three chapters explore the difficulties presented by multiequation models and how to deal with these difficulties. This chapter discusses the simplest model, where there is no reciprocal or simultaneous interaction among endogenous variables. The next two chapters treat increasingly complex systems.

8.4   The General Multiequation Model

Before delving into the statistical implications of estimating multiequation models, we must extend the notation developed for the single equation case. This new notation is designed to make the development and estimation of multiequation systems more general and easier to apply in a larger variety of cases than has been true of most path analysis discussions.[2] Armed with this new notation we can discuss the statistical problems involved in such estimation and suggest possible estimators. This area is, however, extremely complex, and an extensive, sophisticated literature—much of it beyond the scope and level of this book—already exists. Thus this discussion is best thought of as an introduction and road map; frequent references will be made to more advanced discussions elsewhere.

[2]Path models can become quite complex and invoke more sophisticated estimation procedures than OLS with standardized variables. See O. D. Duncan (1975). Many of his examples and procedures parallel our discussion. However, we feel that the structural equation format is more general and easier to apply than the path model format.

For convenience, we begin with a three-equation example. At the outset we must distinguish between two types of variables: endogenous and predetermined. *Endogenous variables*, denoted by Y_{tm} for the mth endogenous variable, are variables whose values in period t are determined by the model; thus, they are analogous to Y_t in the single-equation case. Predetermined variables are variables whose values are determined prior to the current observation (t). There are in turn, two types of predetermined variables. The first, *exogenous variables*, which are denoted X_{tk}, are completely determined outside of the model; examples include policy variables such as the number of days in a school year or underlying population characteristics such as intelligence or the age distribution of a voter group. The second, *lagged endogenous variables*, have values determined prior to the current observation,[3] e.g., $Y_{t-1,m}$.

Equations (8.13)–(8.15) present this stylized model, for the nonsimultaneous case, with each equation "explaining" one of the three endogenous variables:

$$Y_{t1} = \beta_{11}X_{t1} + \beta_{21}X_{t2} + \beta_{31}X_{t3} + \beta_{41}X_{t4} + U_{t1}, \qquad (8.13)$$

$$Y_{t2} - \gamma_{12}Y_{t1} \quad + \beta_{12}X_{t1} + \beta_{22}X_{t2} + \beta_{32}X_{t3} + \beta_{42}X_{t4} + U_{t2}, \qquad (8.14)$$

$$Y_{t3} = \gamma_{13}Y_{t1} + \gamma_{23}Y_{t2} + \beta_{13}X_{t1} + \beta_{23}X_{t2} + \beta_{33}X_{t3} + \beta_{43}X_{t4} + U_{t3}. \qquad (8.15)$$

Here Y_1 is a function of the predetermined variables, and Y_2 and Y_3 are functions of the previous endogenous variables and a set of predetermined variables; as before, X_1 denotes the constant term. In the aspiration model, Y_1 is parental aspiration, Y_2 is occupational aspiration, Y_3 is educational aspiration, intelligence is X_2, and socioeconomic status is X_3. The primary change in the notation at this point is the additional subscript added to each of the coefficients and error terms in order to indicate the appropriate equation in the system. This model and notation can be extended to include as many equations (endogenous variables) and as many predetermined variables in each equation as appropriate for the phenomena being studied.

This system is described as *hierarchical* because the equations can be structured so that higher ordered endogenous variables (higher subscripts) do not appear as explanatory variables in lower ordered equations. In this example, Y_2 and Y_3 do not appear in the equations for Y_1, and Y_3 is excluded from Y_2's equation. This chapter is concerned solely with hierarchical models.

Equations (8.13)–(8.15) are referred to as *structural equations* because they represent the way in which we believe the observed data were generated, i.e., the underlying behavioral and stochastic processes that led to the observed

---

[3]As we proceed, more formal definitions will be required. The essential characteristic of predetermined variables is that they are uncorrelated in the probability limit with the error term in each equation, i.e., $\mathrm{plim}\sum X_{tk}U_t = 0$. At times this assumption will be untenable for lagged endogenous variables.

data. The structural representation corresponds to the theoretical models underlying the analysis and relates to the formulation of the model where a priori information about specification or coefficient values is relevant. Each of the coefficients (the γ_{ij} and the β_{ij}) describes the direct effects of the respective variables. The direct effect, as in the single-equation case, is the change in the endogenous variable that would occur from a change in one of the right-hand side variables while holding all other variables constant. However, contrary to the single equation case, this direct, or first round, effect is generally not the total effect of a change in a right-hand side variable. The model itself hypothesizes that changes in Y_1, or any of the exogenous variables, result in changes in the other endogenous variables. Thus a change in X_2 is hypothesized to affect the Y's directly, as indicated by the coefficients β_{21}, β_{22}, and β_{23}, and indirectly by any influence each endogenous variable has on the other endogenous variables. These secondary or indirect effects transmit the direct effects of a change in an exogenous variable throughout the model. To assess the total effect on the endogenous variables of a change in X_2, we must include all these indirect effects.

The easiest way to calculate the total effects is through transformations into the *reduced form* of the system. The reduced form is a rearrangement of the structural equations so that each endogenous variable appears in only one equation (and as the left-hand side of that equation). For the hierarchical system of Eqs. (8.13)–(8.15) this simply requires successive substitution of Y from the previous equation. [Note that Eq. (8.13) for Y_1 is its own reduced form.] These reduced form expressions are

$$Y_1 = \beta_{11}X_1 + \beta_{21}X_2 + \beta_{31}X_3 + \beta_{41}X_4 + U_1$$

$$= \pi_{11}X_1 + \pi_{21}X_2 + \pi_{31}X_3 + \pi_{41}X_4 + V_1, \tag{8.16}$$

$$Y_2 = (\beta_{12} + \gamma_{12}\beta_{11})X_1 + (\beta_{22} + \gamma_{12}\beta_{21})X_2 + (\beta_{32} + \gamma_{12}\beta_{31})X_3$$

$$+ (\beta_{42} + \gamma_{12}\beta_{41})X_4 + (U_2 + \gamma_{12}U_1)$$

$$= \pi_{12}X_1 + \pi_{22}X_2 + \pi_{32}X_3 + \pi_{42}X_4 + V_2, \tag{8.17}$$

$$Y_3 = \left[\beta_{13} + \gamma_{23}\beta_{12} + (\gamma_{23}\gamma_{12} + \gamma_{13})\beta_{11}\right]X_1 + \left[\beta_{23} + \gamma_{23}\beta_{22} + (\gamma_{23}\gamma_{12} + \gamma_{13})\beta_{21}\right]X_2$$

$$+ \left[\beta_{33} + \gamma_{23}\beta_{32} + (\gamma_{23}\gamma_{12} + \gamma_{13})\beta_{31}\right]X_3 + \left[\beta_{43} + \gamma_{23}\beta_{42} + (\gamma_{23}\gamma_{12} + \gamma_{13})\beta_{41}\right]X_4$$

$$+ \left[U_3 + \gamma_{23}U_2 + (\gamma_{23}\gamma_{12} + \gamma_{13})U_1\right]$$

$$= \pi_{13}X_1 + \pi_{23}X_2 + \pi_{33}X_3 + \pi_{43}X_4 + V_3. \tag{8.18}$$

The coefficients denoted by π are the reduced form coefficients, and the V's are the reduced form error terms. The π's here and throughout the rest of the book are functions of the population parameters in the structural model.

They are *not* standardized coefficients. π_{km} estimates the total change in Y_m expected from a unit change in X_k. It is clear from the expressions for each π that these reduced form coefficients are merely summing the effects attributable to each path in the conventional path model, only the effects are in terms of population parameters rather than standardized or correlational terms. Take the estimated total effect on Y_3 of a change in X_2. This total effect is

$$
\begin{array}{ll}
\beta_{23} & \text{(direct effect)} \\
+ \quad \gamma_{23}\beta_{22} & \text{(indirect effect due to direct changes in } Y_2\text{)} \\
+ \quad \gamma_{13}\beta_{21} & \text{(indirect effect due to direct changes in } Y_1\text{)} \\
+ \gamma_{23}\gamma_{12}\beta_{21} & \text{(indirect effect due to changes in } Y_2 \text{ caused by changes in } Y_1\text{)} \\
\hline
= \quad \pi_{23}, & \text{total effect.}
\end{array}
$$

The reduced form and structural equations are alternative representations of the same behavioral model in that both describe the same phenomena. However, they provide quite different information and present quite different estimation problems. The reduced form equations summarize the entire structural model in terms of the total changes expected in each endogenous variable from a change in any one of the exogenous variables. The structural model on the other hand "explains" how those changes occur and describes the behavioral process underlying the predicted changes. The reduced form equations are completely comparable to the equations we have discussed in the first seven chapters of this textbook. In fact those chapters can be viewed as a discussion of how to estimate reduced form models. Structural equations present additional and more difficult estimation problems.

Why Structural Equations?

If the predetermined variables are independent of the error terms, equation by equation estimation of the reduced form with ordinary least squares yields unbiased and consistent estimates of the reduced form parameters. Additionally, if the other assumptions about the error terms are plausible, these OLS estimates will be minimum variance unbiased estimators. The same cannot be said for the structural parameters. Unfortunately, independence of the error term and included endogenous variables in each structural equation is generally not plausible, and equation by equation estimation of the structural model with ordinary least squares will generally yield undesirable (biased and inconsistent) estimates of the structural parameters. Therefore, why not be satisfied with just estimating the reduced form parameters?

The answer is found in examining the purposes of the model estimation. The structural equations represent the theoretical model hypothesized to underlie the observed data; this is the causal structure assumed to generate the data. The interrelationships of the endogenous variables are often of substantive importance to us since these interrelationships represent an aspect of the behavior in question. Thus, to the extent that understanding the "endogenous portion" of the model is a focal point of interest, the reduced form simply is insufficient. An important subpart of this reason for doing structural estimation is the use of estimated multiequation models to test competing theories. Structural estimation is generally required in order to distinguish the different theoretical models since the theoretical models apply to the structural form rather than the reduced form.

The electoral model discussed in this book provides a good illustration of the need for structural estimation as part of theory testing. We can hypothesize a set of predetermined demographic and social variables that explain both people's issue preferences and evaluations and their party identifications. These variables will be related to people's voting behavior, as many early analysts discovered who found that religion, income, region of the country, and so forth correlated with election results. From the standpoint of considering the electoral process as a means by which individual preferences for public policies are expressed, we must know whether these exogenous variables are more important in explaining evaluations than party identification, whether party attachments are derived from issue evaluations, or whether party affiliations determine voters' evaluations through some psychological screening process. Each structural model predicts correlations between voting and social variables; the important question is the nature of the process generating those correlations. The only way to test these two theories and a model of electoral behavior is through a structural model that treats evaluations, party identifications, and votes as endogenous variables.

A somewhat different way of looking at the preference for structural estimation comes from considering what reduced form coefficients mean. The reduced form coefficients describe the observed relationships between each endogenous variable and the predetermined variables. As long as all of the observed relationships of these variables remain unchanged, the reduced form equations can be used effectively to predict values of the endogenous variables. However, if the historical relationships among any of the structural variables change—i.e., if the underlying structure of the model changes for some reason—the reduced form predictions can be very misleading. In fact, there are often times when we might expect structural changes and, perhaps, are estimating the model solely for the purpose of predicting the effects of such changes. Since each structural parameter tends to appear as part of several different reduced form parameters, a change of but one structural parameter can lead to many different changes in the reduced form parameters. Moreover, even if we know precisely how a given structural parameter changed, we have no way of adjusting the reduced form parameters if we do not also know all the structural parameters.

This chapter and the next two concentrate upon the estimation of structural equations. For the reason sketched above, this estimation is frequently the focal point of interest. To the extent that reduced form estimation is the subject of interest, the previous chapters suffice.

8.5 Estimating Hierarchical Models

We can easily show the difficulties inherent in using ordinary least squares to estimate the two examples and the structural equations shown in Eqs. (8.13)–(8.15). Let us concentrate on Eq. (8.15). If

$$E(r_{Y_1U_3}) = \frac{1}{T}E\sum Y_{t1}U_{t3} = 0 \quad \text{and} \quad E(r_{Y_2U_3}) = \frac{1}{T}E\sum Y_{t2}U_{t3} = 0,$$

the error term implicit in Y_3 is independent of the explanatory variables in the equation, and OLS or path analysis should be appropriate. The reduced form equation for Y_1 and Y_2 and the assumption that the predetermined variables and error terms are independent give

$$\frac{1}{T}E\left(\sum Y_{t1}U_{t3}\right) = \frac{1}{T}E\left[\pi_{11}\sum X_{t1}U_{t3} + \pi_{21}\sum X_{t2}U_{t3} + \pi_{31}\sum X_{t3}U_{t3}\right.$$

$$\left. + \pi_{41}\sum X_{t4}U_{t3} + \sum V_{t1}U_{t3}\right]$$

$$= \frac{1}{T}E\sum V_{t1}U_{t3} = \frac{1}{T}\sum E(U_{t1}U_{t3}) = \sigma_{13},$$

$$\frac{1}{T}E\left(\sum Y_{t2}U_{t3}\right) = \frac{1}{T}E\left[\pi_{12}\sum X_{t1}U_{t3} + \pi_{22}\sum X_{t2}U_{t3} + \pi_{32}\sum X_{t3}U_{t3}\right.$$

$$\left. + \pi_{42}\sum X_{t4}U_{t3} + \sum V_{t2}U_{t3}\right]$$

$$= \frac{1}{T}E\sum (V_{t2}U_{t3}) = \frac{1}{T}\sum E(U_{t2}U_{t3}) + \frac{1}{T}\sum E(\gamma_{12}U_{t1}U_{t3})$$

$$= \sigma_{23} + \gamma_{12}\sigma_{13}.$$

The OLS estimator is unbiased only if the error terms in successive pairs of equations are independent, i.e., only if $\sigma_{13} = \sigma_{23} = 0$. This is the first major difference with estimating structural as opposed to reduced form equations. Not only must one worry about the specification and error term properties of the equation being estimated, but those of the other equations in the system as well.

Multiequation models that are both hierarchical *and* have independent error terms across equations are called recursive systems. OLS is appropriate for estimating recursive models. If the system is not hierarchical (regardless of error term correlations) *or* if the error terms are correlated (regardless of whether the structure is hierarchical) OLS estimates are biased and inconsistent.

How reasonable is it to assume the error terms in each question are independent of each other, i.e., $\sigma_{12} = \sigma_{13} = \sigma_{23} = 0$? The difficulty with making this assumption will be obvious if you recall the discussion for why error terms are added to social science models, and what effects are included in the error terms. A common interpretation of these error terms is that they represent the effects of a large set of omitted small influences on the endogenous variables. Unfortunately, just because they are small and randomly distributed with respect to the exogenous variables does not imply they are uncorrelated across different equations. In the extreme, if the same set of factors are omitted from each equation, the error terms across equations could be perfectly correlated.

In a less extreme case, if the same explanatory factor is excluded from more than one equation, the effect of that factor will be present in more than one error term and will cause the error terms to be somewhat correlated.

Our voting model provides an example. In Chapter 1 we mentioned several influences not included in the equation predicting a person's vote intention. They might be the party and vote intention of a spouse or neighbor, the impressions left by a recent TV appearance of one of the candidates, the receipt of recent campaign literature, the number of female delegates at the convention, or the influence of some issue for which we do not have position and evaluation information. All of these effects, plus others, will cause a person to state more of a Democratic or Republican vote intention than predicted by the systematic model based on the party identification and evaluation variables. These excluded variables may also influence the respondent's answer to the party identification question and their issue evaluations. This clearly leads to systematic positive correlations between the error terms in the equations explaining party affiliations, evaluations, and voting behavior.

Excluded factors are very likely present in the error terms in equations explaining most endogenous variables. Unless the two endogenous variables are quite distant in some causal sense and clearly subject to quite different influences, it will be difficult to assume that error terms in separate equations are independent of each other.

Error terms are also possibly correlated because of measurement difficulties. Errors in measuring endogenous variables are clearly one component of the error terms. In many cases, similar measuring devices are used to obtain several endogenous variables in a system. If these devices introduce similar errors into the endogenous variables, the results are correlated errors. For instance, when similar questions and scales are used to measure several attitudes in a voter survey, aspects of questions, the manner of the interviewer, or the coding of the responses may affect the variables developed from respondents' answers in a similar fashion. The consequences are correlated errors in equations modeling these attitudes. Likewise, any difficulties encountered in creating aggregate data for counties, countries, or voting districts, such as sampling biases, will manifest themselves in all the variables, with the same consequences.

The moral is that there are many reasons for suspecting that the error terms across equations might be correlated, and considerable thought is needed before one concludes that these error terms are not in fact correlated. Similarities in the behavior being related and in the data collection procedure often imply that standard appeals to the law of large numbers in terms of many small omitted factors to justify randomness and independence are likely to fail when viewed across equations.

At this point we have also reached a fairly simple conclusion. If the structural model is nonrecursive, as is likely, OLS assumptions will be violated and OLS will lead to biased and inconsistent parameter estimates for structural equations.

8.6   Hierarchical, Nonrecursive Systems

The remainder of this chapter discusses ways to estimate nonrecursive models where the structural model is hierarchical, but the error terms are correlated. We shall use the structure outlined in Eqs. (8.13)–(8.15), but drop some of the exogenous variables from specific equations (setting some $\beta_{km} = 0$). (This is done for illustrative purposes but, as will be apparent shortly, the specific form of the model dramatically effects its properties.) These new equations are

$$Y_1 = \qquad\qquad \beta_{11}X_1 + \beta_{21}X_2 + \beta_{31}X_3 \qquad\quad + U_1, \qquad (8.19)$$

$$Y_2 = \gamma_{12}Y_1 \qquad \beta_{12}X_1 \qquad\quad + \beta_{32}X_3 + \beta_{42}X_4 + U_2, \qquad (8.20)$$

$$Y_3 = \gamma_{13}Y_1 + \gamma_{23}Y_2 + \beta_{13}X_1 \qquad\qquad\qquad + U_3, \qquad (8.21)$$

and are diagramed in Fig. 8.3 where arrows indicate "causal influences" or nonzero structural coefficients. (The constant term X_1 is omitted from the diagram for simplicity.)

The reduced form expressions for each endogenous variable are

$$Y_1 = \beta_{11}X_1 + \beta_{21}X_2 + \beta_{31}X_3 + U_1 = \pi_{11}X_1 + \pi_{21}X_2 + \pi_{31}X_3 + V_1, \qquad (8.22)$$

$$Y_2 = (\beta_{12} + \gamma_{12}\beta_{11})X_1 + \gamma_{12}\beta_{21}X_2 + (\beta_{32} + \gamma_{12}\beta_{31})X_3$$
$$+ \beta_{42}X_4 + \gamma_{12}U_1 + U_2$$
$$= \pi_{12}X_1 + \pi_{22}X_2 + \pi_{32}X_3 + \pi_{42}X_4 + V_2, \qquad (8.23)$$

$$Y_3 = (\beta_{13} + \gamma_{13}\beta_{11} + \gamma_{23}\beta_{12} + \gamma_{23}\gamma_{12}\beta_{11})X_1 + (\gamma_{13}\beta_{21} + \gamma_{23}\gamma_{12}\beta_{21})X_2$$
$$+ (\gamma_{13}\beta_{31} + \gamma_{23}\beta_{32} + \gamma_{23}\gamma_{12}\beta_{31})X_3 + \gamma_{23}\beta_{42}X_4 + (\gamma_{13} + \gamma_{23}\gamma_{12})U_1$$
$$+ \gamma_{23}U_2 + U_3$$
$$= \pi_{13}X_1 + \pi_{23}X_2 + \pi_{33}X_3 + \pi_{43}X_4 + V_3. \qquad (8.24)$$

Indirect Least Squares

The OLS estimates for the reduced form coefficients are unbiased if the X's and U's satisfy the independence assumption. These coefficients are unbiased

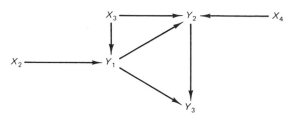

FIGURE 8.3   Simple hierarchical system.

estimates of the parameters in the equation for Y_1 because its structural and reduced form equations are the same due to the absence of endogenous variables. Similarly, from the estimates of Eq. (8.23), we provide an estimate of β_{42}. This leaves just three "unknown" coefficients in Eq. (8.23), γ_{12}, β_{12}, and β_{32}. From the reduced form coefficients π_{12}, π_{22}, and π_{32}, notice that

$$\gamma_{12}\beta_{21} = \pi_{22}, \qquad \beta_{12} + \gamma_{12}\beta_{11} = \pi_{12}, \qquad \text{and} \qquad \beta_{32} + \gamma_{12}\beta_{31} = \pi_{32}.$$

From this information we can estimate γ_{12} using the previous estimate for β_{21} and the estimate for π_{22}. Once γ_{12} is estimated, using the estimates for β_{11} and β_{31} we can estimate β_{12} and β_{32}. Thus by working backward from the reduced form estimates we can identify estimates for all the coefficients in the first two structural equations, regardless of the extent of correlations among the separate error terms.

This procedure is called indirect least squares because it involves the use of ordinary least squares to estimate the reduced form coefficients and then transformation of these to obtain the structural estimates. Unfortunately, while the reduced form estimates may be unbiased, the structural estimates are not. The essence of the difficulty is that the structural coefficients are nonlinear functions of the reduced form coefficients, and distributions of a nonlinear transformation of stochastic but unbiased estimators will not generally be unbiased. [Appendix I shows that in general $E(XY) \neq E(X)E(Y)$.] The indirect least squares estimates are consistent however. That is, as the sample size becomes infinitely large, the estimates converge to the true values and in the probability limit they equal the true values. The explanation for this consistency property can be seen by recalling the discussion and simulation in Chapter 3. The point there was that as the sample size increases, the variance of the distribution of the estimated coefficients decreases, becoming zero in the limit. If we have true estimates for the reduced form coefficients, such as when $T \to \infty$ and the distribution collapses, the nonlinear transformations will give true estimates of the structural coefficients.

The same operation tried on Eq. (8.24), in hopes of getting estimates for the coefficients in Eq. (8.21), presents a dilemma. While we can estimate γ_{23} simply by $\gamma_{23} = \pi_{43}/\beta_{42} = \pi_{43}/\pi_{42}$, we get two separate estimates of γ_{13}: $\gamma_{13} = (\pi_{23} - \gamma_{23}\pi_{22})/\pi_{21}$ and $\gamma_{13}^* = (\pi_{33} - \gamma_{23}\pi_{32})/\pi_{31}$. If we are talking about population values, then $\gamma_{13} = \gamma_{13}^*$. However, the estimates of π are drawn from distributions that at best have the population values as their means and each coefficient is very unlikely to equal this value exactly. Consequently, two estimates based upon sample data are unlikely to equal each other. The greater the variance of the distributions from which the coefficients are drawn, the greater the disparity is likely to be between the two "estimates" of γ_{13}. The two different estimates for γ_{13} also imply more than one estimate for β_{13}, depending upon which value of γ_{13} is used.

Alternatively, we could use the expressions for π_{23} and π_{33} in Eq. (8.24) and the estimates for β_{21}, β_{31}, π_{22}, and π_{32} to get estimates for γ_{13} and γ_{23}. We then get a second estimate for γ_{23} from π_{43}. We call this situation "overidentification." The essence of overidentification is that we have four expressions [one for each reduced form coefficient in Eq. (8.24)] and three unknowns [the parameters in Eq. (8.21)]. Thus any three of the reduced form coefficients from Eq. (8.24) can be used to estimate the structural coefficients. If we know the population values for all reduced form coefficients in Eqs. (8.22)–(8.24), all combinations would give the same structural estimates. But due to sampling variations, each combination of π's from Eq. (8.24) will give different estimates of β_{13}, γ_{13}, and γ_{23}.

As the sample size increases, the variance in each of the estimated reduced form coefficients declines, and the estimates approach their true values for both the reduced form coefficients and the structural coefficients. This implies that any combination of three reduced form coefficients including π_{13} gives consistent estimates of Eq. (8.21). One choice available would seem to be to choose one of the consistent estimates arbitrarily. However, by combining these estimates in some manner, the variance of the structural coefficient estimate can be reduced.

Imagine estimating the mean of the distribution of a random variable W. One observation of W would provide an unbiased estimate of the mean of W with variance equal to the variance of W. A second observation of W could be averaged with this observation to provide another unbiased estimate of the mean; however, in this case, the variance of the mean estimator falls to the variance of W divided by 2. This happens because the deviations of each observation from the true mean tend to cancel. Similarly, we can reduce the variance in the coefficient estimator by combining the two different "observations" of it. Therefore, we might look for an estimation procedure that combines the information contained in the separate estimates from indirect least squares. In effect, this means using all the reduced form coefficients in Eqs. (8.22)–(8.23), not just a subset of them.

The expressions for the reduced form coefficients in Eq. (8.24) suggest a way of approaching this problem. We can rewrite that equation as

$$Y_3 = \beta_{13}X_1 + \gamma_{13}(\pi_{11}X_1 + \pi_{21}X_2 + \pi_{31}X_3)$$
$$+ \gamma_{23}(\pi_{12}X_1 + \pi_{22}X_2 + \pi_{32}X_3 + \pi_{42}X_4) + V_3$$
$$= \beta_{13}Z_1 + \gamma_{13}Z_2 + \gamma_{23}Z_3 + V_3.$$

The parenthetical expression multiplied by γ_{13} is nothing more than the systematic part of the reduced form model for $Y_1, (Y_1 - V_1)$; similarly, the expression multiplied by γ_{23} is the reduced form systematic component for ,Y_2 $(Y_2 - V_2)$. Call these expressions Z_2 and Z_3, respectively. (We are letting $Z_1 = X_1$, the constant term.) If we knew the true reduced form coefficients in the equations for Y_1 and Y_2, we could calculate Z_2 and Z_3 and thereby estimate the coefficients in the structural equation for Y_3 directly by regressing Y_3 on X_1, Z_2, and Z_3. This in effect combines the information in all the reduced form coefficients to give us the structural estimates. We do not, however, know these true reduced form parameters. The best we can do is estimate these coefficients from the reduced form expressions for Y_1 and Y_2; but this might help us. Consider using these values to calculate estimated values $\hat{Z}_2$ and $\hat{Z}_3$ and then using these estimates to estimate the equation for Y_3. This in fact turns out to be a common technique and is referred to as instrumental variables.

Instrumental Variables

The general formulation of the instrumental variable method is only sketched here and a more formal presentation provided as an appendix to this chapter. For simplicity, return briefly to the single equation model of Chapter 5 where the true model is

$$Y = X_1 \beta_1 + X_2 \beta_2 + \cdots + X_K \beta_K + U.$$

Y and X_k represent vectors of observation on endogenous and exogenous variables, respectively, and U is the vector of error terms implicit in the observations on Y. Contrary to the development in Chapter 5, it is now assumed that at least one of the X_k is correlated with the error term, i.e., $E(\Sigma X_{tk} U_t) \neq 0$. This would result in biased and inconsistent OLS estimates. However, if for each such troublesome variable X_k, an alternative variable exists that is uncorrelated with U but still correlated with X_k, it is possible to get consistent estimates of the parameters in the population model. The new variables, referred to as the instrument for X_k, are denoted by Z_k. Any exogenous variable that is uncorrelated with U can serve as its own instrument, i.e., $Z_k = X_k$ if $E(\Sigma X_{tk} U_t) = 0$.

The appendix to this chapter shows precisely how these instrumental variables are used along with the X's and Y to estimate $\beta_1, \beta_2, \ldots, \beta_K$. The appendix also derives the variance of the instrumental variable estimator. The important point is that the higher the correlation between Z_k and X_k (so long as Z_k remains uncorrelated with U), the lower the variance of the estimated coefficients. This makes it clear why one wants to use X_k as its own instrument if it is uncorrelated with U.

This general technique—available for estimating any single equation where a set of explanatory variables are correlated with the error term—can be used for estimating one equation in a hierarchical or (more generally) simultaneous equation system. In fact, multiequation models greatly assist us in choosing instruments. Let us explore this with respect to Eq. (8.21).

In the case of exogenous variables—which are independent of U—the best choice of an instrument is the exogenous variable itself. Thus in Eq. (8.21), X_1 serves as its own instrument [as does X_4 in Eq. (8.20)]). It is then necessary to choose instruments Z_2 and Z_3 that are correlated with Y_1 and Y_2, respectively, and uncorrelated with U_3.

One natural way of forming instruments in such a situation—and, indeed the most common—involves use of the reduced form equations. The true reduced form model (i.e., using the true coefficients) would yield the value of the systematic component of Y_1 and Y_2, and these systematic components would be independent of the error terms both in their structural equation and in the equation for Y_3. We could then estimate Eq. (8.21) by the OLS regression of Y_3 on X_1 and the values of the two systematic components Z_2 and Z_3. However, since we must rely upon an estimate of the reduced form, the desired independence in small samples is generally not obtained. Nevertheless, let us define $\hat{Z}_2$ and $\hat{Z}_3$ for the example of Eq. (8.21) as the estimated systematic component of the reduced form, i.e.,

$$\hat{Z}_2 = \hat{Y}_1 = p_{11}X_1 + p_{21}X_2 + p_{31}X_3, \qquad \hat{Z}_3 = \hat{Y}_2 = p_{12}X_1 + p_{22}X_2 + p_{32}X_3 + p_{42}X_4,$$

where the p's are obtained from the OLS regression of Y_1 and Y_2 on X_1 through X_4. Equation (8.21) is then estimated by the OLS regression of Y_3 on X_1, $\hat{Z}_2$, and $\hat{Z}_3$. Since the estimated reduced form coefficients are functions of U_1 and U_2 (remember the OLS estimator), $\hat{Z}_2$ and $\hat{Z}_3$ will generally not be uncorrelated with U_3 in small samples. Therefore, use of instrumental variables will yield biased estimates of the structural parameters.

Nevertheless, even though the instrumental variables estimators in this case are not unbiased, they do exhibit a weaker desirable property: they are consistent estimators.[4] In other words, as the sample size becomes infinitely large, the variance of the estimator decreases to zero and the probability mass for the estimator converges to the true parameter value. Such convergence to the true parameter is intuitively appealing; but, at the same time, it is a property less desirable than unbiasedness since actual applications involve small samples.[5] The important issue, which will be discussed below, is the behavior of this estimator in small samples. (Remember, OLS estimators of the structural parameters are both biased *and* inconsistent.)

There is a second way of interpreting the instrumental variable estimator.

---

[4]The consistency of these estimators will be discussed later in the general context of simultaneous equations. For the moment, suffice it to note that the reduced form estimates are consistent so that in large samples they converge to the true values and, thus, $\hat{Y}$ converges to the true, systematic part of Y.

[5]This comparison is not quite correct since we must also consider the estimator's variance. An unbiased estimator might have a large variance so that it is inferior in terms of mean square error to be a consistent (but biased) estimator.

Goldberger (1973) shows that the coefficients obtained from instrumental variables are weighted averages (the weights sum to one but are not necessarily positive) of the separate estimates of each overidentified coefficient obtainable from the reduced form coefficients. Each individual estimate's weight varies inversely to its variance. This suggests, and it is the case, that there are other estimators based on the individual estimates. For example, one could simply take the unweighted average of these alternatives. However, these instrumental variable estimates are at least as efficient as any other linear combination of the separate estimates (their asymptotic variance is less than or equal to that of any other estimator that is a linear combination of the alternative individual indirect least squares estimates) as illustrated by Goldberger (1973). There are estimators that are nonlinear combinations of the separate estimates. Except for a brief discussion in Chapter 10, these methods are beyond the scope of this text.

Monte Carlo Simulation

We can use our simulation to illustrate both OLS and instrumental variables (IV) estimation of hierarchical models and the statistical properties of these estimators. A two-equation system is sufficient for demonstration purposes.

The structural equations are

$$Y_1 = \qquad \beta_{11}X_1 + \beta_{21}X_2 + \beta_{31}X_3 \qquad + U_1,$$
$$Y_2 = \gamma_{12}Y_1 + \beta_{12}X_1 \qquad\qquad + \beta_{42}X_4 + U_2,$$

where the following parameter values were chosen: $\beta_{11} = 6.0$, $\beta_{21} = 1.0$, $\beta_{31} = 2.0$, $\beta_{12} = 9.0$, $\beta_{42} = 1.0$, and $\gamma_{12} = 1.0$. Also, the model as shown already includes the prior information that $\beta_{41} = \gamma_{21} = \beta_{22} = \beta_{32} = 0$ and that $X_1 = 1$. By substitution, we see that the reduced form for Y_2 is simply the model used in Chapter 4 [Eq. (4.6)] to simulate the misspecification and multicollinearity problems, i.e.,

$$Y_2 = 15.0 + X_2 + 2X_3 + X_4 + (U_1 + U_2).$$

It should also be clear that γ_{12}, the coefficient of Y_1 in the equation for Y_2, is overidentified in the reduced form; hence we would not get a single estimate for it were we to use indirect estimation.[6] Consequently, it is appropriate to use the instrumental variable procedure. In this case an instrument for Y_1 is obtained by regressing Y_1 on X_2 and X_3 and the constant X_1 and getting estimated values for Y_1. This instrument for Y_1 is then used to estimate γ_{12}.

---

[6]As we did before, write the reduced form in terms of the structural parameters as

$$Y_2 = (\beta_{12} + \gamma_{12}\beta_{11})X_1 + \gamma_{12}\beta_{21}X_3 + \gamma_{12}\beta_{31}X_3 + \beta_{42}X_4 + (\gamma_{12}U_1 + U_2).$$

The first equation can be estimated by OLS, and these values can be used to estimate γ_{12}. Clearly, there are two estimates of γ_{12}.

The simulation itself alters the correlations between U_1 and U_2 (r_{12}) from 0.00 to 0.80 and considers sample sizes ranging from 25 to 200. Additionally, to illustrate the resulting bias in the OLS estimates when the correlation between U_1 and U_2 is not zero, the structural equation for Y_2 is estimated with ordinary least squares.

The results of the simulation (50 replications) with OLS and instrumental variables (IV) estimation of γ_{12} are shown in Table 8.1.[7] Looking at the mean estimates of γ_{12} down each of the OLS columns, there is a noticeable increase in the bias in the ordinary least squares estimator as the correlation between the error terms in the two equations increases. When the error terms are independent of each other, it is a recursive system, and the ordinary least squares estimates are unbiased. However, with each 0.20 increase in the correlation of U_1 and U_2, the bias in the OLS estimator increases. This holds for each sample size. Interestingly, the instrumental variable estimator's mean, even with a small sample size, remains very close to the true value, usually less than one sampling standard deviation from the population value of 1.0. Further, within this experiment, for each case where the correlation of U_1 and U_2 is greater than zero, the bias in the IV estimator is less than that for the OLS estimator. This, of course, is not necessarily the case since both OLS and IV estimators are biased, and we have proven only large sample properties (consistency) for the IV estimator.

A second characteristic of the two estimators is the higher variance of the instrumental variables estimator. (Both estimators' standard deviations are

TABLE 8.1

Mean and Standard Deviation[a] of Estimated γ_{12} by Sample Size, Estimation Procedure, and Correlation of U_1 and U_2 (r_{12})

r_{12}	$N = 25$		50		100		200	
	OLS	IV	OLS	IV	OLS	IV	OLS	IV
0.00	1.014	1.003	1.008	1.010	1.004	0.997	1.010	1.008
	(0.113)	(0.153)	(0.097)	(0.102)	(0.061)	(0.076)	(0.042)	(0.049)
0.20	1.079	1.003	1.083	1.004	1.067	1.002	1.065	1.001
	(0.131)	(0.147)	(0.088)	(0.107)	(0.058)	(0.071)	(0.038)	(0.042)
0.40	1.129	0.980	1.138	0.999	1.142	1.003	1.132	0.995
	(0.104)	(0.153)	(0.085)	(0.103)	(0.057)	(0.071)	(0.037)	(0.051)
0.60	1.221	1.011	1.200	0.984	1.200	0.984	1.199	0.996
	(0.096)	(0.145)	(0.075)	(0.105)	(0.050)	(0.076)	(0.035)	(0.059)
0.80	1.277	1.041	1.283	0.998	1.271	0.994	1.273	1.000
	(0.092)	(0.150)	(0.062)	(0.112)	(0.040)	(0.059)	(0.029)	(0.045)

[a]Observed standard deviations of estimates are given in parentheses.

[7]Some care is required in interpreting these and subsequent Monte Carlo experiments since the theoretical variance is known not to exist for some estimators. In such cases, the experiments demonstrate typical dispersion but cannot be interpreted as yielding the variance.

computed about their sample means, not the population values.) This result should not be surprising, given the discussions in Chapters 3 and 5 about the minimum variance characteristic of the OLS estimator.

The appropriate way to compare the two estimators however is on the basis of their mean squared errors (*MSE*) about the population value. This criterion takes into account differences both in variances and in biases ($MSE = $ variance + bias^2). The sample means and standard deviations in Table 8.1 permit calculations of the mean squared errors for both estimators under the different circumstances. These mean squared errors for the OLS estimator are less than those for the IV estimator when the correlation between error terms is zero. In this case OLS is the best, linear, unbiased estimator and certainly to be preferred to IV. However even for the modest correlation of 0.2, this pattern is reversed. The best the OLS estimator can do is the simulation with only 25 observations. In this instance, the ratio of the two mean squared errors (see Table 8.2) is 0.92, indicating only a slight superiority for IV. As both the sample size and the correlation between the two error terms increase, the mean squared error ratios imply a definite superiority for instrumental variables. Thus for this particular model, the probability of obtaining an estimate within a given distance of the population value is noticeably higher with instrumental variables than with ordinary least squares except with small sample sizes and low correlations among the error terms in the two structural equations.

TABLE 8.2

Ratio of Mean Squared Errors for Simulations in Table 8.1[a]

	Sample size			
r_{12}	25	50	100	200
0.00	1.80	1.10	1.57	1.33
0.20	0.92	0.79	0.64	0.31
0.40	0.87	0.40	0.22	0.14
0.60	0.36	0.25	0.14	0.09
0.80	0.28	0.15	0.05	0.03

[a]Entries are IV mean squared error divided by OLS mean squared error.

One aspect of IV estimators (discussed further in Chapter 10) is not explicitly demonstrated by this simulation. The higher the correlation of the instruments and the variables that they replace, the more efficient is the estimator, i.e., the lower is the mean squared error of the estimated coefficients. This is shown obliquely by the first row of Table 8.2 where $r_{12} = 0$. In that experiment, OLS can be viewed as the limiting case of an IV estimator where the instrument Z_k for each explanatory variable is the explanatory variable itself $Z_k = X_k$. (This is appropriate because X_k is uncorrelated with U_2.) As pointed out above, this "instrumental variable" estimator has a smaller variance than the "imperfect" IV estimator in the table. Thus, we are not indifferent in the choice of instruments. In fact, in the extreme where the instrument is uncorrelated with the explanatory variable, the IV estimator will be undefined.

8.7 Underidentification in Hierarchical Models

There is one more possible situation that we have not covered in structural estimation. In estimating Eq. (8.20), all the coefficients were uniquely derived or identifiable from the reduced form equation. In the case of Eq. (8.21), we obtained more than one estimate for the coefficients β_{13}, γ_{13}, and γ_{23}, but only because of the sampling variance inherent in any estimated coefficient. There are only a finite number of direct estimates of these coefficients however; and the best estimates are a weighted average of the finite alternatives, obtainable perhaps by the instrumental variables procedure. There is also the possibility that we cannot obtain a finite number of estimates of the structural coefficient. This is called underidentification, or the equation is said to be unidentified.

To illustrate this problem, let us assume a fourth equation in our sample system, specified as:

$$Y_4 = \gamma_{14}Y_1 + \gamma_{24}Y_2 + \gamma_{34}Y_3 + \beta_{14}X_1 + \beta_{44}X_4 + U_4. \tag{8.25}$$

This equation still satisfies our hierarchical structure. The new model is diagrammed in Fig. 8.4, again omitting the constant term X_1. The reduced form expression for Y_4 is

$$Y_4 = (\beta_{14} + \gamma_{14}\pi_{11} + \gamma_{24}\pi_{12} + \gamma_{34}\pi_{13})X_1 + (\gamma_{14}\pi_{21} + \gamma_{24}\pi_{22} + \gamma_{34}\pi_{23})X_2$$

$$+ (\gamma_{14}\pi_{31} + \gamma_{24}\pi_{32} + \gamma_{34}\pi_{33})X_3 + (\beta_{44} + \gamma_{24}\pi_{42} + \gamma_{34}\pi_{43})X_4 + V_4$$

$$= \pi_{14}X_1 + \pi_{24}X_2 + \pi_{34}X_3 + \pi_{44}X_4 + V_4. \tag{8.26}$$

Since Y_4 does not appear in the structural equation for Y_1, Y_2, and Y_3, the reduced form equations for those variables will remain unchanged from what they were in the three-equation model. Importantly, none of the structural

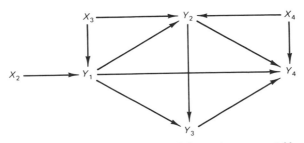

FIGURE 8.4   Hierarchical system with four endogenous variables.

coefficients from Eq. (8.25) appear in the reduced form equations for Y_1, Y_2, or Y_3. (Make sure you understand why this is so.)

Assume we have estimated the reduced form coefficients for these other equations, $\pi_{11}, \pi_{12}, \ldots, \pi_{42}, \pi_{43}$. From scanning Eq. (8.26), however, we find that our four new relationships (for $\pi_{14} \cdots \pi_{44}$) are linear combinations of the *five* unknowns (β_{14}, β_{44}, γ_{14}, γ_{24}, and γ_{34}) and the known reduced form coefficients. Thus,

$$\beta_{14} + \quad \pi_{11}\gamma_{14} + \pi_{12}\gamma_{24} + \pi_{13}\gamma_{34} = \pi_{14},$$

$$\pi_{21}\gamma_{14} + \pi_{22}\gamma_{24} + \pi_{23}\gamma_{34} = \pi_{24},$$

$$\pi_{31}\gamma_{14} + \pi_{32}\gamma_{24} + \pi_{33}\gamma_{34} = \pi_{34},$$

$$\beta_{44} + \quad \pi_{42}\gamma_{24} + \pi_{43}\gamma_{34} = \pi_{44}.$$

Elementary algebra rules remind us that we cannot solve four equations for five unknowns and obtain a finite number of solutions. This is what we call underidentification or the unidentified case.

Looked at from the opposite point of view, there are an infinite number of structures that could have generated our observed data and our estimated reduced form models. On the basis of our data and the a priori information contained in the structural model specification, we have no way to distinguish among the structures that could possibly have generated the data. In this situation, we do not have a statistical estimation problem—it is simply not possible to estimate the structural parameters. Underidentification is a modeling, or conceptual, problem.

It will also be clear that our instrumental variables procedure will fail in this case. Attempting to use brute force simply will not work. Say we use X_1, X_2, X_3, and X_4 to get estimated values for Y_1 to Y_3, yielding instrumental variables that are all linear combinations of the four exogenous variables. When we include these three variables along with X_1 and X_4 in the estimates for Eq. (8.25), we have five variables that are linear combinations of each other, or perfect multicollinearity. In instrumental variables notation $Z = (X_1, X_4, \hat{Y}_1, \hat{Y}_2, \hat{Y}_3)$ and $\hat{Y}_1$, $\hat{Y}_2$, and $\hat{Y}_3$ are linear combinations of X_1 through X_4.

Importantly, recursive models are always identified. The a priori information introduced by the maintained hypotheses of the hierarchial form of the endogenous variable *and* the independence of error terms across equations is sufficient for identification of each equation.

The identification problem is pervasive throughout the field of parameter estimation. The key is merging outside information with the observed data to obtain structural parameter estimates. Our approach to this issue—which will occupy a large part of the next chapter—is to decide how much a priori information is required. The previous examples, and indeed, even the discussion of single equation estimation provide a brief glimpse of issues to be considered. However, before going into the identification problem, two examples of nonrecursive, hierarchical systems are appropriate.

8.8 Nonrecursive Hierarchical Models: Two Examples

The aspiration and voting models discussed at the beginning of this chapter offer excellent opportunities to illustrate the considerations required in estimating even simple nonrecursive structural models. The models, as hypothesized in Eqs. (8.3)–(8.5) and (8.1) and (8.2), are unidentified unless the assumption of independence among the error terms in the separate equations is valid, in which case OLS (or path analysis) is the appropriate procedure. What can we do if the error terms are likely to be correlated, which is a more reasonable assumption?

Taking the case of the aspiration model first, it is easy to see that the occupation and education aspiration equations cannot be identified from the reduced form equations as the model is presently formulated. Rewriting the model with the general notation developed in Section 8.4, we have

$$P = \beta_{11}I + \beta_{21}S \qquad\qquad + U_1, \qquad (8.3a)$$
$$O = \beta_{12}I + \beta_{22}S + \gamma_{12}P \qquad + U_2 \qquad (8.4a)$$
$$E = \beta_{13}I + \beta_{23}S + \gamma_{13}P + \gamma_{23}O + U_3. \qquad (8.5a)$$

This model completely parallels the general model shown and discussed in Eq. (8.13)–(8.15), with two fewer exogenous variables. The expressions relating the reduced form coefficients to the structural parameters are obtained from Eq. (8.16)–(8.18):

$$\pi_{11} = \beta_{11}, \qquad \pi_{12} = \beta_{12} + \pi_{11}\gamma_{12}, \qquad \pi_{13} = \beta_{13} + \pi_{11}\gamma_{13} + \pi_{12}\gamma_{23},$$

$$\pi_{21} = \beta_{21}, \qquad \pi_{22} = \beta_{22} + \pi_{21}\gamma_{12}, \qquad \pi_{23} = \beta_{23} + \pi_{21}\gamma_{13} + \pi_{22}\gamma_{23}.$$

It should be obvious from these equations that only the coefficients in the parental aspiration equation, β_{11} and β_{21}, are identified unless there is further information about the values for some of the other coefficients. This information must be in the form of specified values for certain coefficients. If we knew that one of the coefficients in Eq. (8.4a) equalled zero, for example, then the information in π_{12} and π_{22} would be sufficient to estimate the remaining coefficients in that equation. In the case of Eq. (8.5a), we would need to specify values for two of the coefficients on the basis of prior knowledge if the equation is to be identified. Since we have no reason or prior information to support any such assertions or assumptions which would identify the model, we are left with a structure that is impossible to estimate. The only way out of this impasse is to add information to the model in the form of additional variables and hypothesized relationships which may serve to identify the model. This in fact is what Duncan, Haller, and Portes (1968) do and what we shall do in the next chapter, by adding the effect of peer influences on a person's occupational and educational aspirations. While this considerably complicates the model and adds to the estimation problems, it also adds enough information to identify each equation.

The presidential voting equations outlined in the model of Eqs. (8.1) and (8.2) is also unidentified. However, each equation in the complete model from which this simple illustration is drawn is identified, and the estimation of the vote equation offers a straightforward application of instrumental variables.

The full voting model includes people's issue positions, evaluations, and party affiliations, in addition to the intended vote, as endogenous variables, plus a large set of exogenous variables.[8] (The full specification is shown in Figure 9.1.) The exogenous variables, including income, religion, age, education, parental party affiliation, and other demographic variables, are included in the equations explaining issue positions and party identifications and can be used to create the instruments for evaluations and affiliations needed to estimate the vote equation. The vote equation follows the specification shown at the end of Chapter 4, which includes the slope dummy variable for party identification among voters with indifferent evaluations. The three explanatory variables in the vote equation, evaluations (E), party identification (P), and party identification for indifferent evaluations (IP), are all endogenous to the model. Consequently, instruments for all three are created by regressing them against the exogenous variables and using the estimated equations to predict values for each variable. These predicted values are then used to estimate the vote equation. This two-step procedure is referred to as two-stage least squares (TSLS). (In fact, an easier way to compute instrumental variables estimates is to use a TSLS program when available.)

The vote equation estimated with two stage least squares (IV) is

$$V = -0.07 + 0.90\text{Eval} + 0.31\text{Party} + 0.26\text{Indiff} \times \text{Party}$$

These estimates can be compared to the earlier OLS estimates,

$$V = 0.07 + 0.39\text{Eval} + 0.61\text{Party} + 0.12\text{Indiff} \times \text{Party}.$$

The instrumental variable estimates differ greatly in magnitude from the ordinary least squares estimates. The previously discussed reasons why the error terms in the respective equations are correlated suggests that the coefficient estimates differ because OLS is inappropriate for this model. (Chapter 10 discusses at some length an additional problem that leads to systematic correlations between the explanatory variables and error terms in the vote equation. This is the already mentioned difficulty in accurately measuring evaluations and party affiliations and is referred to as the errors in the variables problem. That chapter also points out that instrumental variables, as used here, are an appropriate way to deal with this problem.)

[8] These exogenous variables are age < 25, age > 54, education < 8 yr, college degree, education in years, income, east, south, rural, central city, father's party identification, race, religion, union. For a discussion of the full model, see Jackson (1975).

8.9 Conclusion

This chapter serves only as an introduction to multiequation systems and to the difficulties encountered in trying to estimate these more complex, but richer models. Some of the reasons for considering such models should be fairly obvious. Additional ones will become apparent in the course of the next two chapters. The estimation problems, even for the simple models discussed here, are far from trivial and depend on the specification of the model and the assumptions made about the nature of the error terms in the individual equations. The basic cause of these difficulties lies in the fact that we can no longer automatically assume that the error terms are distributed independently of the explanatory variables. The discussions in Chapters 2–5 clearly indicate that this independence assumption is fundamental to ordinary least squares estimation.

The discussions in this chapter leave us in the following position. If the multiequation system is recursive (hierarchical structure *and* error terms independent across equations), ordinary least squares estimation is appropriate and yields consistent estimates. If the system is hierarchical but not recursive (error terms not independent across equations), the appropriate estimation technique depends upon whether each equation is unidentified, just identified, or overidentified. In the unidentified case, no estimation of the structural parameters is possible. In the just identified case, all consistent estimation techniques provide identical estimates, and these estimates are most easily seen from the method of indirect least squares. (Computationally, since all estimators are equivalent in this case, it is often easier to use instrumental variables than to solve through indirect least squares.) Finally, in the overidentified case, there is a wide menu of possible consistent estimators, and we develop the most common of these—instrumental variables. In each of these cases, ordinary least squares provides biased *and* inconsistent estimators. As both the simulation and the examples illustrate, OLS estimates may be very misleading, relative to estimates obtained with procedures not assuming recursiveness.

The problems outlined here, in simple form for the hierarchical model as well as the estimation techniques described for that model, are common to all work with structural equations. It becomes more difficult to recognize and deal with them as the models become more complex, but they remain fundamentally the same. In the next two chapters, we shall discuss some of these more elaborate structural models and estimation procedures. It is important to keep in mind however that these are basically extensions of the problems discussed in this chapter. If these problems are not understood in the context of the simple hierarchical model, it would be useful to review this chapter once more.

APPENDIX 8.1

Instrumental Variables Estimator

The development of the instrumental variables estimator is quite simple, however it is best seen in the matrix notation presented in Chapter 5. We assume that the true model is

$$Y = X\beta + U, \tag{8.1a}$$

and that at least one of the K explanatory variables is correlated with the error term U. We denote by Z the set of instrumental variables that are correlated with X, but not with U, in the probability limit such that

$$\text{plim}(T^{-1}Z'X) = S_{ZX} \neq 0, \quad \text{plim}(T^{-1}Z'Z) = S_{ZZ} \neq 0, \quad \text{and} \quad \text{plim}(T^{-1}Z'U) = 0.$$

We maintain the assumption that $E(UU') = \sigma^2 I$. From this it is clear that any explanatory variable not correlated with U can serve as its own instrument.

The instrumental variables estimator b^* is

$$b^* = (Z'X)^{-1}Z'Y = (Z'X)^{-1}Z'(X\beta + U) = \beta + (Z'X)^{-1}Z'U. \tag{8.1b}$$

Of course, if all variables are their own instruments, this is simply the ordinary least squares estimator. It is easy to see that b^* is a consistent estimator of β. [Since $\text{plim}(T^{-1}Z'X)^{-1}\text{plim}(T^{-1}Z'U) = S_{ZX} \cdot 0 = 0$.] The variance of b^* in this case is asymptotically

$$\text{asy}\, E\,(b^* - \beta)(b^* - \beta)' = T^{-1}\text{plim}\left[T(Z'X)^{-1}Z'UU'Z(X'Z)^{-1}\right]$$

$$= T^{-1}\text{plim}(T^{-1}Z'X)^{-1}\text{plim}(T^{-1}Z'UU'Z)\text{plim}(T^{-1}X'Z)^{-1}$$

$$= T^{-1}\sigma^2 S_{ZX}^{-1}S_{ZZ}S_{ZX}^{-1'} \quad [\text{since plim}(T^{-1}Z'UU'Z)$$

$$= \sigma^2 \text{plim}(T^{-1}Z'Z)]. \tag{8.1c}$$

Loosely, b^* is better—has lower variance—the higher the correlations between Z and X. For a given set of X's, the variance expression is minimized when $Z = X$ if Z is independent of U (see Johnston, 1972, pp. 278–281).

REVIEW QUESTIONS

1. Consider the system

$$Y_1 = \qquad\qquad\qquad + \beta_{21}X_2 + \beta_{31}X_3 \qquad\qquad + U_1,$$

$$Y_2 = \gamma_{12}Y_1 \qquad\qquad\qquad\qquad + \beta_{42}X_4 + U_2,$$

$$Y_3 = \gamma_{13}Y_1 + \gamma_{23}Y_2 + \qquad\qquad \beta_{33}X_3 + \beta_{43}X_4 + U_3,$$

$$Y_4 = \gamma_{14}Y_1 + \gamma_{24}Y_2 + \gamma_{34}Y_3 \qquad\qquad\qquad + U_4.$$

(All variables are written as deviations about their respective mean.)

   (a)   What coefficients are unidentified, just identified, and overidentified if all error terms are correlated?

   (b)   What information about additional variables in the model or about error term covariances is necessary to identify the structure completely?

2. Given the following multiequation system and estimated reduced form equations, estimate the coefficients of the structural model. (All variables are in deviation form and the constant term is omitted.)

Structural:

$$Y_1 = \qquad \beta_{21}X_2 + \beta_{31}X_3 \qquad\qquad + U_1,$$

$$Y_2 = \gamma_{12}Y_1 \qquad\qquad + \beta_{32}X_3 + \beta_{42}X_4 \qquad + U_2,$$

$$Y_3 = \gamma_{13}Y_1 + \gamma_{23}Y_2 \qquad\qquad\qquad + \beta_{53}X_5 + U_3.$$

Reduced form:

$$Y_1 = 0.521X_2 + 0.247X_3,$$

$$Y_2 = 0.194X_2 + 0.448X_3 + 0.123X_4,$$

$$Y_3 = 0.209X_2 + 0.216X_3 + 0.040X_4 + 0.171X_5.$$

3. Given the following variance–covariance matrix and using the IV estimator where appropriate, estimate the identified equations in Problem 1. Compare these with the OLS estimates.

X_2 31.395						
X_3 11.621	29.841					
X_4 13.186	12.234	28.442				
Y_1 21.465	17.446	4.375	25.996			
Y_2 18.001	18.570	17.141	27.018	38.169		
Y_3 25.869	27.879	25.136	30.425	31.030	63.885	
Y_4 29.516	29.356	25.332	30.136	32.574	47.983	58.467

9 | Structural Equations: Simultaneous Models

9.1 Introduction

The specialized examples of hierarchical systems in the previous chapter are only the beginning. The analysis of fully simultaneous systems is more important in social science research. The distinguishing feature of simultaneous models is that the causal structure is no longer hierarchical, meaning that at least one equation contains a higher ordered endogenous variable as an explanatory variable. This general form, where each of the endogenous variables can interact with all the other endogenous variables, is much more useful.

Even though it is conceptually and philosophically possible to talk about causal ordering in a time-dependent manner, very few observed variables capture this ordering. In experimental research it is possible to control which variables are changed, and in which order, and observe the resulting changes in the other endogenous variables. This experimental method is not available in most social sciences. Neither is it normally possible to obtain continuous observations on all the variables in a system, which would permit one to observe the sequence of changes even though the variables were not controlled. In cross-sectional studies, where data on many individuals are collected at only one point in time, the observations on any variables subject to reciprocal influences will appear simultaneous because no time ordering is observed.

Even in longitudinal panel and time series data the measured variables are likely to exhibit the same problems. Consider the macroeconomic model's consumption function estimated with aggregate data. This model hypothesizes that consumption (C) is determined by income (Y), $C = \beta_0 + \beta_1 Y$, and that consumption expenditures are an important part of aggregate income by the identity $Y \equiv C + I + G$, where I is investment and G is governmental expenditures. If the time it takes income changes to affect consumption and work their way back into increased incomes is less than the period between observations, then the observed variables must be considered to be simultaneously related. Many times lagged income is used in estimating the consumption function. However, with data observed over long time periods, such as a year, this may be an incorrect specification since people are consuming from the current year's income. Similarly, in a panel study of political behavior, attitudes and behavior—such as party affiliations—are measured at two points in time, maybe years, months, or weeks apart. If both attitudes and behavior differ at each time, one must be very careful about estimating the magnitudes of the reciprocal influences and the influences of the exogenous variables that are part of the system because the interactions between attitudes and behavior may take place over shorter periods than the time between interviews. In this case, the observed attitude and behavior variables will have affected each other and the measured variables will be simultaneously related. As with the consumption function example, if the period of interaction between attitudes and behavior is less than the sampling period, proper specification requires that the current values of the endogenous variable be used as explanatory variables and not the values at the previous interview. (In a dynamic model the lagged values will be included but along with the current values for some variables.) Consequently, many structural models of social behavior will contain simultaneous relationships if properly specified.

Both examples discussed in the previous chapter are more usefully formulated as simultaneous equation models. The discussion of alternative theories of voting behavior can be properly tested only by specifying a reciprocal relationship between evaluations and party identification. One theory holds that evaluations are derived from people's party identifications by the way they filter information and perceive candidates' positions. (This is the theory expressed in Fig. 8.1.) The alternative theory holds that party affiliations are based on evaluations of the parties' and candidates' positions on different issues and that voters will identify with the party that best represents their views on public policy issues.

It is quite likely that reality encompasses some combination of these alternatives, and that true behavior is best modeled by a reciprocal relationship between the evaluation and party identification variables. The relative merits of each theory are tested by the magnitude of the effects as indicated by the two coefficients in this reciprocal relationship. In the context of one survey it certainly will not be possible to observe the independent changes in evaluations and party affiliations required to estimate the magnitudes of these two influences separately. Consequently, the hierarchical model estimated in Chapter 8 is replaced by a more complete model that includes equations simultaneously modeling both evaluations and party identifications as functions of each other and additional appropriate explanatory variables. The model, displayed here in Fig. 9.1 [and discussed in more detail in Jackson

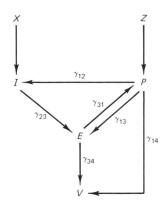

FIGURE 9.1   *Simultaneous voting model.* $X = $ *(education, race, region, union, religion, income),* $Z = $ *(father's party, region and place of residence, age over 54, union, religion, income).*

(1975)], includes a set of exogenous variables that are also expected to determine evaluations and party affiliations. (These are the variables used in Chapter 8 to create the instruments in the vote equation.) The variable I measures the respondent's position on each of the seven separate issues summarized in the evaluation variable. In the full study, separate issue and evaluation equations are estimated for each issue.

The coefficients denoted by γ's in Fig. 9.1 are the key to the structural model; they represent the relationships among the endogenous variables. The test of the different voting theories centers on the magnitudes of γ_{12} and γ_{13}, compared to the size of γ_{31}. The key to estimating this model is the hypothesis that some of the exogenous variables determine party identifications but not issue positions (and thus evaluations), while others affect issue positions and thus evaluations but not party identifications. The specification of these variables and their importance in estimating simultaneous equation models is a major theme in this chapter.

The complete aspiration model investigated by Duncan, Haller, and Portes (1968) is properly simultaneous for a different set of reasons. In their case, they are not trying to test or evaluate alternative theories of aspiration levels. Their primary interest is in estimating the influence of peers on an individual's aspirations and in comparing the magnitude of peer influences to those of exogenous environmental and personal variables. Since friendship is a reciprocal relationship, it is impossible to imagine that a peer can influence the respondent without in turn being influenced. Duncan, Haller, and Portes also collected measures for the variables used in the previous chapter for a best friend of each of the respondents. They then specify a model with simultaneous relationships between respondent's and friend's occupational and educational aspirations. Note that we now have some exogenous variables excluded from each equation. In this case, it is the cohort's exogenous and parental aspiration variables. Thus we exclude the possibility that the friend's intelligence, parental aspirations, or socioeconomic status influence the respondent's aspirations, except through the interrelationships among cohort's aspirations. (This specification differs from that of the Duncan, Haller, and Portes model in that they include friend's socioeconomic level as an explanatory variable in the respondent's aspirations. We see no a priori reason to continue this specification.) The new model is shown in Eqs. (9.1)–(9.6) and diagramed in Fig. 9.2; the subscript 1 on each variable refers to the respondent, while 2 refers to peer:

$$P_1 = \beta_{11}I_1 + \beta_{21}S_1 \qquad\qquad\qquad\qquad\qquad\qquad\qquad + U_1, \tag{9.1}$$

$$P_2 = \qquad\quad \beta_{32}I_2 + \beta_{42}S_2 \qquad\qquad\qquad\qquad\qquad\quad + U_2, \tag{9.2}$$

$$O_1 = \beta_{13}I_1 + B_{23}S_1 \qquad\quad + \gamma_{13}P_1 \qquad\qquad + \gamma_{43}O_2 \qquad + U_3, \tag{9.3}$$

$$O_2 = \qquad\quad \beta_{34}I_2 + \beta_{44}S_2 \qquad + \gamma_{24}P_2 + \gamma_{34}O_1 \qquad\qquad + U_4, \tag{9 4}$$

$$F_1 = \beta_{15}I_1 + \beta_{25}S_1 \qquad\quad + \gamma_{15}P_1 \qquad + \gamma_{35}O_1 \qquad + \gamma_{65}E_2 + U_5, \tag{9.5}$$

$$E_2 = \qquad\quad \beta_{36}I_2 + \beta_{46}S_2 \qquad + \gamma_{26}P_2 \qquad + \gamma_{46}O_2 + \gamma_{56}E_1 \quad + U_6. \tag{9.6}$$

The primary interest in this model concerns the importance of peer's

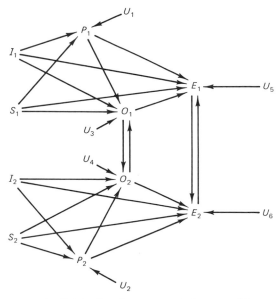

FIGURE 9.2   Simultaneous peer influence model.

aspirations (indicated by the coefficients γ_{43}, γ_{34}, γ_{65}, and γ_{56}) relative to the other variables in the model. If these peer coefficients are large, they indicate that aspirations are largely derived through the operation of peer influences, and not directly from the effects of family and personal characteristics.

The considerations involved in estimating the equations in these two examples, and in similar models, parallel those of the hierarchical models in Chapter 8. However, the potential problems are more numerous, harder to visualize, and less tractable. The very first consideration is to verify that the structure is identified. The mathematics required for this verification is more difficult than in the hierarchical case. We shall spend considerable time discussing the identification problem and how one proceeds to ascertain whether the hypothesized structure is identified. Once this is accomplished, presuming it is, the structure can be estimated in ways quite similar to those used in the previous chapter.

We first want to give one of the classic illustrations of the identification problem and to show how to cope with it. Following this we discuss the identification and estimation problems, and their resolution, in more general terms.

9.2   Identification in Simultaneous Systems: An Example

In discussing identification for the hierarchical case, the issue was seen to be whether or not different structures—i.e., different parameter values in the structural model—could produce the same set of reduced form parameters. The key to identification is knowledge about the specification of the equations, in particular whether specific parameters or certain error covariances are zero. This information must come from outside the data and cannot be deduced from the observational information. Such a priori information may come from theory about the causal relationships or may come from knowledge of the data and the potential errors in the system.

The approach that we take to the identification problem is consideration of how much such outside information is necessary. We begin with a simple story.

Suppose we wish to estimate a demand function for an agricultural commodity, say wheat.[1] We posit that quantity demanded is inversely related to its price, as in Fig. 9.3 and the equation

$$Q_t^d = \alpha_1 + \alpha_2 P_t, \qquad \text{where} \quad \alpha_2 < 0. \tag{9.7}$$

If Equation (9.7) is nonstochastic, any two observations such as points 1 and 2 in Fig. 9.3a are sufficient to estimate the two parameters (α_1 and α_2). If the demand function is stochastic, as in Fig. 9.3b and the equation

$$Q_t^d = \alpha_1 + \alpha_2 P_t + U_{t1}, \tag{9.7a}$$

---

[1]Identification of a demand function is perhaps the classic example of the identification problem. It was first discussed by Working (1927). This example can also be found in Lawrence Klein (1962).

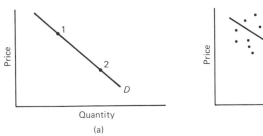

FIGURE 9.3   *(a) Nonstochastic demand, (b) stochastic demand.*

we can use an ordinary least squares estimator to obtain unbiased and consistent estimates of α_1 and α_2.

However, the story does not end here. It is clearly reasonable to think that the demand for wheat is related to price. However, at the same time one would expect (because of relationships between costs of production and amount produced) the price and the quantity supplied to be positively related, as in Fig. 9.4a and 9.4b and the equations

$$P_t = \beta_1 + \beta_2 Q_t^s, \tag{9.8}$$

$$P_t = \beta_1 + \beta_2 Q_t^s + U_{t2}, \qquad \text{where} \quad \beta_2 > 0. \tag{9.8a}$$

This relationship is referred to as the supply curve. When we observe actual data, we observe sales of wheat at a given price where the amount sold is both the quantity demanded and the quantity supplied. Thus, we do not need the superscripts on Q.

Now what is the situation? In the nonstochastic case only one point is observed (point 1 in Fig. 9.4a). Because the data come from actual transactions, the price paid by the consumer is the price received by the producer, and the quantity purchased equals the quantity supplied. We have no way of observing what consumers might have been willing to pay for different quantities of wheat or what farmers might have been willing to supply for different prices. These underlying demand and supply relationships are not revealed. The only thing that we know is that both the supply and demand curves must go through point 1. But, this knowledge does not help us in estimating the slopes of either the demand or the supply curve.

Looked at from the opposite point of view, there are an infinite number of

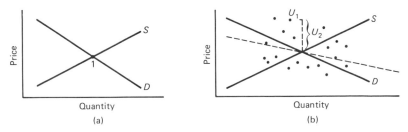

FIGURE 9.4   *(a) Nonstochastic supply and demand, (b) stochastic supply and demand.*

demand and supply curves that could go through point 1 and, thus, an infinite number of underlying supply and demand relationships that could generate exactly the same observed data about sales and price. On the basis of the observed data, we have no way of choosing one of the infinite number of possible structures.

We are no better off in the stochastic case. The observed data still relate to the intersection of the supply and demand curves. Because of the random deviations (such as U_1 and U_2 in Fig. 9.4b), the actual observations provide several distinct points. Further, there may be some pattern to these points within a given sample. Indeed, the dashed line in Fig. 9.4b represents the ordinary least squares estimates of observed quantity regressed on observed price. Because it slopes downward, is it a demand curve? Clearly not. It is a mixture of the demand and the supply curve, and no amount of observational information will allow us to choose the true structural parameters from all the potential parameters that could have generated the observational data. Both the demand and the supply equation are unidentified.

Suppose, however, that the supply curve is known to shift for different observations while the demand curve remains stable. This can arise because of weather variations—as measured by, say, rainfall—affecting crop yields. If, for given inputs of seed, fertilizer, time, etc., the wheat crop yield is larger in years of higher rainfall, the cost of producing a given quantity of wheat falls, and the supply curve for wheat for high rainfall years would lie to the right and below that for low rainfall years. (That is, the cost and "supply price" for any amount of wheat is less in the high rainfall year.) This is depicted in Fig. 9.5 and the equations

$$P_t = \beta_1 + \beta_2 Q_t + \beta_3 R_t, \tag{9.9}$$

$$P_t = \beta_1 + \beta_2 Q_t + \beta_3 R_t + U_{t2}, \tag{9.9a}$$

where different supply curves correspond to different rainfalls ($R_1 < R_2 < R_3 < R_4$). From Fig. 9.5, we see that the observed quantities (points 1–4) trace out the demand curve; using any two observations (corresponding to different rainfalls), it is now possible to estimate the two demand parameters, and the demand curve is identified. [The stochastic case in Eq. (9.9a) is completely analogous to this: variations in rainfall shift the supply curve so that the

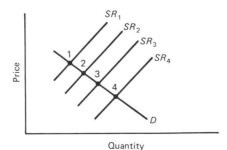

Quantity

FIGURE 9.5 Shifting supply curve.

observations tend to trace out a demand curve, albeit with error in the stochastic case.]

Note also that the observational data still provide no help in unraveling the *supply* parameters. There remain an infinite number of supply structures that could have generated the observational data.

Let us look at this situation in the same way that we looked at the hierarchical case. If R is independent of U_1 and U_2 and the expected values of U_1 and U_2 are zero, we can obtain consistent estimates of the reduced form that is represented in[2]

$$Q_t = \pi_{11} + \pi_{21} R_t + V_{1t}, \tag{9.10}$$

$$P_t = \pi_{12} + \pi_{22} R_t + V_{2t}. \tag{9.11}$$

In terms of the structural parameters in Eqs. (9.7a) and (9.9a)

$$\pi_{11} = (\alpha_1 + \alpha_2 \beta_1)/(1 - \alpha_2 \beta_2), \qquad \pi_{22} = \beta_3/(1 - \alpha_2 \beta_2),$$
$$\pi_{21} = \alpha_2 \beta_3/(1 - \alpha_2 \beta_2), \qquad V_1 = (U_1 + \alpha_2 U_2)/(1 - \alpha_2 \beta_2),$$
$$\pi_{12} = (\beta_1 + \beta_2 \alpha_1)/(1 - \alpha_2 \beta_2), \qquad V_2 = (U_2 + \beta_2 U_1)/(1 - \alpha_2 \beta_2).$$

From this it follows that

$$\alpha_2 = \pi_{21}/\pi_{22} \quad \text{and} \quad \alpha_1 = \pi_{11} - \alpha_2 \pi_{12} = \pi_{11} - (\pi_{21}/\pi_{22})\pi_{12}.$$

However, there is no way to unravel β_1, β_2, and β_3 from the reduced form parameters. Thus, we conclude that Eq. 9.7 [(9.7a)] is identified, but that Eq. (9.9) [(9.9a)] is not identified.

One further variant is useful. If quantity demanded is a function of not only price but also income (Y), Eqs. (9.7) and (9.7a) become

$$Q_t = \alpha_1 + \alpha_2 P_t + \alpha_3 Y_t \quad \text{and} \quad Q_t = \alpha_1 + \alpha_2 P_t + \alpha_3 Y_t + U_{t1}.$$

Income changes now tend to shift the demand curve relative to the supply curve and to trace out different points on the supply curve. Thus, the income variable does for the supply curve what the rainfall variable did for the demand curve. The supply curve in this expanded model is now identified, and we can estimate its parameters (β_1, β_2, and β_3) on the basis of the new reduced form parameters.[3] (Exercise: write out the new structural and reduced form equations and solve for β_1, β_2, and β_3.)

The central point of this discussion has been that the introduction of specific a priori information—information that is supplied before looking at the data and which cannot be deduced from the data—allows identification and estimation of the structural parameters. It is important to understand, however, just what the necessary information was for the previous examples. The demand equation was identified by introducing the information that rainfall, a factor affecting supply, did *not* affect demand. It was this *exclusion restriction* that enabled identification of the demand parameters. To see this,

---

[2]In simultaneous equation estimation, we concentrate upon consistent estimators, and thus we shall begin to cast the discussion in those terms. For example, in the above discussion, the necessary assumption is really that $\text{plim}_{T \to \infty} \Sigma R_t U_{t1}/T = \text{plim}_{T \to \infty} \Sigma R_t U_{t2}/T = 0$.

[3]When all of the equations of a model are identified, we say that the model is identified.

note that the inclusion of rainfall did not identify the supply equation and, further, that including the rainfall variable in the demand equation would have left it unidentified. An alternative way of specifying this restriction is to say that a given parameter is known a priori to be zero.[4] For example, rainfall is assumed to have a zero coefficient in the demand equation, i.e., to be excluded from the demand equation. Further, these restrictions cannot be deduced from the observed data; they require additional information based on theory or previous empirical work.

The preceding discussion has concentrated upon the distinction between unidentified and identified equations. This, as we have pointed out, is clearly the crucial distinction because, if an equation is unidentified, there is simply nothing further to do.

The example developed so far is the exactly, or just, identified case. Once we estimate the reduced form coefficients, we get only one estimate for each structural coefficient. To illustrate the overidentified case, consider that supply might also be influenced by the mean temperature during the growing season, denoted as T. Adding T to Eqs. (9.9) and (9.9a) leaves the demand equation overidentified, in the sense that we get more than one, but a finite number of, estimates of some coefficients. We leave it as an exercise to check this by finding alternative ways of estimating the demand equation coefficients from the reduced form coefficients of this expanded model.

This example should serve as an illustration of the identification problem and what is required for a model to be identified. In the discussion of the general problems of identification and of how to ascertain whether an equation is identified, it will be useful to refer to this example.

9.3 Identification in Simultaneous Models

The conclusions from the previous chapter and the supply and demand example vividly point out the importance of identification. Identification must logically precede any estimation of equations in a simultaneous system and merits considerable attention. This—as with the other topics in this and the next chapter—is the subject of a large literature, much of which is beyond the scope of this book. Therefore, the development here concentrates on the most common situations and is by no means complete.[5]

There is one important concept to keep in mind during all discussions of the identification problem and subsequent attempts to estimate structural

[4]Exclusion restrictions are by far the most common restriction. However, other types of information can also be used to achieve identification. For example, a precise (nonzero) value of a coefficient will suffice anytime that a zero restriction does. Likewise specifying that two coefficients must equal each other is also a form of prior information. Or, the recursive model, which we shall come back to, achieves identification through restrictions on the error covariances across equations. A menu of alternative restrictions can be found in Fisher (1966).

[5]For a complete, but quite difficult, discussion of identification, see Fisher (1966). The discussion of identification in this chapter relies heavily upon that work.

equation models: identification is a modeling problem, *not* a statistical one. If an equation is unidentified, there is an infinite number of possible coefficients, or true structures, which give the same relationships among observed variables, regardless of the presence or absence of error terms. No matter how many times we replicate an experiment or how many observations we have, there is no way to distinguish observationally among these alternative structures. There is no way statistically to overcome this difficulty.

After determining that an equation is identified and after collecting data for the variables in the model, we can turn to the statistical problems of estimating it. Many alternative estimators have been proposed for simultaneous equations. However, if the structure is exactly identified, these estimators will be equivalent. If an equation is overidentified, the statistical problem is not so easy. Because the equation is overidentified, we get more than one estimate for some structural coefficients. Again, these estimates differ only because of the stochastic nature of our estimates of the relations among observed variables. (Without this stochastic element, our several estimates would be equal.)

At this point, we can see the statistical question. How do we use these several estimates of the same coefficient to get a better estimate of the true value? All of the different estimation methods for structural equations amount to alternative ways of combining these several estimates into a better, single estimate. This is the statistical problem.

The immediate question now is to develop some general rules for determining whether an equation is under-, exactly, or overidentified. To do this for the general case we need to resort to the matrix notation developed in Chapter 5.

Matrix Notation

In Chapter 5, we saw that it is convenient to write the single equation model in matrix form and to develop the estimation techniques using matrix algebra. The matrix representation allows us to write compactly the T equations of the observed data, i.e., the equations for each of the T observations of the model.

Matrix notation also leads to considerable simplification of the discussion of multiequation systems. By convention, we write the structural equations for each observation t of a general system of M equations with K exogenous variables in implicit form as:

$$\gamma_{11} Y_{t1} + \gamma_{21} Y_{t2} + \cdots + \gamma_{M1} Y_{tM} + \beta_{11} X_{t1} + \cdots + \beta_{K1} X_{tK} + U_{t1} = 0,$$

$$\vdots$$

$$\gamma_{1M} Y_{t1} + \gamma_{2M} Y_{t2} + \cdots + \gamma_{MM} Y_{tM} + \beta_{1M} X_{t1} + \cdots + \beta_{KM} X_{tK} + U_{tM} = 0,$$

for $t = 1, 2, \ldots, T$. It is also convenient for most purposes to normalize the system of equations by setting one coefficient in each equation equal to -1. Typically, we set $\gamma_{mm} = -1$ for $m = 1, \ldots, M$. Thus, we can think of the mth equation as "explaining" the observed values of Y_m. In most instances, our theories are not dependent upon the units of measure—we are, for example, usually indifferent between measuring income in dollars or thousands of dollars.[6] There is not only an interpretive gain from this normalization, but there also is a statistical gain. Through specification of one coefficient in this manner, one less coefficient must be estimated. It is also easier to address identification when the model is placed in this form. This normalization convention will be used throughout this chapter.

In matrix form, we may write this system as

$$Y_t.\Gamma + X_t.\beta + U_t. = 0$$

where Y_t. is the $1 \times M$ row vector of the tth observation of all endogenous variables, X_t. is the $1 \times K$ row vector of the tth observation of all exogenous variables, U_t. is the $1 \times M$ row vector of the tth observation of error term for each equation, Γ is the $M \times M$ matrix of coefficients for the endogenous variables, β is the $K \times M$ matrix of coefficients for the exogenous variables, M is the number of equations (and number of endogenous variables), and K is the number of exogenous variables. It is instructive to display the matrices corresponding to the aspiration example in Eqs. (9.1)–(9.6). (Note that we have put these into implicit form and have applied the normalization rule.)

$$[P_{1_t} \quad P_{2_t} \quad O_{1_t} \quad O_{2_t} \quad E_{1_t} \quad E_{2_t}] \begin{bmatrix} -1 & 0 & \gamma_{13} & 0 & \gamma_{15} & 0 \\ 0 & -1 & 0 & \gamma_{24} & 0 & \gamma_{26} \\ 0 & 0 & -1 & \gamma_{34} & \gamma_{35} & 0 \\ 0 & 0 & \gamma_{43} & -1 & 0 & \gamma_{46} \\ 0 & 0 & 0 & 0 & -1 & \gamma_{56} \\ 0 & 0 & 0 & 0 & \gamma_{65} & -1 \end{bmatrix}$$

$$+ [I_{1_t} \quad S_{1_t} \quad I_{2_t} \quad S_{2_t}] \begin{bmatrix} \beta_{11} & 0 & \beta_{13} & 0 & \beta_{15} & 0 \\ \beta_{21} & 0 & \beta_{23} & 0 & \beta_{25} & 0 \\ 0 & \beta_{32} & 0 & \beta_{34} & 0 & \beta_{36} \\ 0 & \beta_{42} & 0 & \beta_{44} & 0 & \beta_{46} \end{bmatrix}$$

$$+ [U_{1_t} \quad U_{2_t} \quad U_{3_t} \quad U_{4_t} \quad U_{5_t} \quad U_{6_t}]$$

$$= [0 \quad 0 \quad 0 \quad 0 \quad 0 \quad 0]. \qquad (9.12)$$

It is important to contrast this formulation with the single equation formulation. Here the matrix equation (9.12) describes a *single* observation of the whole system of equations; in the single equation case, the matrix formulation described *all* observations of one equation.

---

[6]In almost all cases of interest, this normalization will not affect our discussions. Occasionally, there are situations where one particular scale of coefficients is appropriate (because of a priori information dependent upon scale), but it is beyond our development.

It will often be useful to expand this description of the system to include all the observations. We do this by defining the matrices

$$Y = \begin{bmatrix} Y_{1\cdot} \\ Y_{2\cdot} \\ \cdot \\ \cdot \\ \cdot \\ Y_{T\cdot} \end{bmatrix} = \begin{bmatrix} Y_{11} & Y_{12} & \cdots & Y_{1m} & \cdots & Y_{1M} \\ Y_{21} & Y_{22} & \cdots & Y_{2m} & \cdots & Y_{2M} \\ \cdot & \cdot & & \cdot & & \cdot \\ \cdot & \cdot & & \cdot & & \cdot \\ \cdot & \cdot & & \cdot & & \cdot \\ Y_{T1} & Y_{T2} & \cdots & Y_{Tm} & \cdots & Y_{TM} \end{bmatrix} = \begin{bmatrix} Y_1 & Y_2 & \cdots & Y_M \end{bmatrix},$$

$$X = \begin{bmatrix} X_{1\cdot} \\ \cdot \\ \cdot \\ \cdot \\ X_{T\cdot} \end{bmatrix} = \begin{bmatrix} X_{11} & \cdots & X_{1K} \\ \cdot & & \cdot \\ \cdot & & \cdot \\ \cdot & & \cdot \\ X_{T1} & \cdots & X_{TK} \end{bmatrix} = \begin{bmatrix} X_1 & X_2 & \cdots & X_K \end{bmatrix},$$

$$U = \begin{bmatrix} U_{1\cdot} \\ \cdot \\ \cdot \\ \cdot \\ U_{T\cdot} \end{bmatrix} = \begin{bmatrix} U_{11} & \cdots & U_{1M} \\ \cdot & & \cdot \\ \cdot & & \cdot \\ \cdot & & \cdot \\ U_{T1} & \cdots & U_{TM} \end{bmatrix} = \begin{bmatrix} U_1 & U_2 & \cdots & U_M \end{bmatrix}.$$

The matrix X in this formulation is the same as the $T \times K$ matrix of observations on the explanatory variables used in the multivariate model in Chapter 5 except that it includes all of the predetermined variables and not just those that appear in a specific equation. Analogously, the matrix Y is the $T \times M$ matrix of observations for all endogenous variables. The coefficient matrices Γ and β describe not only the magnitude of the relationships within given equations but also which variables appear in each equation. Thus, if $\gamma_{im} = 0$, the ith endogenous variable does not appear in the mth equation; similarly, if $\beta_{km} = 0$, the kth predetermined variable does not appear in the mth equation. From this we may write all observations of all equations as

$$Y\Gamma + X\beta + U = 0. \tag{9.13}$$

This notation should be completely understood before proceeding. For the variable matrices (Y, X, and U), a row corresponds to a given observation of all variables, and a column corresponds to all observations of a single variable. For the coefficient matrices, rows relate to coefficients of specific variables in all equations, and columns relate to the different variables in specific equations.

Equation (9.13) is the structural model for the hypothesized system. For purposes of estimation and consideration of the identification problem, as well as some applications, the reduced form equations are required. As in Chapter 8, these equations summarize the total effects of each exogenous variable on the endogenous variables. However, attempts to derive the reduced form model one equation at a time, as done with the hierarchical system, are quite cumbersome for complicated systems such as the aspiration model. Fortunately, the matrix notation greatly simplifies derivation of the reduced form.

If Γ is nonsingular,[7] then

$$Y_{t\cdot} = -X_{t\cdot}\beta\Gamma^{-1} - U_{t\cdot}\Gamma^{-1} = X_{t\cdot}\Pi + V_{t\cdot} \qquad \text{for} \quad t = 1,\ldots,T, \qquad (9.14)$$

or

$$Y = -X\beta\Gamma^{-1} - U\Gamma^{-1} = X\Pi + V, \qquad (9.15)$$

where $\Pi = -\beta\Gamma^{-1}$ and $V = -U\Gamma^{-1}$. Π is a $K \times M$ matrix of reduced form coefficients relating each endogenous variable to all the exogenous variables. For each observation, each endogenous variable now appears in one and only one equation. Each column of Π expresses the causal relationships with the exogenous variables for a different endogenous variable. Note that the individual Π coefficients assess the total expected change in the endogenous variable from a unit change in the relevant exogenous variable. For example, π_{km} measures the total effect of a unit change in X_k on Y_m.

Stop at this point. Understanding the notion is essential to following the subsequent discussion, and a skimming of the previous discussion will probably lead to confusion within a few pages. As a check, if you knew the values of the third observation for all of the exogenous variables (i.e., $X_{31}, X_{32}, \ldots, X_{3K}$), into what row or column of X would the data you have fit? Given that, what row or column of Π would you use to predict the third value of the second endogenous variable (i.e., Y_{32})? If these questions gave you trouble, go back to the beginning of this section.

Identification

Identification is a question of how much prior information is needed in specifying a model. The previous supply and demand model was quite simple —simpler than many models of interest in the social sciences. And, while it was easy to see what was needed for identification in that case, it is not so easy to see it as the structures get more complicated, as more equations and more variables are entered. Therefore, we shall now develop more general rules and procedures for determining whether an equation is underidentified, just identified, or overidentified. As we go through this, we shall also discuss estimation schemes.

The general approach to the identification problem follows the discussion

---

[7]It is not very restrictive to assume that Γ is nonsingular. If Γ is singular, it implies that the coefficients in each equation are a linear combination of the coefficients in the other questions. There are two possibilities. First, the model may be incomplete in that there are more endogenous variables than equations. Secondly, in the terminology of simultaneous equations, one of the equations may be redundant. The most likely occurrence of a redundant equation is the case where there is an identity relationship in the model. The assumption of nonsingularity of Γ simply implies that we agree to eliminate redundant equations before expressing the model and we agree to consider only complete models.

of the previous examples. If we look at a particular equation in the simultaneous system and if we know the reduced form parameters, is it possible to obtain uniquely the structural parameters? In general, without the introduction of a priori information, it is easy to see that this is not possible. Consider the relationship between the reduced form parameters and the structural parameters:

$$\Pi = -\beta\Gamma^{-1} \quad \text{or} \quad \Pi\Gamma = -\beta.$$

The question is simply, Is there more than one set of structural coefficients (β, Γ) that can yield the same Π? Knowing that Γ is nonsingular and applying the normalization rule that $\gamma_{mm} = -1$ is insufficient.

Concentrating upon the relationship between the reduced form parameters and the structural parameters is a natural way to view identification since the reduced form represents our observational information, i.e., the information about the joint distributions of the X's and Y's. The identification problem is then a question of whether a different form of the structural model could have generated the observed relationship.

Before entering into the mathematics of this, it is useful to give a numerical example. Assume that the known reduced form parameters (Π) in a two-equation system with three exogenous variables are given by

$$\Pi = \begin{bmatrix} 1 & -2 \\ 3 & -1 \\ 5 & -4 \end{bmatrix}$$

These reduced form coefficients are compatible with, among others, the following very different structural coefficients:

$$\Gamma^* = \begin{bmatrix} -1 & 2 \\ 3 & -1 \end{bmatrix} \quad \text{and} \quad \beta^* = \begin{bmatrix} +7 & -4 \\ +6 & -7 \\ +17 & -14 \end{bmatrix}$$

or

$$\Gamma^{**} = \begin{bmatrix} -1 & -3 \\ 7 & -1 \end{bmatrix} \quad \text{and} \quad \beta^{**} = \begin{bmatrix} +15 & 1 \\ +10 & 8 \\ +33 & 11 \end{bmatrix}.$$

(Confirm that these structures are both compatible with the known reduced form coefficients.) Without additional information, we have no way of distinguishing between these alternative structures or, in fact, between these and the infinite number of alternative ways of generating Π.

The formal way of stating this problem is to consider the true coefficients in a system β and Γ. We can take any nonsingular, $M \times M$ matrices F and F^* and generate alternative structures: $\beta^* = \beta F, \Gamma^* = \Gamma F$, and $\beta^{**} = \beta F^*, \Gamma^{**} = \Gamma F^*$. The identification problem arises because[8] $\beta\Gamma^{-1} = \beta^*\Gamma^{*-1} = \beta^{**}\Gamma^{**-1} = -\Pi$. Thus from observing Π, we have no way of discriminating between

---

[8]This formulation considers only linear transformations of the structures by putting the problem in terms of an F matrix. This is the only transformation that must be considered since a nonlinear transformation would add other variables to the model. Thus, with additional variables there would be no difficulty in distinguishing the true structure from the alternative structure.

the alternative structures. The additional information required specifics values for or relationships among certain elements of β and Γ. Each such restriction must be maintained in the alternative structures β^* and Γ^*. Thus, if $\gamma_{11} = -1$, $\gamma_{11}^* = \gamma_{11}^{**} = -1$; or if $\gamma_{1m} = 0$, then $\gamma_{1m}^* = \gamma_{1m}^{**} = 0$. These restrictions on β and Γ impose conditions on the elements of F, so that β^* and Γ^* maintain the restrictions. Identification is possible if β and Γ are sufficiently restricted so that only the identity matrix $F = I$ maintains the restrictions.

For most purposes, we are interested in identifying a single equation in the multiequation system, and this section will deal exclusively with that case. Because we are no longer dealing with hierarchical models, it is notationally convenient (and leads to no loss of generality) to consider the first equation. It is also useful to divide the endogenous variables into two groups: first, the M_1 included endogenous variables (those whose coefficients are not restricted to zero a priori) and, second, the M_2 excluded endogenous variables (those for which $\gamma_{m1} = 0$ a priori). Similarly, the exogenous variables are arranged so that the K_1 included variables come before the K_2 excluded variables. Note that $M_1 + M_2 = M$ and $K_1 + K_2 = K$. This division of the variables can be seen schematically in Fig. 9.6. The shaded areas in Γ and β represent parameters for the other equations in the system and thus are not of direct interest in identification of the first equation.

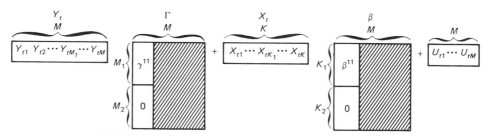

FIGURE 9.6   *Schematic structure of simultaneous equation system.*

Returning to the relationship between the structural parameters and the reduced form parameters, we note that the system relationship also holds equation by equation, as

$$\Pi \gamma^1 = -\beta^1, \tag{9.16}$$

where γ^1 and β^1 are the first columns of Γ and β. Next, partition Π so that it is conformable to the above partitioning of γ^1 and β^1. See Fig. 9.7, where π^{11} and π^{21} refer to coefficients in the reduced form equations for the M_1 endogenous variables included in the equation being estimated. π^{11} refers to the coefficients on the exogenous variables included in this first equation, and π^{21} refers to the coefficients on the excluded exogenous variables. [Partitioned multiplication is covered in the matrix algebra appendix (II) and should be referred to if this operation is not clear.] The shaded areas of Π are multiplied

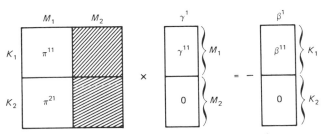

FIGURE 9.7   *Relationship of reduced form and structural parameters.*

by the zero elements in γ^1 and therefore do not enter into these calculations. Multiplying the partitioned matrices yields two matrix equations:

$$\pi^{11}\gamma^{11} = -\beta^{11},\tag{9.17}$$

$$\pi^{21}\gamma^{11} = 0.\tag{9.18}$$

The key to identifying the first equation in the model is Eq. (9.18). If we can solve for γ^{11} in Eq. (9.18), Eq. (9.17) will yield estimates of β^{11}. Equation (9.18) is a set of K_2 homogenous linear equations with M_1 unknown values (γ^{11}). It can be solved for a unique solution (up to a scale factor) if and only if the rank of π^{21} is $M_1 - 1$. In fact, we are interested only in the solution up to a scale factor because we are also going to apply the normalization rule that $\gamma_{11} = -1$. The necessary and sufficient condition for identification of the first equation is

> Rank condition for identification:
> rank of π^{21} equals $M_1 - 1$.

However, the rank condition is quite opaque. It is difficult to see what this implies about the model; and, furthermore, calculating the rank of even medium sized matrices can be difficult. Therefore, it is instructive to recast the identification criterion into the form of necessary (but not sufficient) conditions that relate to the size requirements of the submatrices of Π.

π^{21}, the matrix of interest at this point, is a $K_2 \times M_1$ matrix; i.e., its dimensions are dictated by the number K_2 of excluded predetermined variables and the number M_1 of included endogenous variables. If π^{21} is to have rank $M_1 - 1$, the number of rows (the number of excluded exogenous variables) must be at least $M_1 - 1$ (the number of included endogenous variables less one). Thus,

> Order condition for identification:
> $K_2 \geqslant M_1 - 1.$
(9.19)

The order condition is a counting rule. For an equation to be identified, it is necessary, but not sufficient, that the number of exogenous variables excluded from the equation equal or exceed the number of endogenous variables included as explanatory variables. (Remember that an additional endogenous variable, the one being "explained," is included with a coefficient set to -1.) This rule is not infallible because it does not ensure that the rank of $\pi^{21} = M_1 - 1$. It does however provide an initial reading about whether an equation is identified.

There are alternative ways of stating this rule that might be useful in understanding the basic idea. If we add M_2 (the number of excluded endogenous variables) to both sides of (9.19), we have

$$M_2 + K_2 \geqslant M - 1. \tag{9.20}$$

In other words, the *total* number of excluded variables must be at least as great as the number of equations minus one.

The order conditions are generally easy to check, and such a check is a natural first step in considering simultaneous equations. However, since they are only necessary conditions, they may fail in certain circumstances. It is possible for K_2 to equal or exceed $M_1 - 1$, but for some rows to be linear combinations of each other. Then the matrix will not be of full rank and will in fact have rank less than $M_1 - 1$. We can illustrate several such situations.

The most obvious case is the exclusion of one variable from all equations. This implies that the variable does not systematically affect any of the endogenous variables, hence its true reduced form coefficients are zero in all equations. A row of zeros in π^{21} clearly reduces its rank by one. Although this example seems farfetched, it means that we cannot use exogenous variables that are not included in the specification of at least one of the structural equations to help identify the model.

A second case of satisfying the order but not the rank condition occurs when two equations have exactly the same set of excluded variables. This results in π^{21} having at least two rows that are linearly related once we do the multiplication $\beta \Gamma^{-1}$. (We suggest you try a couple of simple examples to see this. Remember that we are talking about the population values of β, Γ, and Π when we are talking about identification. Due to sampling variations it is very unlikely that the estimated reduced form coefficients are linearly related even when the true values are.)

These cases provide some examples of situations where the simple counting rules are misleading, i.e., where the order conditions might hold while the rank conditions fail. However, checking the rank condition is more difficult than counting the included and excluded variables, and it is particularly difficult when the rank conditions are stated in terms of the reduced form parameters.

Sometimes it is easier to examine the identification question in terms of the structural equations rather than through the reduced form coefficients. The way to do this is to ascertain whether the restrictions imposed on β and Γ are sufficient to restrict F to the identity matrix, where F is the $M \times M$ nonsingular matrix in the expressions $\beta^* = \beta F$ and $\Gamma^* = \Gamma F$. In terms of a single equation, all we need to do is restrict the appropriate column of F—it does not matter if the other columns are not sufficiently restricted until we try to estimate those equations. Whether these restrictions are sufficient to obtain identifications depends upon the specification of the other equations in the model.

To show this, we re-create Fig. 9.6, only now we show some of the coefficients for the excluded variables (Fig. 9.8). The equivalent rank condition[9] to the one previously discussed is that the rank of the matrix $\begin{bmatrix} \gamma^{22} \\ \beta^{22} \end{bmatrix}$ equals $M-1$. In this form, it is easy to see how the two previous examples affect the rank condition for identification by introducing a row or column dependency in either β^{22} or γ^{22}. By putting the identification condition in terms of the structural coefficients, it is also more feasible to check the rank conditions when there is doubt.

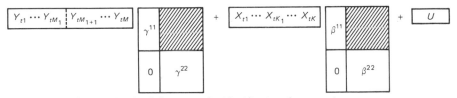

FIGURE 9.8   Submatrices for identification of structural parameters.

At this point, it is possible to relate the mathematical conditions for identification to our previous examples and terminology:

(a)   If rank $(\pi^{21}) < M_1 - 1$ or rank $\begin{bmatrix} \gamma^{22} \\ \beta^{22} \end{bmatrix} < M-1$, the equation in question is *unidentified*.

(b)   If rank $(\pi^{21}) = M_1 - 1$ *and* $K_2 = M_1 - 1$, the equation in question is *just identified*.

(c)   If rank $(\pi^{21}) = M_1 - 1$ *and* $K_2 > M_1 - 1$, the equation in question is *overidentified*.

Case (a) can arise because $K_2 < M_1 - 1$ or because of a dependency relationship such that the rank condition fails. In either situation, there is no way of obtaining meaningful estimates of the structural parameters. The remainder of this chapter focuses on estimation when the structural parameters are identified.

In the just identified case, the restrictions on the structure are needed in order to use the observational data to estimate the structural parameters.

[9]The proof of this equivalence is given in Kmenta (1971) and Wonnacott and Wonnacott (1970). It essentially relies on the fact that $\gamma^{22}F_1 = 0$ and $\beta^{22}F_1 = 0$, where F_1 indicates the second through the Mth elements in the first column of F.

Thus, if a restriction is needed for identification, the validity of the restriction cannot be tested with the data. Overidentified restrictions can, however, be tested. For example, if by assuming that $\beta_{k1}=0$, the first equation is over-identified, we can test this restriction by including X_k (but not all other excluded variables) in the estimation of the first equation. If the equation is exactly identified by assuming $\beta_{k1}=0$ this assumption cannot be tested.

An Identification Problem

Consider the system

$$-Y_1+\gamma_{21}Y_2+\gamma_{31}Y_3 \qquad +\beta_{11}X_1+\beta_{21}X_2+\beta_{31}X_3+\beta_{41}X_4 \qquad =U_1,$$
$$\tag{9.21}$$

$$\gamma_{12}Y_1- \quad Y_2+\gamma_{32}Y_3+\gamma_{42}Y_4+\beta_{12}X_1 \qquad \beta_{32}X_3 \qquad =U_2,$$
$$\tag{9.22}$$

$$\gamma_{13}Y_1+\gamma_{23}Y_2- \quad Y_3 \quad +\beta_{13}X_1 \qquad +\beta_{43}X_4 \qquad =U_3,$$
$$\tag{9.23}$$

$$\gamma_{34}Y_3- \quad Y_4+\beta_{14}X_1 \qquad +\beta_{44}X_4+\beta_{54}X_5=U_4.$$
$$\tag{9.24}$$

Equation (9.21) has one excluded exogenous variable ($K_2=1$) and three included endogenous variables. Thus, $K_2=1<M_1-1=2$, and Eq. (9.21) is unidentified. For Eq. (9.22), $K_2=3=M_1-1=4-1$, and the equation is just identified. Equations (9.23) and (9.24) illustrate the overidentified case since $K_2=3>M_1-1=2$ and $K_2=2>M_1-1=1$, respectively. These latter three cases are the subject of the following sections. (A useful exercise is to show that Eqs. (9.22)–(9.24) also satisfy the rank conditions.[10])

Identification of the Aspiration Model

Ascertaining whether the aspiration model specified in Eqs. (9.1)–(9.6) is identified illustrates many of the points made in this discussion of the identification problem. Because of the symmetry of the model, we need only discuss Eqs. (9.1), (9.3), and (9.5). Any conclusions we reach about these equations will hold for the other three. From the discussion in the previous chapter, it should be clear that so long as the exogenous intelligence and socioeconomic status variables (I and S) are independent of the error terms in the model, the equation for parental aspirations (P_1) is identified *and* can be estimated with OLS. We can also see immediately that the equation for educational aspirations (E_1) is unidentified. This equation has four included endogenous variables (P_1, O_1, E_1, and E_2) and only two excluded endogenous variables (I_2 and S_2). Thus $K_2=2<M_1-1=3$, and the necessary condition

[10]Note that equations in systems that are identities do not affect identification. Identities can be eliminated from consideration of identification.

for identification is not fulfilled. What about the equation for occupational aspirations (O_1)? It satisfies the necessary condition since $K_2 = 2 = M_1 - 1 = 2$. However, the same exogenous variables included in the parental aspiration equation are included in this equation. This illustrates an equation that satisfies the necessary but not the sufficient condition and is thus unidentified. To see this, we form the matrix corresponding to the rank conditions

$$\text{rank}\begin{bmatrix} \gamma^{22} \\ \beta^{22} \end{bmatrix} = \text{rank}\begin{bmatrix} \gamma_{24} & 0 & \gamma_{26} & 0 & -1 \\ 0 & -1 & \gamma_{56} & 0 & 0 \\ 0 & \gamma_{65} & -1 & 0 & 0 \\ \beta_{34} & 0 & \beta_{36} & 0 & \beta_{32} \\ \beta_{44} & 0 & \beta_{46} & 0 & \beta_{42} \end{bmatrix} < 5.$$

Columns correspond to Eqs. (9.4), (9.5), (9.6), (9.1), and (9.2) respectively, while rows correspond to the excluded endogenous variables (P_2, E_1, E_2) and the excluded exogenous variables (I_2, S_2) in Eq. (9.3). Note that the rank is less than 5 since one column is identically equal to zero. Thus, the equation is unidentified.

We can, however, identify γ_{43}, γ_{34}, γ_{65}, γ_{56}, γ_{35}, and γ_{46}. Thus we can estimate the magnitude of the peer influences, the major objective of the investigation. To do this, we need to rewrite the equations for occupational and educational aspirations, substituting the expressions for P_1 and P_2 in Eqs. (9.1) and (9.2) for P_1 and P_2 in the subsequent equations:

$$O_1 = \quad \gamma_{43}O_2 \qquad\qquad + (\beta_{13} + \gamma_{13}\beta_{11})I_1 + (\beta_{23} + \gamma_{13}\beta_{21})S_1 + U_3 + \gamma_{13}U_1,$$

$$O_2 = \gamma_{34}O_1 \qquad\qquad + (\beta_{34} + \gamma_{24}\beta_{32})I_2 + (\beta_{44} + \gamma_{24}\beta_{42})S_2 + U_4 + \gamma_{24}U_2,$$

$$E_1 = \gamma_{35}O_1 \qquad + \gamma_{65}E_2 + (\beta_{15} + \gamma_{15}\beta_{11})I_1 + (\beta_{25} + \gamma_{15}\beta_{21})S_1 + U_5 + \gamma_{15}U_1,$$

$$E_2 = \quad \gamma_{46}O_2 + \gamma_{56}E_1 \qquad + (\beta_{36} + \gamma_{26}\beta_{32})I_2 + (\beta_{46} + \gamma_{26}\beta_{42})S_2 + U_6 + \gamma_{26}U_2,$$

or

$$O_1 = \quad \gamma_{43}O_2 \qquad\qquad + \beta_{13}^*I_1 + \beta_{23}^*S_1 \qquad\qquad + U_3^*, \quad (9.3a)$$

$$O_2 = \gamma_{34}O_1 \qquad\qquad + \beta_{34}^*I_2 + \beta_{44}^*S_2 + U_4^*, \quad (9.4a)$$

$$E_1 = \gamma_{35}O_1 \qquad + \gamma_{65}E_2 + \beta_{15}^*I_1 + \beta_{25}^*S_1 \qquad\qquad + U_5^*, \quad (9.5a)$$

$$E_2 = \quad \gamma_{46}O_2 + \gamma_{56}E_1 \qquad + \beta_{36}^*I_2 + \beta_{46}^*S_2 + U_6^*. \quad (9.6a)$$

The equations for O_1 and E_1 (and O_2 and E_2) now satisfy both the necessary and sufficient conditions for identification, O_1 being overidentified and E_1 being exactly identified. The only structural coefficients not identified are the separate influences of parental aspirations and intelligence and socioeconomic status. These influences are aggregated in the coefficients for the exogenous variables, β_{13}^*, β_{23}^*, β_{34}^*, and so on. (We shall discuss a method for identifying these structural coefficents at the end of this chapter.)

9.4   Estimating Identified Models

Exact Identification

When the equation in question is just identified, all consistent simultaneous equation estimation techniques (but not ordinary least squares) are equivalent. Thus, it is not necessary to dwell upon the estimation method. Perhaps the easiest to see conceptually is the method of indirect least squares. If the reduced form parameters are estimated by ordinary least squares, the structural coefficients can be estimated by taking the appropriate combination of the reduced form coefficients. Since there is only one possible combination, there are no questions about how this should be done.

Ordinary least squares estimators of the reduced form coefficient are unbiased and consistent since the exogenous variables are independent of the reduced form error terms. The estimates of the structural coefficients will *not* however be unbiased since the transformation from reduced form to structural estimates does not preserve unbiasedness. The structural estimates will be consistent. This is not true of OLS estimates of the structural parameters since the explanatory variables and structural errors are not independent, even in large samples.

We illustrate indirect least squares by considering Eq. (9.22). In this case X_2, X_4, and X_5 are omitted ($K_2 = 3$) and Y_1 to Y_4 are all included ($M_1 = 4$). Equation (9.18) is

$$\begin{bmatrix} \pi_{21} & \pi_{22} & \pi_{23} & \pi_{24} \\ \pi_{41} & \pi_{42} & \pi_{43} & \pi_{44} \\ \pi_{51} & \pi_{52} & \pi_{53} & \pi_{54} \end{bmatrix} \begin{bmatrix} \gamma_{12} \\ -1 \\ \gamma_{32} \\ \gamma_{42} \end{bmatrix} = \begin{bmatrix} 0 \\ 0 \\ 0 \end{bmatrix} \qquad (9.25)$$

or

$$\pi_{21}\gamma_{12} + \pi_{23}\gamma_{32} + \pi_{24}\gamma_{42} = \pi_{22},$$

$$\pi_{41}\gamma_{12} + \pi_{43}\gamma_{32} + \pi_{44}\gamma_{42} = \pi_{42},$$

$$\pi_{51}\gamma_{12} + \pi_{53}\gamma_{32} + \pi_{54}\gamma_{42} = \pi_{52}.$$

Now, as long as the matrix

$$\begin{bmatrix} \pi_{21} & \pi_{23} & \pi_{24} \\ \pi_{41} & \pi_{43} & \pi_{44} \\ \pi_{51} & \pi_{53} & \pi_{54} \end{bmatrix}$$

is nonsingular, the estimated structural coefficients can be obtained from

$$\begin{bmatrix} \gamma_{12} \\ \gamma_{32} \\ \gamma_{42} \end{bmatrix} = \begin{bmatrix} \pi_{21} & \pi_{23} & \pi_{24} \\ \pi_{41} & \pi_{43} & \pi_{44} \\ \pi_{51} & \pi_{53} & \pi_{54} \end{bmatrix}^{-1} \begin{bmatrix} \pi_{22} \\ \pi_{42} \\ \pi_{52} \end{bmatrix}$$

by using the OLS estimates of the reduced form parameters. Once we have γ^{11} (γ_{12}, γ_{32}, and γ_{42}) estimated, we can calculate b_{12} and b_{32} from Eq. (9.17).

With Π estimated by OLS, this is the indirect least squares estimator of the structural coefficients.

The equivalence between indirect least squares and other methods in this situation implies that other techniques might be more convenient in actual application. For example, two-stage least squares (discussed below) is frequently available in packaged computer programs, thus making it computationally advantageous over indirect least squares. Further, while the asymptotic distributions of the estimated coefficients are difficult to derive from indirect least squares, such distributions are readily available from other techniques.

The more interesting issue is how to handle the overidentified case. With overidentification there are alternative ways to estimate the structural parameters, and, thus, alternative estimators have been developed for application to different situations.

Overidentification and Two-Stage Least Squares (TSLS)

The overidentified situation is where there are more excluded exogenous variables than included explanatory endogenous variables, $K_2 > M_1 - 1$. This gives the same result as the overidentified hierarchical model. There can be more than one estimate for a structural coefficient due to the deviations of the estimated reduced form coefficients about their true values. We can illustrate this by considering Eq. (9.23), the structural equation for Y_3. In this case X_1 and X_4 are the only included exogenous variables, while Y_4 is an excluded endogenous variable, so that Eq. (9.18) is

$$\begin{bmatrix} \pi_{21} & \pi_{22} & \pi_{23} \\ \pi_{31} & \pi_{32} & \pi_{33} \\ \pi_{51} & \pi_{52} & \pi_{53} \end{bmatrix} \begin{bmatrix} \gamma_{13} \\ \gamma_{23} \\ -1 \end{bmatrix} = \begin{bmatrix} 0 \\ 0 \\ 0 \end{bmatrix} \quad \text{or} \quad \begin{matrix} \pi_{21}\gamma_{13} + \pi_{22}\gamma_{23} = \pi_{23}, \\ \pi_{31}\gamma_{13} + \pi_{32}\gamma_{23} = \pi_{33}, \\ \pi_{51}\gamma_{13} + \pi_{52}\gamma_{23} = \pi_{53}. \end{matrix}$$

Substituting the OLS estimates of the reduced form parameters could give three different estimates of the structural coefficients γ_{13} and γ_{23} depending upon which pair of equations we use.

Just as in the hierarchical case, we need to find a method for combining the information contained in all possible estimates for each coefficient. Again this is something more complicated than a simple average of the different estimates. There are many procedures available at this point,[11] but the most common one is equivalent to the instrumental variable procedure discussed in Chapter 8.

The instruments for each included endogenous variable are obtained from the reduced form equations. Each included endogenous variable is regressed against the complete set of exogenous variables, and the predicted values of Y

---

[11]For an extended discussion of other procedures see Goldberger (1964), Johnston (1972), or Kmenta (1971).

from those equations are subsequently used to estimate the structural equation—hence the name two-stage least squares. This procedure can be accomplished with any OLS program that calculates and stores $\hat{Y}$ or the residuals.

The computations actually do not need to be done in two steps however. Let b be the vector of estimated coefficients for included exogenous variables and c be the vector of estimated coefficients on included endogenous variables. A computationally more efficient way to compute the two-stage least squares estimated coefficients for a single equation is

$$
\begin{bmatrix} b \\ c \end{bmatrix} = \begin{bmatrix} X_1'X_1 & X_1'Y_1 \\ Y_1'X_1 & Y_1'X(X'X)^{-1}X'Y_1 \end{bmatrix}^{-1} \begin{bmatrix} X_1'Y_m \\ Y_1'X(X'X)^{-1}X'Y_m \end{bmatrix}, \quad (9.26)
$$

where X_1 and Y_1 are matrices of observations on included exogenous and endogenous variables, respectively, X is the matrix of observations on all exogenous variables, and Y_m is the vector of observations on the endogenous variable being "explained," whose coefficient we have set to -1, $\gamma_{mm} = -1$.

The two-stage least squares estimator in Eq. (9.26) is simply the instrumental variables estimator discussed in Chapter 8 and presented in the appendix to that chapter (also see Goldberger, 1964, pp. 331–333). Consequently, these estimates are consistent, in that as the sample size becomes very (infinitely) large, the distribution of estimated coefficients collapses about the true value.

The equivalence between instrumental variables and two-stage least squares means the estimated asymptotic variance-covariance matrix of the estimated coefficients is Eq. (8.1c) from the appendix to Chapter 8. Note that this expression estimates the variance obtained from a large sample about the population value of the estimates. The population parameter is not necessarily the mean for the given sample, but the population value that is approached as the sample size becomes infinitely large. The expression for the coefficients' estimated variance-covariance matrix in the two-stage least squares model, denoted as S_{bc}, is

$$
S_{bc} = s_m^2 \begin{bmatrix} X_1'X_1 & X_1'Y_1 \\ Y_1'X_1 & Y_1'X(X'X)^{-1}X'Y_1 \end{bmatrix}^{-1}, \quad (9.27)
$$

where s_m^2 is the estimated variance of the error term in the structural equation, denoted as U_m.

The estimate s_m^2 is computed from the residuals in the same fashion as it is in the ordinary least squares case. In this case the residual $e_{t,m}$ is

$$
e_{t,m} = Y_{t,m} - \sum_{k=1}^{K_1} X_{t,k} b_k - \sum_{j=1}^{M_1} Y_{t,j} c_j.
$$

Note that these residuals are computed using the observed values of the explanatory variables, not the calculated values used to estimate the equation.[12] The estimated error variance is

$$s_m^2 = \frac{1}{T - (K_1 + M_1)} e'e = \frac{1}{T - K_1 - M_1}(Y_m - X_1 b - Y_1 c)'(Y_m - X_1 b - Y_1 c).$$

$$(9.28)$$

The expression for the sum of squared residuals $e'e$ can also be used to calculate a statistic that is analogous to the R^2 value calculated for the ordinary least squares model. This expression, which is often also referred to as an R^2, is simply $1 - (e'e/\Sigma(Y_m - \bar{Y}_m)^2)$. However, due to the method of calculating the structural coefficients, there is no guarantee that the TSLS R^2 value has a lower bound of zero.

9.5 Simultaneous Equations: The Voting and Aspiration Examples

Estimation of the voting and aspiration models illustrates the treatment of simultaneous systems and the application of two-stage least squares. We begin with the aspiration example because the model is more straightforward and the discussion of its identification is already completed. We then discuss the estimation of the complete voting model, which includes equations for voters' positions and evaluations on each issue, as well as their party affiliations.

The parameters shown in Eqs. (9.3a)–(9.6a) can be appropriately estimated with two-stage least squares since we have shown those more limited equations to be identified. There are four exogenous variables, the two intelligence and two status variables, which are used to create the instruments for the endogenous variables included in each equation. Table 9.1 shows the TSLS and OLS estimates for these four equations.

The most notable difference in the two estimation methods is the much larger estimates of peer influences with two-stage least squares. The TSLS estimates of these coefficients are close to 40% and 65% higher than the OLS estimates for the occupation and education equations, respectively. Intuitively one might expect the TSLS estimates to be lower than the OLS estimates because of the positive correlation between error terms and explanatory variables introduced by the simultaneous relationship between respondents' and peers' aspirations. However, the next chapter presents a plausible explanation why that correlation might be negative.

The structure of the voting model shown in Fig. 9.1 is more complicated

---

[12]Computationally, TSLS can be accomplished by using two passes of an OLS regression program. After the first stage estimation of the reduced form for each of the included endogenous variables, the predicted Y's can be printed and subsequently read back into the OLS program for a second estimation. This section shows, however, that even though the parameter estimates are correct, the R^2, standard error of estimate, and estimated standard errors of the coefficients will be incorrect.

TABLE 9.1

Estimates of Limited Occupational and Educational Aspiration Equations[a]

		Respondent			Friend	
		TSLS	OLS		TSLS	OLS
Parental aspirations						
Intelligence	(β_{11})		0.11	(β_{32})		0.14
			(0.03)			(0.03)
Status	(β_{21})		−0.01	(β_{42})		−0.08
			(0.03)			(0.03)
Occupational aspirations						
Intelligence	(β_{13}^{*})	0.643	0.698	(β_{34}^{*})	0.819	0.870
		(0.124)	(0.115)		(0.129)	(0.112)
Status	(β_{23}^{*})	0.348	0.408	(β_{44}^{*})	0.348	0.394
		(0.124)	(0.113)		(0.122)	(0.107)
Occupational Aspirations—Peer	(γ_{43})	0.409	0.296	(γ_{34})	0.350	0.257
		(0.104)	(0.050)		(0.124)	(0.048)
Educational aspirations						
Intelligence	(β_{15}^{*})	0.050	0.032	(β_{36}^{*})	0.036	0.050
		(0.039)	(0.011)		(0.067)	(0.010)
Status	(β_{25}^{*})	0.053	0.043	(β_{46}^{*})	0.027	0.035
		(0.023)	(0.010)		(0.033)	(0.010)
Occupational Aspirations	(γ_{35})	0.020	0.046	(γ_{46})	0.051	0.042
		(0.060)	(0.005)		(0.079)	(0.005)
Educational Aspirations—Peer	(γ_{65})	0.197	0.120	(γ_{56})	0.197	0.121
		(0.264)	(0.046)		(0.270)	(0.041)

[a] Estimated standard errors appear in parentheses.

than that of the aspiration model. Much of this complexity results from one of the model's objectives, which is to compare two alternative theories of voting behavior. One theory holds that party identifications are central to the voting process and are primarily determined by forces exogenous to elections. According to this theory, party affiliations are the primary determinant of voters' stated positions on different issues and of their evaluations of the parties, as well as of actual vote intentions. The alternative explanation holds that people's issue positions are cental and that these determine evaluations that in turn dominate the party affiliation and voting decisions. The structure outlined in Fig. 9.1 includes explanatory variables to represent both theories. The entire structure, once both approaches are considered, is clearly simultaneous. Party identifications and evaluations directly influence each other, as well as being simultaneously related through the issue position variables. The complete model includes separate issue position and evaluation equations for each of the seven issues included in the model. However, for the sake of brevity, we shall discuss only the party identification equation.

The party affiliation equation is estimated both with ordinary least squares and with two-stage least squares to indicate the differences in the estimates

TABLE 9.2
Party Choice Equation

Variable	TSLS	OLS	Variable	TSLS	OLS
Evaluation	0.771	0.440	Age > 54	−0.045	−0.045
	(0.109)	(0.025)		(0.020)	(0.018)
South	0.085	0.104	Income	−0.004	−0.007
	(0.024)	(0.022)		(0.003)	(0.002)
East	−0.074	−0.061	Catholic	0.075	0.079
	(0.024)	(0.022)		(0.023)	(0.022)
Central City	0.048	0.068	Jew	0.010	0.080
	(0.030)	(0.027)		(0.065)	(0.056)
Rural	0.020	0.029	Union	0.076	0.113
	(0.024)	(0.022)		(0.026)	(0.021)
Father Democrat	0.265	0.299	Constant	−0.021	0.161
	(0.025)	(0.021)		(0.067)	(0.031)
Father independent	0.190	0.221	R^2	0.36	0.45
	(0.030)	(0.027)			

provided by each method. These estimates are shown in Table 9.2. There are considerable differences in the size of the evaluation coefficient between the two techniques. The TSLS estimate of the influence of evaluations on party affiliations is much greater than the OLS estimate. This result is again counterintuitive, in that if evaluations and party affiliations are simultaneously related in the manner hypothesized, we should expect that OLS would overestimate the true effect and that the TSLS coefficient would be lower. The likely explanations for this result rest with the difficulties in accurately measuring issue positions and evaluations and are similar to explanations for the TSLS and OLS differences both in the aspiration model and in the voting equation in Chapter 8.

The restructuring of the voting model into a nonrecursive multiequation model and the use of two-stage least squares (instrumental variables) has produced a substantially different picture than one gets from the single equation model estimated with ordinary least squares (see the path analysis in Chapter 8). Both the instrumental variables estimates of the voting equation and the equivalent TSLS estimation of the party identification equation imply that issues and people's evaluations of the parties' positions, relative to their own positions, are important determinants of both party affiliations and voting behavior. These results illustrate how substantive conclusions can be influenced by the choice of statistical technique.

9.6   Identification through Assumptions about Error Terms

We have not yet completed the discussion of ways to identify the coefficients in structural equations. So far we have used only the prior information known about coefficients in the model, namely, which ones are believed to be zero. This does not exhaust the possible ways to identify coefficients, although it is surely the most commonly used method. If certain conditions are met, it is possible to use the covariances of the error terms in the reduced form equations to aid in the identification process. We shall briefly point out how this can be done, show its application to the unidentified equation (9.21) and to the remaining coefficients in the aspiration model, and discuss why this is not a common procedure.

In Eq. (9.15) the error terms in the reduced form equation are written as $V = - U\Gamma^{-1}$. Assuming that $(1/T)E(U_i'U_j) = \sigma_{ij}$ for each pair i,j of structural equations means that we can write the expected variance–covariance matrix for the error terms from these reduced form equations as

$$\Sigma_V = E\left[\frac{1}{T}(Y'Y - \Pi'X'X\Pi)\right]$$

$$= E\left[\frac{1}{T}(V'V)\right] = \Gamma^{-1\prime}E\left(\frac{1}{T}U'U\right)\Gamma^{-1} = \Gamma^{-1\prime}\Sigma_U\Gamma^{-1}.$$

Because the ordinary least squares estimates of the reduced form equations are consistent and unbiased (so long as we can satisfy the requirement that the U's are uncorrelated with the exogenous variables), we shall have consistent estimates of the reduced form equation errors and can use them to estimate Σ_V from our observational information. If we are able to specify any of the elements of Σ_U, then we can use this information along with the estimates for the elements of Σ_V to identify some of the coefficients in Γ. We do this by rewriting the above expression as $\Gamma'\Sigma_V\Gamma = \Sigma_U$ and then trying to solve the individual equations from this expression for unknown values of Γ.

The most common prior information one might have is an argument that certain pairs of structural error terms are uncorrelated. This means that specific off-diagonal entries of Σ_U are zero, and that we can relate given expressions involving Γ and Σ_V to these zero values.

It should be clear at this point that even if we are to assume that all pairs of error terms are uncorrelated, it will not be sufficient to identify a complete set of structural coefficients on the endogenous variables. If there are M structural equations and M endogenous variables, there will be $M(M-1)/2$ pairs of error terms and thus at most $M(M-1)/2$ equations to set equal to a known value.[13] However there are a possible $M(M-1)$ structural coefficients to identify. Some of these may be specified to be zero by exclusions of certain exogenous variables from the structural equation. This information is summarized in Eq. (9.18), $\pi^{21}\gamma^{11} = 0$, for each structural equation. The object

---

[13]Since the variance–covariance matrix is symmetric, we consider only half of it. $M(M-1)/2$ is merely the number of off-diagonal elements in the top of an $M \times M$ matrix. We can now see why the restriction that Σ_U is diagonal is sufficient to identify the hierarchical model. In that model, Γ is triangular, meaning all elements below the main diagonal are zero, so there are at most $M(M-1)/2$ unknown coefficients in Γ (after setting the M elements on the main diagonal equal to -1).

now is to combine these two sources of information, $\pi^{21}\gamma^{11} = 0$ and $\Gamma'\Sigma_V\Gamma = \Sigma_U$, to identify as many of the structural coefficients as possible.

If one or more of the structural equations is already identified, it means that those columns of Γ are "known," and that any expressions for Σ_U involving that column may be helpful in identifying some other structural coefficients. If we denote one of the elements of Σ_U by σ_{ij}, meaning the covariance between the errors in the structural equations for Y_i and Y_j, then

$$\sigma_{ij} = \gamma_i'\Sigma_V\gamma_j \qquad (9.29)$$

where γ_i and γ_j are the i and j columns of Γ. If the elements of γ_i are already identified and we can assume $\sigma_{ij} = 0$, Eq. (9.29) becomes a linear expression in terms of the unknown γ_j. It may be possible to combine this information with other expressions involving γ_j from the derivation of the reduced form coefficients [Eq. (9.18)] to identify γ_j. In cases where neither γ_i nor γ_j are identified, we are dealing with a series of nonlinear simultaneous equations; this becomes more difficult and is the subject of the next chapter.

There is an important difference between identification with information about the error structure and identification from information only about the reduced form parameter matrices. With identification through the error term, identification is no longer an issue of a single equation but now explicitly involves more than one structural equation, because we must know something about γ_i in order to identify γ_j. This dependence upon other structural equations may also present estimation difficulties. If γ_i is badly estimated for any reason, such as the equation "explaining" Y_i being misspecified, γ_j consequently will also be badly estimated.

We shall illustrate how this procedure might operate in practice by going back to the underidentified equation in the first illustration of Eqs. (9.21)–(9.24). In this case we had four structural equations, of which the first was unidentified. In this case we had only one excluded exogenous variable and two included endogenous variables. Equation (9.18) yields one expression, $\pi_{52}\gamma_{21} + \pi_{53}\gamma_{31} = \pi_{51}$, which does not provide unique estimates for γ_{21} and γ_{31}. However, information about the error structure, such as the covariance of U_1 and U_2 being zero ($\sigma_{12} = 0$), can be used to improve the situation. Because the equation for Y_2 is identified, we can write Eq. (9.29) as

$$\gamma_2'\Sigma_V\gamma_1 = Q_2\gamma_1 = \begin{bmatrix} q_{21} & q_{22} & q_{23} \end{bmatrix} \begin{bmatrix} -1 \\ \gamma_{21} \\ \gamma_{31} \end{bmatrix} = \sigma_{12} = 0,$$

where $Q_2 = \gamma_2'\Sigma_V$. (We have already included the restriction that $\gamma_{41} = 0$.) We can compute the elements of Q because we were previously able to identify values for $\gamma_2' = [\gamma_{12} - 1 \quad \gamma_{32} \quad \gamma_{42}]$. This now gives us two linear expressions in the two unknowns γ_{21} and γ_{31},

$$\pi_{52}\gamma_{21} + \pi_{53}\gamma_{31} = \pi_{51} \qquad \text{[from Eq. (9.18)]},$$

$$q_{22}\gamma_{21} + q_{23}\gamma_{31} = q_{21} \qquad \text{[from Eq. (9.29)]}.$$

We can solve this set of equations for γ_{21} and γ_{31}. This then makes our system completely identifiable, as long as $\sigma_{12} = 0$.

Alternatively, had we believed that $\sigma_{13} = 0$, we could use the equations involving γ_{13} and γ_{23} and Σ_V to get the second identifying equation. If both covariances equal zero, we again have an overidentified system. (If we also assume that $\sigma_{23} = 0$, then all the equations are overidentified.) It should be possible to use the alternative estimates of each coefficient, obtained from each set of overidentified restrictions, to improve our estimates. This is analogous to the way two-stage least squares effectively weights each estimate of the structural coefficients obtained from restrictions on the coefficients to obtain a single estimate of each coefficient. However, techniques to incorporate overidentifying information about the structure of the variance and covariance of the error terms are quite complicated and, except for the general discussion in the next chapter, beyond the scope of this book.

The difficulty with relying on assumptions about the error term covariances to identify the model is discussed in the context of recursive and nonrecursive hierarchical models in Chapter 8. In most situations it is difficult to justify the assumption that the error terms are uncorrelated. We pointed out in Chapter 8 that if any of the excluded factors comprising the error terms are common to both equations or if there are similar difficulties in measuring the endogenous variables being explained by each equation, then the error terms will not be independent. Either or both circumstances would seem to be present in most social science models, making it hazardous to count on this approach to solve the identification problem.

The assumption of uncorrelated structural error terms may be plausible in the model of peer influences on aspirations, however, and it will permit us to identify the remaining coefficients in Eqs. (9.1)–(9.6). We may assume that the error terms in the equation for parents' aspirations for each respondent are uncorrelated with the error terms in the peers' occupational and educational aspiration equations, $\sigma_{14} = \sigma_{16} = 0$. From the symmetric structure of the peers' model we also assume that $\sigma_{23} = \sigma_{25} = 0$.

These assumptions are justified on the grounds that the omitted systematic influences on the respondent's (peer's) parents' aspirations, P_1 (P_2), are unlikely to affect the peers' (respondents') educational and occupational aspirations systematically. These omitted influences in the equation for P_1 (P_2) are possible influences in the respondents' (peers') aspirations, so it is difficult to argue that σ_{13}, σ_{15}, σ_{24}, and σ_{26} also equal zero.

This specification that $\sigma_{14} = \sigma_{16} = \sigma_{23} = \sigma_{25} = 0$ is sufficient to permit identification and estimation of the remaining structural coefficients from the covariances of the residuals from the reduced form estimations of the parental aspiration and occupational and educational aspiration equations. To see how this is done, note that from Eq. (9.12) we can partition the matrix of coefficients on the endogenous variables as

$$\Gamma = \begin{bmatrix} -I & \vdots & \gamma^{12} \\ - & \top & - \\ 0 & \vdots & \gamma^{22} \end{bmatrix} \quad \text{where} \quad \gamma^{12} = \begin{bmatrix} \gamma_{13} & 0 & \gamma_{15} & 0 \\ 0 & \gamma_{24} & 0 & \gamma_{26} \end{bmatrix}.$$

We have already estimated γ^{22} and are trying to estimate γ^{12}. We can partition Σ_U as

$$\Sigma_U = \left[\begin{array}{c|c} \Sigma^{11} & \Sigma^{12} \\ \hline \Sigma^{12'} & \Sigma^{22} \end{array} \right],$$

where with the assumption that $\sigma_{14} = \sigma_{16} = \sigma_{23} = \sigma_{25} = 0$,

$$\left[\begin{array}{c|c} \Sigma^{11} & \Sigma^{12} \end{array} \right] = \left[\begin{array}{cc|cccc} \sigma_{11} & \sigma_{12} & \sigma_{13} & 0 & \sigma_{15} & 0 \\ \sigma_{12} & \sigma_{22} & 0 & \sigma_{24} & 0 & \sigma_{26} \end{array} \right].$$

From the relationship between reduced form and structural errors, $\Gamma'\Sigma_V\Gamma = \Sigma_U$, we arrive, in partitioned form, at

$$\left[\begin{array}{c|c} -I & 0 \\ \hline \gamma^{12'} & \gamma^{22'} \end{array} \right] \left[\begin{array}{c|c} \omega^{11} & \omega^{12} \\ \hline \omega^{21'} & \omega^{22} \end{array} \right] \left[\begin{array}{c|c} -I & \gamma^{12} \\ \hline 0 & \gamma^{22} \end{array} \right] = \left[\begin{array}{c|c} \Sigma^{11} & \Sigma^{12} \\ \hline \Sigma^{12'} & \Sigma^{22} \end{array} \right], \tag{9.29}$$

where ω^{ij} are the conformably partitioned elements of Σ_V. Multiplication of this expression gives the needed relationships

$$\left[\begin{array}{c|c} \omega^{11} & -\omega^{11}\gamma^{12} - \omega^{12}\gamma^{22} \end{array} \right] = \left[\begin{array}{c|c} \Sigma^{11} & \Sigma^{12} \end{array} \right].$$

For estimation, define S as the observed reduced form error matrix using the estimated reduced form coefficients: $S = (1/T)[Y'Y - p'X'Xp]$. We can estimate ω^{11} by S^{11} and ω^{12} by S^{12}; γ^{22} is composed of previously identified parameters whose estimates are denoted as c_{ij}. Substituting estimated values into Eq. (9.29)., where appropriate, we can relate the unknown parameters of γ^{12} to estimated sample values:

$$-\left[\begin{array}{cc} s_{11} & s_{12} \\ s_{12} & s_{22} \end{array} \right] \left[\begin{array}{cccc} \gamma_{13} & 0 & \gamma_{15} & 0 \\ 0 & \gamma_{24} & 0 & \gamma_{26} \end{array} \right]$$

$$-\left[\begin{array}{cccc} s_{13} & s_{14} & s_{15} & s_{16} \\ s_{23} & s_{24} & s_{25} & s_{26} \end{array} \right] \left[\begin{array}{cccc} -1 & c_{34} & c_{35} & 0 \\ c_{43} & -1 & 0 & c_{46} \\ 0 & 0 & -1 & c_{56} \\ 0 & 0 & c_{65} & -1 \end{array} \right]$$

$$= \left[\begin{array}{cccc} \sigma_{13} & 0 & \sigma_{15} & 0 \\ 0 & \sigma_{24} & 0 & \sigma_{26} \end{array} \right].$$

This yields estimators of γ^{12}:

$$c_{24} = \frac{s_{14} - c_{34}s_{13}}{s_{12}}, \qquad c_{26} = \frac{s_{16} - s_{15}c_{56} - s_{14}c_{46}}{s_{12}},$$

$$c_{13} = \frac{s_{23} - s_{24}c_{43}}{s_{12}}, \qquad c_{15} = \frac{s_{25} - s_{26}c_{65} - s_{23}c_{35}}{s_{12}}.$$

With these estimates for γ^{12} and the earlier estimates for $\beta_{13}^*, \beta_{23}^*, \ldots, \beta_{46}^*$ we can estimate $\beta_{13}, \beta_{23}, \ldots, \beta_{46}$. These estimates are compared to the OLS estimates of the same coefficients in Table 9.3. The results of the two

TABLE 9.3
Structural and OLS Estimates of Aspiration Coefficients

	Occupation : Respondent			Occupation : Peer	
	Structural	OLS		Structural	OLS
Parental Aspiration (γ_{13})	−0.34	0.51	(γ_{24})	−0.44	0.49
Intelligence (β_{13})	0.68	0.64	(β_{34})	0.88	0.80
Status (β_{23})	0.35	0.42	(β_{44})	0.31	0.43

	Education : Respondent			Education : Peer	
	Structural	OLS		Structural	OLS
Parental Aspiration (γ_{15})	−0.62	0.51	(γ_{26})	−1.09	0.63
Intelligence (β_{15})	0.56	0.28	(β_{36})	0.51	0.43
Status (β_{25})	0.53	0.43	(β_{46})	0.19	0.39

estimations are substantially different. The OLS estimates suggest that parental aspirations have a substantial positive impact on their offspring's occupational and educational aspirations. The structural estimates predict exactly the opposite, that parental aspirations have a substantial negative influence on a son's aspirations.

One obvious weakness of this approach is its sensitivity to the accuracy of the estimates of the other structural coefficients. Only slight differences in the estimates of γ_{34} and γ_{43} give substantially different estimates of γ_{13} and γ_{24}. For example, if we use the OLS estimates for γ_{34} and γ_{43}, the estimates of parental aspirations influences become 0.08 and 0.19 for respondents and peers, respectively (instead of −0.34 and −0.44). The problem of evaluating structural estimates obtained from analyzing the reduced form residual covariances is compounded because there are no estimated standard errors for these coefficients. Thus we do not know if the estimate $\hat{\gamma}_{13} = -0.34$ has a large standard error, thereby making the estimate unreliable and preventing strong conclusions about the perverse influence of parental aspirations. What we would like, and what is described in more detail in the next chapter, is a method that simultaneously estimates all the structural coefficients.

9.7 Alternative Estimators

The previous discussions have considered exclusively the most popular estimators for simultaneous equations: indirect least squares for the exactly identified case and two-stage least squares for the overidentified case. However, there are other estimators that have been developed for simultaneous equation estimation. This section summarizes some leading alternatives and discusses the choice among them.

Let us remember the development of ordinary least squares. In the course of that development, we noted that there are many alternative estimators, even many alternative estimators that are unbiased. The choice of OLS is then based upon another criterion: efficiency. Among all linear, unbiased estimators (under the assumptions of Chapter 3), OLS estimators have the smallest variance. Similarly, when some of the assumptions were relaxed in Chapter 6, we opted for the generalized least squares estimator on efficiency grounds. Thus, it seems natural to consider relative efficiencies here when choosing between alternative estimators. As one might guess, however, from the complexity of the discussions in the chapter, the choice among simultaneous equation estimators is more complicated.

First, the relevant class of estimators is not that of unbiased estimators but instead the class of consistent estimators. This is important because the only properties that can be proved generally for simultaneous equation estimation are large sample properties—i.e., properties that hold as the sample size gets infinitely large. This is unfortunate on two counts: (1) we often are faced in practice with estimation for quite small samples; and (2) on theoretical grounds we can generally talk only about large sample efficiency—relative asymptotic variances. Thus the choice on the basis of theoretical information is less obvious.[14]

Secondly, other complications also enter the picture. Various estimators are characterized by differing "robustness"; in other words, they tend to hold up more or less well as the model and data deviate from the underlying assumptions. Related to this is the sensitivity of the estimator to specification errors in various parts of the model, or the preciseness of the model specification that is required.

Finally, in the case of simultaneous equations, the issue of computational costs must be introduced. Clearly, the cost of computing has dropped significantly in recent years, and one would expect this cost decline to continue. Nevertheless, some simultaneous equation estimators that require solutions of large nonlinear systems remain quite expensive, and programs for their solution are not readily available.

Within these guidelines, we now offer a cursory summary of leading alternative estimators and then a discussion of what we know about small sample properties.

Limited Information/Full Information

Simultaneous equation estimators are divided into two groups: limited information and full information. Limited information methods are used for estimation of a single equation in simultaneous equation systems and rely upon neither the restrictions imposed in other equations nor the parameter estimates of other equations. Two-stage least squares is a limited information method. Full information methods estimate the entire system simultaneously

---

[14]It is even more complicated than this. Some different estimators have the same asymptotic variance even though their small sample properties are different.

and, thus, incorporate information about specification of all the equations. Chapter 10 deals extensively with a full information model.

Full information methods tend to increase the efficiency of the estimators by virtue of incorporating more information in the estimation process. On the other hand, this asymptotic efficiency gain is not free. These methods are considerably more expensive computationally (although some computer routines for full information methods are becoming quite commonly available and the price for applications continues to fall). Perhaps more importantly, though, full information estimators appear to be more sensitive to specification error than limited information methods.

Small Sample Properties

The small sample distributions of simultaneous equation estimators are generally unavailable. (Small sample distributions have been solved for only very specific cases.) Yet most interest in the properties of estimators centers on the small sample properties, not the large sample properties of consistency and asymptotic variance.

The approach followed has been to do Monte Carlo studies of alternative estimators with small samples. From these, some insights can be gained as to the characteristics of different estimators. A series of studies has involved a similar set of estimation techniques and allows for some generalization in different situations.[15] One useful finding is that the asymptotic properties of these estimators are often quite useful in a small sample context.

The results of these simulations generally support the arguments in favor of full information procedures if *all* the equations are specified correctly. The difficulty is that if one, or a subset, of the equations is misspecified, the estimates for even the correctly specified equations are affected, and the full information procedures lose their advantage. Among the limited information procedures, two-stage least squares estimators generally seem to be the most robust and have the least small sample bias as well as the smallest variance among consistent estimators. Of course OLS estimates usually have the least variance, but they are both inconsistent and biased. This, we believe, accounts for the popularity of TSLS, particularly when combined with its computational ease and efficiency.

9.8   Summary and Conclusions

This chapter extends the analysis of multiequation systems begun in Chapter 8. In this chapter, completely simultaneous (as opposed to hierarchical) models are considered.

---

[15]These studies generally involve at least the following techniques: OLS, TSLS, limited information maximum likelihood, three-stage least squares, and full information maximum likelihood. The last two are full information methods with three-stage least squares being a generalization of TSLS. A summary of Monte Carlo studies can be found in Johnston (1972).

An essential first step in considering simultaneous equations is ascertaining whether or not the structural parameters are identified. This must precede any estimation.

Identification is the introduction of a priori information into the analysis so that structural parameters can be deduced from the observational information. The reduced form, which describes each endogenous variable as a function of only exogenous variables, represents this observational information. Thus, the identification problem is equivalent to considering how much a priori information is needed to establish that a particular structural model, and no other one, generated the reduced form parameters.

Identification can be obtained in a variety of ways. The most common way of identifying an equation is through restrictions on the values of structural parameters. These restrictions are often exclusion restrictions, i.e., a priori statements that certain variables do not appear in a structural relationship. If one considers only exclusion restrictions, a necessary condition for identification of an equation is that the number of excluded exogenous variables must be equal to or greater than the number of included endogenous variables less one. (The rule includes the number of endogenous variables *less one* because the coefficient on one endogenous variable in each equation is set equal to −1.) An alternative way of obtaining identification is through a priori knowledge of the values for individual coefficients or the equality of two or more coefficients.

Also, less commonly, if one can describe parts of the error variance–covariance matrix across equations, such information can be used to aid in identification. This is, for example, how identification is achieved in recursive models. Identification with this last type of information also differs from identification through exclusion restrictions because use of the error variance–covariance matrix explicitly considers identification of more than one equation at a time.

Identification is a logical precondition to estimation. If the model is unidentified, there is no way to estimate the structural coefficients. If a model is just identified (there is exactly the minimum number of restrictions needed for identification), all consistent estimators of the model yield the same results. Therefore, there is no reason to consider alternative estimators; only one, such as indirect least squares, is needed. However, when the model is overidentified (there is more prior information than is needed for identification), there are alternative ways of using the observational information to estimate some of the structural parameters. The variety of estimation methods for simultaneous equations merely represents alternative methods of incorporating the a priori information in an overidentified model into the estimation.

Perhaps the most common estimation technique for simultaneous equations is two-stage least squares. This is equivalent to an instrumental variables estimation where the excluded exogenous variables from a given equation become instruments for the included endogenous variables. The first stage creates instruments, regressing each included (right-hand side) endogenous variable on the set of *all* exogenous variables. The second stage uses these instruments to obtain consistent estimates of the structural coefficients.

There are a number of alternative estimation schemes. Some of these estimate the model's parameters one equation at a time (limited information methods) and some estimate all of the parameters at the same time (full information methods). Full information methods provide estimators with smaller mean square errors when the whole model is correctly specified. However, if parts of the model are incorrectly specified, limited information methods are often superior. The popularity of limited information methods partially reflects such a concern about the accuracy of specification for an entire model and partially reflects differences in computational costs.

APPENDIX 9.1

Variances and Covariances for Peer Influence Data [a]

	A_1	A_2	O_1	O_2	E_1	E_2	I_1	S_1	I_2	S_2
A_1	10.210									
A_2	1.129	9.449								
O_1	8.636	2.951	159.542							
O_2	3.379	7.702	67.110	158.533						
E_1	10.890	2.678	98.250	51.400	154.878					
E_2	3.254	10.360	50.080	97.710	55.380	147.016				
I_1	3.135	1.672	27.690	17.460	26.810	16.810	28.442			
S_1	0.855	1.562	22.360	19.190	27.530	18.480	6.468	29.841		
I_2	1.341	3.457	20.390	33.940	19.150	33.860	9.642	6.761	28.950	
S_2	0.340	−0.758	20.740	25.470	16.800	27.920	5.558	8.295	8.894	31.395

[a]Data from Duncan *et al.* (1968) with variances very kindly provided by Professor Haller and Stephen Miller. $N = 329$.

REVIEW QUESTIONS

1. Consider the following system (again all variables are in deviations about their means and the constant term is omitted):

$$- Y_1 \qquad\qquad + \gamma_{41} Y_4 + \beta_{21} X_2 \qquad\qquad\qquad = U_1,$$

$$\gamma_{12} Y_1 - \quad Y_2 + \gamma_{32} Y_3 \qquad\qquad + \beta_{32} X_3 + \beta_{42} X_4 \qquad\qquad = U_2,$$

$$\gamma_{13} Y_1 + \gamma_{23} Y_2 - \quad Y_3 \qquad\qquad + \beta_{33} X_3 + \beta_{43} X_4 \qquad\qquad = U_3,$$

$$\gamma_{24} Y_2 + \gamma_{34} Y_3 - \quad Y_4 + \beta_{24} X_2 \qquad\qquad\qquad + \beta_{54} X_5 + \beta_{64} X_6 = U_4.$$

(a) Write this system in matrix form, showing the complete coefficient matrices.
(b) Which coefficients are under-, exactly, and overidentified if all error terms are correlated?

(c) From the purely computational standpoint of getting estimates for the coefficients, will TSLS fail if used on the second and third equations? If not, are the estimates interpretable? Why not?

2. Given the following simultaneous equation system and reduced form estimates, estimate the structural coefficients.

Structural:

$$-Y_1 \qquad +\gamma_{31}Y_3 + \beta_{21}X_2 \qquad +\beta_{41}X_4 = U_1,$$

$$-\quad Y_2 + \gamma_{32}Y_3 \qquad +\beta_{32}X_3 \qquad = U_2,$$

$$\gamma_{13}Y_1 + \gamma_{23}Y_2 - \quad Y_3 \qquad +\beta_{43}X_4 = U_3.$$

Estimated reduced form:

$$Y_1 = 0.413X_2 + 0.150X_3 + 1.275X_4,$$

$$Y_2 = 0.084X_2 + 0.713X_3 + 0.806X_4,$$

$$Y_3 = 0.141X_2 + 0.188X_3 + 1.493X_4.$$

10 | Estimating Models with Erroneous and Unobserved Variables

10.1 Introduction

The first seven chapters proceeded on the basic assumption that all explanatory variables are measured accurately and can be held fixed for successive replications. Relaxing the assumption of fixed exogenous variables has little impact as long as the exogenous variables and error terms are independent. Chapters 8 and 9 further weakened this argument by considering cases where important explanatory variables are determined within the system being modeled. In simultaneous equation cases, it is not possible to maintain the assumption of fixed explanatory variables measured without error because we assume stochastic explanations for some variables. However, we maintained the assumption of error-free measure for all exogenous variables. We shall now consider situations where some variables that are presumably exogenous to the model, or at least predetermined with respect to the equation being estimated, are either not measured accurately, are subject to stochastic effects, or are simply not measured at all. We shall also consider the situation where one or more of the endogenous variables are not directly observed.

The problems of erroneous and unmeasured variables are becoming more apparent in the social sciences as researchers explore empirically new areas and subjects. Many of these new research areas concern theoretical variables where the measures are known to be faulty or where measures do not exist at all. Important examples are studies of people's status attainment or earnings, which include "ability" as one explanatory variable, and models of political attitudes and their effects on behavior. No one has yet observed or measured ability, and the measurement problems with the issue and party variables in the voting model have been discussed at several points. Researchers are then faced with the problem of discarding present data and waiting until adequate measures are developed, which may be a long time, or of using statistical procedures that can deal with these inadequate or missing measures. There are a number of variables, supposedly related to ability, which under the proper conditions can be used to estimate status and earnings models. In the case of the voting model, with certain assumptions the instrumental variable model can overcome some of the measurement difficulties. The use of new statistical models may not be possible in all situations because it places very large burdens on the researcher with regard to the specification of the model and the amount of prior information required about parts of the model. However, it is important to investigate these techniques and their assumptions and requirements so that this strategy can be followed where appropriate.

The voting and aspiration models discussed in the two previous chapters provide good illustrations of both the erroneous and missing variable problems, respectively. The discussions of both the party affiliation and evaluation variables throughout the earlier chapters point out the errors introduced by the way these variables are constructed from the survey questions. The limited response categories for the questions mean that many people with different true party affiliations or with different true issue positions will all class their partisanship as weak (or strong) and their issue positions as weak (strong) agreement or disagreement. At the same time, the wording of the questions and the terms weak and strong may imply different things to different respondents, so that some voters with the same party attachments or issue positions may erroneously class themselves differently. Certainly one definition of erroneous variables is where observations with different true values are measured the same and others with the same true values are measured differently. Besides these errors, the aggregate evaluation variable used to estimate the voting and party affiliation equations contains some additional errors because it is a simple average of people's binary evaluations on all issues where they had an evaluation. This process does not allow for the fact that one party or the other may be greatly preferred on some issue and only slightly preferred on others or that some issues are more important to the voter than others. Theoretical models of issue based voting behavior hold that both a distance metric and a measure of issue saliency are necessary variables. Neither is available for the 1964 election, so we are left with either dealing with erroneous variables or not doing the analysis.

The peer influence on aspirations model, as completely specified by Duncan, Haller, and Portes (1968), contains an important variable that was not or could not be measured at the time. They define this variable as "ambition." In the complete model, it is ambitions that are determined by the exogenous variables and which are simultaneously related among peers. Each boy's specific occupational and educational plans are derived from his ambitions. This final model is shown in Fig. 10.1 and the equations

$$P_1 = \beta_{11}I_1 + \beta_{21}S_1 \qquad\qquad\qquad\qquad\qquad\qquad + U_1, \quad (10.1)$$

$$P_2 = \qquad\qquad \beta_{32}I_2 + \beta_{42}S_2 \qquad\qquad\qquad\qquad + U_2, \quad (10.2)$$

$$Y_1 = \beta_{13}I_1 + \beta_{23}S_1 \qquad\quad + \gamma_{13}P_1 \qquad\qquad + \gamma_{43}Y_2 + U_3, \quad (10.3)$$

$$Y_2 = \qquad\qquad \beta_{34}I_2 + \beta_{44}S_2 \quad + \gamma_{24}P_2 + \gamma_{34}Y_1 \qquad + U_4, \quad (10.4)$$

$$O_1 = \qquad\qquad\qquad\qquad\qquad\qquad Y_1 \qquad\quad + \varepsilon_1, \quad (10.5)$$

$$O_2 = \qquad\qquad\qquad\qquad\qquad\qquad\qquad Y_2 + \varepsilon_2, \quad (10.6)$$

$$E_1 = \qquad\qquad\qquad\qquad\qquad\qquad \alpha_1 Y_1 \qquad + \varepsilon_3, \quad (10.7)$$

$$E_2 = \qquad\qquad\qquad\qquad\qquad\qquad\qquad \alpha_2 Y_2 + \varepsilon_4. \quad (10.8)$$

(The variables are considered as deviations from their means so we can omit the constant term.) Y_1 and Y_2 are the unobserved ambition variables of the respondent and friend, respectively, and are scaled to match the occupation aspiration variable.[1]

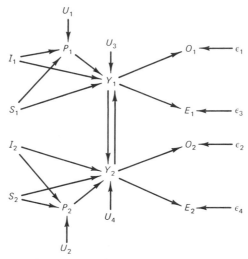

FIGURE 10.1 Simultaneous peer influence model with unobserved ambition variable.

This postulates a quite different model than that estimated in Chapter 9. Besides the presence of the ambition variables, occupational and educational aspirations are no longer causally related but both are derived from the ambition variable. The model also hypothesizes that all of the influence of the exogenous variables on occupational and educational aspirations is exerted through ambitions. The question is, How does one estimate such a model when important endogenous variables are not measured?

---

[1] This is simply an arbitrary scaling, as is measuring income in dollars, thousands of dollars, pounds, or whatever.

We shall see that attempts to conceptualize and estimate models with erroneous variables and with unobserved variables have many points in common. We start with a discussion of the simple errors in the explanatory variables problem and show a simple remedy for it. The errors in variables problem is a convenient introduction to the general unobserved variable model. One way to conceptualize the errors in the variables problem is to treat the true, or theoretically desired, measure as one variable and the observations as a second variable that equals the true variable plus whatever error is present. This procedure in effect adds an additional equation to the model explicitly stating the relationship between the true, but unobserved, variable and the measured one. This additional equation converts even a single equation model into a structural equation system involving observed and unobserved variables. Throughout this chapter the general strategy is to state explicitly the relationships among all observed and unobserved variables as a structural equation model. The questions then become: What information is required in order to identify the structure? and How is the information provided by the observed variables best used in conjunction with a priori knowledge to estimate the parameters of the model?

In some situations we can use the instrumental variables method discussed in the previous chapters to estimate the model. In other cases we must resort to other techniques, some of which are almost ad hoc in nature and depend upon the specification of the particular model. One technique we discuss has been used by psychologists and some sociologists under the general name factor analysis. We shall show that it can be an appropriate technique for estimating certain multiequation models where specific relationships are hypothesized between unmeasured variables and a set of observed variables. Some of the more elaborate applications of path analysis are also helpful in dealing with the problems encountered in estimating models with unobserved variables.

The more elaborate methods discussed toward the end of this chapter, the sophisticated path analyses and factor analyses, share a common structure in the general development of the estimators, and this commonality distinguishes them from the derivation of the instrumental variables and the estimation procedures discussed so far in this book. Instead of concentrating on using the individual observations to develop individual estimator equations, as we did with OLS and TSLS, we switch to the analysis of the correlations or covariances among the observed variables and ask, What values of the parameters in the underlying structural model could be expected to produce this observed variance–covariance matrix?[2] We briefly encountered this approach at the end of the previous chapter in showing how the

---

[2] It is possible to reformulate the OLS method into an analysis of covariances among the observed variables. However, it is easier and more enlightening statistically to develop OLS and its variants through the observation by observation format. The distinction here is more a pedagogical one than a fundamental difference in methodology.

covariances among the residuals in the reduced form equations could be used to identify and estimate simultaneous equation models. This is a subtle but important development because it opens up the possibility of estimating more complicated structural equation models.

Most of the problems discussed in this chapter have only recently begun to receive serious attention, and it is impossible to outline any "conventional wisdom." Further, many of the methods involve quite sophisticated statistical operations and discussions which are beyond the scope of this text. We hope to provide an understanding of the types of models being discussed and the general approach to these methods and to demonstrate their application to some simple problems. Anyone contemplating serious use of these methods should investigate more advanced discussions.[3]

10.2   Erroneous Explanatory Variables

The easiest place to begin our discussion is the simple case where we suspect that the explanatory variables in a single-equation model are erroneous measures of the true variables. We want to show the consequences of using erroneous explanatory variables without correcting for the errors and then present a solution to this problem.

We postulate a true model

$$Y = X\beta + U, \tag{10.9}$$

where Y is $T \times 1$ and X is $T \times K$, but for X consider measures that are not X but variables that are X plus some random error. We denote these observed variables as Z_1, and express their relationship with X as

$$Z_1 = X + \varepsilon_1, \tag{10.10}$$

where ε_1 is $T \times K$. Equation (10.10) simply says that each observed variable Z_{1_k} is the corresponding true value X_k plus a random error component ε_{1_k}. If any of the true explanatory variables are measured without error, then the corresponding column of ε_1 is a vector of zeros, and $Z_{1_k} = X_k$. (We can consider the first column of both X and Z_1 to be a vector of ones used to estimate the constant term in the model, meaning the first column of ε_1 is zero. Alternatively, if we presume that each variable is measured as deviations about its mean, we can drop the constant term from consideration.)

The errors represented by ε_1 are assumed to be the random variation introduced in trying to measure X. In the case of our party identification variable, the ε_1's correspond to the difference between the person's true party affiliation and the value assigned to the group in which this person classes himself or herself. Thus ε_1 includes both the grouping effects and the errors attributable to person's differing perceptions of what the category definitions mean.

---

[3] See Goldberger and Duncan (1973), Griliches (1974), and literature cited therein.

The structure of these ε_1's is critical to the development of estimation procedures. Important assumptions must be satisfied in order to estimate the model. One set of plausible assumptions may be that the errors denoted by ε_1 have mean zero and are distributed independently both of the errors in the original equation and of the true value of X. Symbolically,

$$E(\varepsilon_1) = 0, \qquad E(\varepsilon_1 U') = 0, \qquad E(\varepsilon_1 X') = 0. \tag{10.11}$$

We also continue to assume that X and U are independent of each other, as we have through most of the book.

Consider estimating β by regressing Y on Z_1; the implicit model is

$$Y = Z_1 \beta + (U - \varepsilon_1 \beta) = Z_1 \beta + V. \tag{10.12}$$

[This is derived from $X = Z_1 - \varepsilon_1$ and Eq. (10.9).] This model does not satisfy the basic assumption of ordinary least squares regression. Z_1 and V are correlated because both are functions of ε_1. We can see this by observing that[4]

$$\frac{1}{T} E(Z_1' V) = \frac{1}{T} E\left[(X + \varepsilon_1)'(U - \varepsilon_1 \beta) \right] = \frac{1}{T} E(X' U - X' \varepsilon_1 \beta + \varepsilon_1' U - \varepsilon_1' \varepsilon_1 \beta)$$

$$= \frac{1}{T} E(-\varepsilon_1' \varepsilon_1 \beta) = -\frac{1}{T} E(\varepsilon_1' \varepsilon_1) \beta = -\Sigma_{\varepsilon_1} \beta. \tag{10.13}$$

where $(1/T)E(\varepsilon_1'\varepsilon_1) = \Sigma_{\varepsilon_1}$ and is the variance–covariance matrix of ε_1.[5] We can see that this estimate for β will be both biased and inconsistent. Regardless of the sample size, $\Sigma_{\varepsilon_1}\beta$ will in general not equal zero. Thus, the OLS extimator using Z_1 will yield a biased and inconsistent estimate of β.

If we are considering a simple bivariate model so that $Y_t = \beta_1 + \beta_2 X_t + U_t$ and $Z_{1_t} = X_t + \varepsilon_{1_t}$, the correlation between V and Z_1 is negative or positive, depending upon whether the effect of X on Y, β_2, is positive or negative. Consequently, the estimate of β_2 obtained by regressing Y on Z_1 will be systematically biased toward zero. Thus, if β_2 is positive, we shall underestimate it; while if β_2 is negative, our estimates will be biased upward. In the multivariate case this assessment of the effects is more complicated because it depends upon the matrix product of $\Sigma_{\varepsilon_1}\beta$. Moreover, in the multivariate case, even when one variable, say X_k, is measured without error, the OLS estimator

---

[4]If we consider the OLS estimator, $b = (Z_1'Z_1)^{-1}Z_1'Y$, by substitution we have $b = \beta + (Z'Z)^{-1}Z_1'V$. The text expression evaluates the expected value of the last portion $Z_1'V$. The conclusion is that this term will not equal zero, leading to biased estimators of *all* of the parameters. (Even if $\varepsilon_{1_k} \equiv 0$, the estimate of the kth parameter will generally be biased.) Further, even if the samples get larger, the estimators remain asymptotically biased and inconsistent.

[5]This statement assumes that all individual error terms associated with any one Z_{1_k} have the same variance and that the covariance between error terms for each pair of Z_1's is the same for all observations. We can indicate these assumptions as

$$E(\varepsilon_{1_{tk}} \varepsilon_{1_{sm}}) = \begin{cases} \sigma_k^2, & s = t \text{ and } k = m, \\ \sigma_{km}, & s = t, \\ 0, & s \neq t, \end{cases} \qquad \text{or} \qquad E(\varepsilon_{1_k}\varepsilon_{1_m}') = \sigma_{km}I.$$

of β_k will generally be biased and inconsistent when there are measurement errors in the remaining exogenous variables.

We can provide an intuitive explanation for the result in the bivariate case. The model we are implicitly estimating when we regress Y on Z_1 is a model relating Y to Z_1. However, the variable Z_1 consists of a systematic component X and a random component ε_1. The greater the variance of ε_1, the more the variable Z_1 resembles a random variable rather than the systematic variable X. We do not expect Y to be related to a random variable and would expect the coefficient on such a random variable to have an expected value of zero. The regression of Y on Z_1 may accurately estimate the effect Z_1 has on Y, but this effect approaches zero as Z_1 approaches being a random variable. This in fact is what we observe.

One straightforward way to estimate β in Eq. (10.9) is to develop a structural model relating all the variables and then determine if we can estimate that model. For the problem at hand, the approach is to develop equations that explain X on the basis of some other exogenous variables that can be measured accurately. This converts our problem into a three-equation, hierarchical system which we can estimate with the procedures developed in Chapter 8. We denote the set of variables explaining X as W, where W is a $T \times K^*$ matrix of observations on these additional exogenous variables, with $K^* \geqslant K$, and A is a $K^* \times K$ matrix of coefficients. This gives the system

$$X = W\text{A} + U^*, \tag{10.14}$$

$$Z_1 = X + \varepsilon_1 = W\text{A} + (U^* + \varepsilon_1), \tag{10.15}$$

$$Y = Z_1 \beta + (U - \varepsilon_1 \beta). \tag{10.16}$$

Because X is not observed, we can work only with Eqs. (10.15) and (10.16). These equations are heirarchical, but nonrecursive because the two error terms $U^* + \varepsilon_1$ and $U - \varepsilon_1 \beta$ are clearly correlated because of the presence of ε_1 in both terms. (This is in addition to any correlation that might exist between U^* and U.) If we can assume that ε_1, U^*, and U are distributed independently of W and each other, then we can use the *instrumental variables estimator* described previously (which is essentially two-stage least squares) to estimate Eq. (10.16) as long as W contains as many or more variables than X and Z. (Why?) [We can estimate Eq. (10.15) directly using OLS. Why?] This instrumental variables method uses the OLS estimates of A in Eq. (10.15) to obtain estimated values for Z_1, $\hat{Z}_1 = W\hat{\text{A}} = W(W'W)^{-1}W'Z_1$. Following the two-stage least squares procedure, these estimated values for Z_1, which are purged of the error term $U^* + \varepsilon_1$, are used as the explanatory variables in the second stage estimation of Eq. (10.16).

This is precisely the logic behind the instrumental variables estimation of the voting equation in Chapter 8. In that equation we anticipated that both the evaluation and party identification variables used to explain voting decisions contained errors. The complete voting behavior model in Fig. 9.1

specifies a set of exogenous variables hypothesized to effect both party affiliations and issue positions, and thus evaluations. These postulated relationships correspond to Eq. (10.14) for both the party identification and evaluation variables. The exogenous variables were then used to construct instruments for both explanatory variables and the voting equation estimated with the instrumental variables. The IV and OLS estimates are quite different. The explanation offered for these differences is that the instrumental variables estimates correct for the inconsistencies in OLS.

The strategy of developing equations to explain the true variables as functions of well-measured exogenous variables is only one method for estimating Eq. (10.16). This procedure is particularly appropriate to the voting equation and similar models because we treat party identification and issue evaluations as variables endogenous to the electoral process and already have a set of exogenous variables hypothesized to éxplain them. We may not always be able to do this; it may be hard to hypothesize the relationships indicated by Eq. (10.14). The method we shall develop for the general unobserved variable model is an alternative for these cases where an equation explaining the true variable cannot be specified.

10.3 Unobserved Variables

We can easily see the parallel between the errors in the explanatory variables and the unobserved variables cases. Equations (10.14)–(10.16) include a specific equation for the true variable X and then proceed as if it is simply an unobserved variable explaining Z_1. The general method used to estimate models with unobserved variables is to develop additional measures for X to use along with Z_1. These additional measures are denoted as Z_2. In the instrumental variables method above, $Z_2 = W\hat{A} = W(W'W)^{-1}W'Z_1 = W(W'W)^{-1}W'(X + \varepsilon_1)$. In the more general case, Z_2 is any set of variables assumed to be systematically related to X. This hypothesized relationship is then specified as an additional structural equation.

The fundamental problem we face with models containing erroneous or unobserved variables is the absence of information about the true variables and how they covary with the observed variables. This deficiency quickly leads to unidentified models. There are an infinite combination of values for the structural coefficients and error variances that could produce the variances and covariances observed among the measured variables. What we want to show with the simple errors in variables case is the nature of this identification problem and how additional measures of the missing variable, such as Z_2, can be used along with assumptions about the specification of the structural model and the relationships among the error terms to identify the model.

Considerable prior information is needed to identify and estimate models with unobserved variables. The problem is analogous to the one faced in the preceding chapter, only systems with unobserved variables require substantially more prior information in the form of restrictions on coefficients, variances, and covariances. We substitute information in the form of restrictions on terms for information usually provided by the variances and covariances of the observed variables. One should examine these assumptions carefully and consider them skeptically when trying to estimate a complicated model with erroneous and unobserved variables.

The easiest way to see the unidentified nature of the model in Eqs. (10.9) and (10.10) in the absence of prior information or restrictions is to again consider the bivariate case, where

$$Y_t = \beta_1 + \beta_2 X_t + U_t,$$ (10.17)

$$Z_{1_t} = X_t + \varepsilon_{1_t}.$$ (10.18)

We shall treat all variables as deviations about their respective means and denote them with lowercase letters; this eases the mathematical exposition considerably:

$$y_t = Y_t - \overline{Y} = \beta_2 \left(X_t - \overline{X} \right) + \left(U_t - \overline{U} \right) = \beta_2 x_t + u_t,$$ (10.17a)

$$z_{1_t} = \left(Z_{1_t} - \overline{Z}_1 \right) = \left(X_t - \overline{X} \right) + \varepsilon_{1_t} = x_t + \varepsilon_{1_t}.$$ (10.18a)

[We have already assumed that $E(\varepsilon_{1_t}) = \overline{\varepsilon}_1 = 0$ in Eq. (10.11).] We have two observed variables z_1 and y; the statistical information we observe is their sums of squares and cross products. We write this information as

$$S = \frac{1}{T} \begin{bmatrix} \sum y^2 & \sum yz_1 \\ \sum yz_1 & \sum z_1^2 \end{bmatrix} = \begin{bmatrix} \mathrm{var}(Y) & \mathrm{cov}(YZ_1) \\ \mathrm{cov}(YZ_1) & \mathrm{var}(Z_1) \end{bmatrix} = \begin{bmatrix} V_y & C_{yz_1} \\ C_{yz_1} & V_{z_1} \end{bmatrix}.$$

(10.19)

The terms in Eq. (10.19) contain all the observable information we have, given our two variables z_1 and y. The model in Eq. (10.17) and (10.18) can be used to derive the expected values for these moments. We denote the expected value of S by Σ; $E(S) = \Sigma$. The equation

$$E(S) = \Sigma = E \begin{bmatrix} \beta_2^2 V_x + V_u + 2\beta_2 C_{xu} & \beta_2 V_x + \beta_2 C_{x\varepsilon_1} + C_{u\varepsilon_1} + C_{xu} \\ \beta_2 V_x + \beta_2 C_{x\varepsilon_1} + C_{u\varepsilon_1} + C_{xu} & V_x + 2C_{x\varepsilon_1} + V_{\varepsilon_1} \end{bmatrix}$$

(10.20)

displays Σ as a function of the unknown parameters of the model by substituting Eq. (10.17a) and (10.18a) into (10.19) and taking the expected value of both sides. If we retain the assumptions made previously about u [i.e., $E(uu') = \sigma_u^2 I$, $E(u) = 0$, and that x and u are independent] and add similar assumptions about ε_1 [i.e., $(1/T)E(\varepsilon_1\varepsilon_1')=\sigma_{\varepsilon_1}^2 I$, and that ε_1 is independent of x and u], this expression becomes

$$E(S)=\Sigma=\begin{bmatrix} \beta_2^2 V_x + \sigma_u^2 & \beta_2 V_x \\ \beta_2 V_x & V_x + \sigma_{\varepsilon_1}^2 \end{bmatrix}. \tag{10.21}$$

The observed variances and covariances [Eq. (10.19)] should on average equal the population variances and covariances in Eq. (10.21). They will not be exactly equal because the elements in Eq. (10.19) are based on sample information. However, given that we have only one set of observations on z_1 and y, Eq. (10.19) represents our best guess of the appropriate values for these cross products, and it seems reasonable to use the observed cross products to estimate the unknown population parameters in Eq. (10.21).

Using the observed data to estimate the elements of Σ in Eq. (10.21) is equivalent to OLS when x is measured without error. When there is no error in x, $\sigma_{\varepsilon_1}=0$, and the upper right-hand corner of the matrix equation obtained from setting S equal to Σ provides an estimate of β_2, namely, $b_2=C_{yz}/V_x$ (since $z_1=x$). This is just the OLS estimator for the bivariate model. (In fact, the OLS estimator could have been developed in just this manner.)

However, when x is observed with error, the nature of the identification problem becomes immediately obvious. We can view the matrix equation found from setting S equal to $E(S)$ in Eq. (10.21) as providing three equations. These correspond to the distinct elements in S, i.e., V_y, V_{z_1}, and C_{yz_1}. However, there are four unknowns, β_2, V_x, σ_u, and σ_{ε_1}. We cannot obtain a unique solution based upon our observational information to this system of estimating equations.

We have already introduced some strong assumptions about the errors in the model. We first repeated the assumptions made in Chapter 3, that U has an expected value of zero and is independent of X, and that each U_t has the same variance σ_u^2. We introduced further assumptions that the errors introduced into Z_1 in obtaining the measure for X are completely random and not systematically correlated with either X or U and that each ε_{1_t} has the same variance and is independent of the other ε_{1_t}'s, i.e., $E(\varepsilon_1)=0$ and $E(\varepsilon_1\varepsilon_1')=\sigma_{\varepsilon_1}^2 I$.

These latter assumptions about ε_1 imply that whatever causes the measurement error is equally present in all observations and not associated with certain values of the explanatory variable X or any of the contributors to the error term U. In terms of the party identification variable, for example, this assumption means that all respondents must be equally likely to deviate from their true party affiliation in categorizing themselves according to the schema in the questionnaire. If certain party identifiers are likely to understate (or overstate) consistently the strength of their party affiliation, it will lead to correlations between true party affiliation and the error term. Or if people whose stated vote intentions deviate from those predicted by their true evaluations and party affiliations in a given way are also likely to misperceive or misstate their party identification, it leads to correlations between ε_1 and U. Finally, if some people are more likely to misperceive their party affiliation than others, even though not in any particular direction, the variance of ε_1 will be greater for some people than others. All of these situations result in errors in measuring X, in this case party affiliation, errors that do not satisfy the stated assumptions. Be certain to consider these assumptions carefully prior to starting analysis and when evaluating alternative variables as measures for X.

Unfortunately, even with these assumptions (or restrictions) the model remains unidentified. There are four unknown terms: β_2, V_x, σ_u^2, and $\sigma_{\varepsilon_1}^2$ which are related to only three observed values V_y, V_{z_1} and C_{yz_1}.

There are a variety of ways to proceed at this point. One method would be to assign a value arbitrarily to one of the remaining unknowns. For example, one might argue that the unknown variable x could be scaled in such a way that its variance equals one, $V_x = 1$. This gives

$$b_2 = C_{yz_1} = \frac{1}{T} \sum yz_1.$$

Alternatively, one might assume that the relationship between y and x is exact, so that $\sigma_u^2 = 0$. This leads to

$$b_2 = V_y / C_{yz_1} = \sum y^2 / \sum yz_1,$$

or the reciprocal of b_2 in the regression of z_1 on y.

Such procedures may impose more constraints on the model than one can legitimately justify. In the case of the first assumption, it would generally take unreasonably strong beliefs about the specified variables and the data collection procedures to believe that the variable X exists with a variance exactly equal to one or that the data set could be replicated for any other sample that maintains that restriction. Similarly, in most social science work, it is unreasonable to assume that the "natural experiment" generating the data adequately controls for all factors, other than x, influencing y and that they can be ignored and treated as excluded from the data. Thus neither of the above restrictions seems appropriate for general application. How then to proceed?

One answer is to find Z_2, a set of additional measures for X. (Again we will use $z_{2_t} = Z_{2_t} - \overline{Z}_2$.) This procedure gives us a third observable variable z_2. Now instead of three observational variances and covariances, we have six observed values. The additional three are V_{z_2}, C_{z_2y}, and $C_{z_1z_2}$, so that the observed covariance matrix is

$$S = \begin{bmatrix} V_y & C_{yz_1} & C_{yz_2} \\ C_{yz_1} & V_{z_1} & C_{z_1z_2} \\ C_{yz_2} & C_{z_1z_2} & V_{z_2} \end{bmatrix}. \tag{10.22}$$

Assume that z_2 is related to x in the following way:

$$z_{2_t} = (Z_{2_t} - \bar{Z}_2) = \alpha_2(X_t - \bar{X}) + (\varepsilon_{2_t} - \bar{\varepsilon}_2) = \alpha_2 x_t + (\varepsilon_{2_t} - \bar{\varepsilon}_2).$$

Thus, z_2 functions as an alternative measure of x. If similar assumptions made previously about ε_1 hold for ε_2, namely,

$$E(\varepsilon_2) = 0, \qquad E(\varepsilon_2\varepsilon_2') = \sigma_{\varepsilon_2}^2 I \qquad E(\varepsilon_2' u) = 0 \qquad E(\varepsilon_2' x) = 0$$

and, additionally,

$$E(\varepsilon_2'\varepsilon_1) = 0,$$

then the expected value of Eq. (10.22) is

$$E(S) = \begin{bmatrix} \beta_2^2 V_x + \sigma_u^2 & \beta_2 V_x & \beta_2 \alpha_2 V_x \\ \beta_2 V_x & V_x + \sigma_{\varepsilon_1}^2 & \alpha_2 V_x \\ \beta_2 \alpha_2 V_x & \alpha_2 V_x & \alpha_2^2 V_x + \sigma_{\varepsilon_2}^2 \end{bmatrix}. \qquad (10.23)$$

We now have six relationships (corresponding to the unique elements of S) and six unknowns β_2, α_2, V_x, σ_u^2, $\sigma_{\varepsilon_1}^2$ and $\sigma_{\varepsilon_2}^2$. From this set of relationships, β_2 is identified and can be estimated by observing that

$$E(C_{yz_2})/E(C_{z_1 z_2}) = \beta_2 \alpha_2 V_x / \alpha_2 V_x = \beta_2$$

so that

$$b_2 = \frac{C_{yz_2}}{C_{z_1 z_2}} = \frac{\frac{1}{T}\sum yz_2}{\frac{1}{T}\sum z_1 z_2}.$$

For this two variable case, this is exactly the formula for instrumental variable estimates given in the appendix to Chapter 8, $b = (Z_2'Z_1)^{-1}Z_2'Y$.

The advantage of starting with the instrumental variable format is that we can derive the asymptotic variances of the estimated coefficients. From the appropriate formula from the appendix to Chapter 8, we find that this asymptotic variance is[6]

$$\text{var}(b_2) = (\sigma_u^2 + \beta_2^2 \sigma_{\varepsilon_1}^2) \frac{\sum z_2^2}{\left[\sum z_1 z_2\right]^2} = (\sigma_u^2 + \beta_2^2 \sigma_{\varepsilon_1}^2) \frac{\text{var}(z_2)}{\left[\text{cov}(z_1 z_2)\right]^2}.$$

We can investigate this relationship further by noting that asymptotically $\text{var}(z_2) = \alpha_2^2 \text{var}(x) + \sigma_{\varepsilon_2}^2$ and that $\text{cov}(z_1 z_2) = \alpha_2 \text{var}(x)$ under the assumption that the error terms are independent of each other and of X. This gives the

[6] The correspondence to Chapter 8 develops from the definition of $V_t = U_t - \beta_2 \varepsilon_{1_t}$ in Eq. (10.12) so that $\text{var}(V) = \sigma_U^2 + \beta_2^2 \sigma_{\varepsilon_1}^2$.

asymptotic variance of b_2:

$$\text{var}(b_2) = \left(\sigma_u^2 + \beta_2^2 \sigma_{\varepsilon_1}^2\right) \left\{ \frac{\alpha_2^2 \text{var}(x) + \sigma_{\varepsilon_2}^2}{\left[\alpha_2 \text{var}(x)\right]^2} \right\}$$

$$= \left(\sigma_u^2 + \beta_2^2 \sigma_{\varepsilon_1}^2\right) \left\{ \frac{1}{\text{var}(x)} + \frac{\sigma_{\varepsilon_2}^2}{\alpha_2^2 \left[\text{var}(x)\right]^2} \right\}. \tag{10.24}$$

If we accept σ_u^2, $\text{var}(x)$, β_2, and $\sigma_{\varepsilon_1}^2$ as being fixed, so that the only decisions are about a choice of z_2, we want a z_2 which is strongly related to x_1 so that α_2 is large, and measured with as small a variance as possible, so that $\sigma_{\varepsilon_2}^2$ is small. These two considerations should guide the choice of z_2.

This analysis is easily extended to the multivariate model discussed in Section 10.2. It is simply a matter of replacing the terms in Eqs. (10.17)–(10.22) with the symbols for the appropriate matrices from Eqs. (10.9) and (10.10). We have now reintroduced the constant term, so that the first variable in Z identically equals one, $Z_{1_t} = 1$ for all t. Now the observed cross-products matrix is

$$S = \frac{1}{T} \begin{bmatrix} Y'Y & Y'Z_1 \\ Z_1'Y & Z_1'Z_1 \end{bmatrix} = \begin{bmatrix} \sum Y_t^2 & \sum Y_t & \sum Y_t Z_{1_{t2}} & \cdots & \sum Y_t Z_{1_{tK}} \\ \sum Y_t & T & \sum Z_{1_{t2}} & \cdots & \sum Z_{1_{tK}} \\ \vdots & \vdots & \vdots & & \vdots \\ \sum Z_{1_{tK}} Y_t & \sum Z_{1_{tK}} & \sum Z_{1_{tK}} Z_{1_{t2}} & \cdots & \sum Z_{1_{tK}}^2 \end{bmatrix}.$$

$$\tag{10.25}$$

The expected value of this matrix, from Eq. (10.9) and (10.10), is

$$\Sigma = \frac{1}{T} E \begin{bmatrix} \beta'X'X\beta + U'U + 2\beta'X'U & \beta'X'X + \beta'X'\varepsilon_1 + U'\varepsilon_1 + U'X \\ X'X\beta + \varepsilon_1'X\beta + \varepsilon_1'U + X'U & X'X + X'\varepsilon_1 + \varepsilon_1'X + \varepsilon_1'\varepsilon_1 \end{bmatrix}.$$

With the same assumptions shown in Eq. (10.11) and used above, we have

$$\Sigma = \begin{bmatrix} \beta'\Sigma_X\beta + \sigma_U^2 & \beta'\Sigma_X \\ \Sigma_X\beta & \Sigma_X + \Sigma_{\varepsilon_1} \end{bmatrix}, \tag{10.26}$$

where Σ_X is $1/T$ times the cross-product matrix for the implicit values of the true variables for the sample and Σ_{ε_1} is the variance–covariance matrix of errors from the model relating Z_1 to X.

We again assume that our matrix of observed cross products in Eq. (10.25) is our best estimate of Σ in Eq. (10.26), so we set

$$S = \frac{1}{T}\begin{bmatrix} Y'Y & Y'Z_1 \\ Z_1'Y & Z_1'Z_1 \end{bmatrix} = \begin{bmatrix} \beta'\Sigma_X\beta + \sigma_U^2 & \beta'\Sigma_X \\ \Sigma_X\beta' & \Sigma_X + \Sigma_{\varepsilon_1} \end{bmatrix}, \qquad (10.27)$$

which corresponds to Eq. (10.21). As with that equation, this model is unidentified. There are $1 + K + K(K+1)/2$ separate relationships from the observed cross products [the main diagonal plus the upper (or lower) triangle of S]. [S is a $(K+1)\times(K+1)$ matrix since it includes the K exogenous variables and Y.] The number of unknowns equals this number (σ_U^2, β, and Σ_X), plus the number of unknown terms in Σ_{ε_1}. If there were no unknown elements in Σ_{ε_1} (i.e., $\varepsilon_1 = 0$), the model would be identified, and this would simply be the OLS estimator.[7]

At this point we can use additional measures of X, denoted as Z_2, to resolve the identification problem. We hypothesize that Z_2 is systematically related to X in the following manner:

$$Z_2 = X\mathbf{A} + \varepsilon_2, \qquad (10.28)$$

where Z_2 is a $T \times K$ matrix of observations on the additional measures for X and $\mathbf{A}$ is a $K \times K$ matrix of coefficients relating Z_2 to X. (For any X measured without error we assume the appropriate element of $\mathbf{A}$ is one with zeros in the remainder of the column and a vector of zeros in the corresponding column of ε_2.) With the presence of Z_2, our matrix of observed cross products is expanded to

$$S = \frac{1}{T}\begin{bmatrix} Y'Y & Y'Z_1 & Y'Z_2 \\ Z_1'Y & Z_1'Z_1 & Z_1'Z_2 \\ Z_2'Y & Z_2'Z_1 & Z_2'Z_2 \end{bmatrix}. \qquad (10.29)$$

The expected value of this expression, from Eqs. (10.9), (10.10), and (10.28) and the further assumptions that $E(\varepsilon_2) = 0$, $E(\varepsilon_2'u) = 0$, $E(\varepsilon_2'X) = 0$, and $E(\varepsilon_2'\varepsilon_1) = 0$ is

$$\Sigma = \begin{bmatrix} \beta'\Sigma_X\beta + \sigma_U^2 & \beta'\Sigma_X & \beta'\Sigma_X\mathbf{A} \\ \Sigma_X\beta & \Sigma_X + \Sigma_{\varepsilon_1} & \Sigma_X\mathbf{A} \\ \mathbf{A}'\Sigma_X\beta & \mathbf{A}'\Sigma_X & \mathbf{A}'\Sigma_X\mathbf{A} + \Sigma_{\varepsilon_2} \end{bmatrix}. \qquad (10.30)$$

To get the estimate for β, we note from Eq. (10.30) that

$$\beta = (\mathbf{A}'\Sigma_X)^{-1}(\mathbf{A}'\Sigma_X\beta) \qquad (10.31)$$

if $\mathbf{A}'\Sigma_X$ is nonsingular. This nonsingularity condition holds if both $\mathbf{A}$ and Σ_X

---

[7] It is easy to see how to develop the OLS estimator from this analysis of covariance structures. If $\varepsilon_1 = 0$, then $Z_1 = X$, and Eq. (10.27) implies that $Z_1'Z_1 = \Sigma_X$ and $Z_1'Y = \Sigma_X\beta = (Z_1'Z_1)\beta$. This gives the estimator $b = (Z_1'Z_1)^{-1}Z_1'Y = (X'X)^{-1}X'Y$, assuming $Z_1'Z = X'X$ is nonsingular.

are nonsingular, meaning that: (a) no true variable is a linear combination of other true variables; and (b) the *systematic* part of each Z_2 variable is not a linear combination of the systematic components of the other variables in Z_2. This last condition rules out situations where two or more of the variables in Z_2 differ only in their random errors ε_2 but not in their relationships with X. If these conditions hold, we can estimate β by using $Z_2'Z_1$ from Eq. (10.29) as an estimate of $A'\Sigma_X$ and $Z_2'Y$ as an estimate of $A'\Sigma_X\beta$, and

$$b = (Z_2'Z_1)^{-1}(Z_2'Y). \tag{10.32}$$

This is simply the instrumental variables estimator shown in the appendix to Chapter 8. At this point we shall not try to estimate A and Σ_X. Subsequent work in this chapter deals with the problems involved with that task and the procedures for dealing with them.

Monte Carlo Simulation of Erroneous Variables

We want to use our Monte Carlo simulation to illustrate the consequences of using erroneous explanatory variables and how appropriate use of an additional measure for the true variables can correct the problems. We use the same population model we have been using all along, $Y = 15 + X_2 + 2X_3 + U$. [This corresponds to Eq. (10.9).] Now instead of observing the real X_3 that generates the values for Y, we shall assume that we observe a different variable Z_1 that is X_3 plus an error term. We write this model for Z_1 as $Z_1 = X_3 + \varepsilon_1$, which corresponds to Eq. (10.18). (We assume that we observe X_2 without error.) We choose values of 4, 16, and 36 for the variance of ε_1 in separate simulations. [This corresponds to correlations between true and measured variables ranging from about 0.8 ($\sigma_{\varepsilon_1}^2 = 4$) to about 0.4 ($\sigma_{\varepsilon_1}^2 = 36$).] Our expectation is that the larger the variance of ε_1, the more we shall underestimate the coefficient on X_3. For each simulation 100 replications are run with 200 observations, and the average values for b_3 in each one are shown in Table 10.1. These results are quite consistent with expectations. The mean estimated b_3 coefficients range from 1.31 to 0.36, quite a distance below the true value 2.00.

TABLE 10.1

Means and Standard Deviations of Estimated Coefficients for Erroneous Variables (OLS)
($\beta_2 = 1.00$, $\beta_3 = 2.0$, $T = 200$)

	Variance of $\varepsilon_1 = \sigma_{\varepsilon_1}^2$		
	4.0	16.0	36.0
$\bar{b}_3$	1.306	0.647	0.355
$\hat{\sigma}_{b_3}$	(0.234)	(0.165)	(0.122)
$\bar{b}_2$	0.992	0.978	0.971

The correction for this problem is the additional variable Z_2 which is correlated with X and thus Z_1 but not with ε_1. This variable is modeled as

$$Z_2 = X_3\alpha + \varepsilon_2.$$

We can then apply the instrumental variables model with Z_2 serving as the instrument for Z_1. The constant term and X_2 function as their own instruments since they are presumed to be observed without error. This instrumental variables estimator is $(Z_2^{*\prime}Z_1^*)^{-1}Z_2^{*\prime}Y$, where $Z_2^* = (1\ X_2\ Z_2)$ and $Z_1^* = (1\ X_2\ Z_1)$. This simulation is run with values for the variance of ε_2 equal to 16 and 144 and values for α of 2 and 5, for each of the previous values for the variances of ε_1. The correlations between X and Z_2 vary from about 0.25 to 0.90. The simulated results are shown in Table 10.2, indicating the mean of the 100 estimated coefficients and the observed standard deviation of these values. Again, these results are consistent with those expected from the model. All are apparently consistent, and close to the true value of 2.0. The exceptions are the two cases where we expect very large variances to the estimated coefficient because both Z_1 and Z_2 contain a large error component. The observed standard deviations of the estimated coefficients follow the expected pattern. If we take the variance of ε_1 as fixed, the better Z_2 measures X, i.e., the smaller the variance of ε_2 and/or the larger the value of α, the smaller the variance in our estimated coefficients. And, of course, the smaller the variance of ε_1, for a given set of characteristics of Z_2, the smaller the variance of the estimated coefficient. These results then illustrate the possibility of getting consistent estimates with the instrumental variables, or unobserved variables, model and demonstrate that within the set of consistent estimators, the better the observed variables do at approximating the true or unobserved variable, the smaller the variance of our estimates.

TABLE 10.2

Instrumental Variable Estimator ($\beta_3 = 2.0$)

	$\alpha = 2$		$\alpha = 5$	
Var(ε_2)	16	144	16	144
Variance of $\varepsilon_1 = 4$				
$\bar{b}_3$	1.999	2.034	1.990	2.004
$\hat{\sigma}_{b_3}$	(0.343)	(0.697)	(0.299)	(0.366)
Variance of $\varepsilon_1 = 16$				
$\bar{b}_3$	2.003	2.082	1.989	2.011
$\hat{\sigma}_{b_3}$	(0.408)	(0.835)	(0.344)	(0.441)
Variance of $\varepsilon_1 = 36$				
$\bar{b}_3$	2.027	2.291	2.000	2.042
$\hat{\sigma}_{b_3}$	(0.520)	(0.983)	(0.418)	(0.574)

The derivation of the instrumental variables estimator indicates that it is still biased even though it is consistent. However, the Monte Carlo experiments indicate that the bias of instrumental variables for sample sizes of 200 is quite small. Further, as indicated in Table 10.3, on a mean square error criterion, the IV estimator dominates the least squares estimator, at least in these experiments.

TABLE 10.3

Mean Square Error—Alternative Estimators with Errors in Variables (b_3)

		Instrumental variables			
		$\alpha = 2$		$\alpha = 5$	
$\sigma_{\varepsilon_1}^2$	OLS	$\sigma_{\varepsilon_2}^2 = 16$	$\sigma_{\varepsilon_2}^2 = 144$	$\sigma_{\varepsilon_2}^2 = 16$	$\sigma_{\varepsilon_2}^2 = 144$
4	0.536	0.118	0.487	0.090	0.134
16	1.858	0.166	0.704	0.118	0.195
36	2.721	0.271	1.051	0.175	0.331

Finally, in our experiments, X_2 and X_3 are virtually uncorrelated ($r_{23} = -0.1$). Therefore, the OLS and IV estimator for β_2 are very close to being unbiased. This finding of unbiasedness for b_2 does not have to be true. It will not hold if X_2 and Z_1 (X_3) are correlated within the sample.

The Ambition–Aspiration Model

Rewriting the equations in the ambition–aspiration model enables us to estimate the ambition equations with the instrumental variables method just described. Substituting the parental aspiration equations (10.1) and (10.2) into the ambition equations (10.3) and (10.4) (the same substitution used in Chapter 9), we get

$$Y_1 = \beta_{13}^* I_1 + \beta_{23}^* S_1 + \gamma_{43} Y_2 + U_3^* \quad \text{and} \quad Y_2 = \beta_{34}^* I_2 + \beta_{44}^* S_2 + \gamma_{34} Y_1 + U_4^*.$$

We cannot observe, or measure, the ambition variables Y_1 and Y_2 directly, but from Eqs. (10.5) and (10.6) for occupational aspirations O_1 and O_2, we have the relationships $Y_1 = O_1 - \varepsilon_1$ and $Y_2 = O_2 - \varepsilon_2$, which give

$$Y_1 = O_1 - \varepsilon_1 = \beta_{13}^* I_1 + \beta_{23}^* S_1 + \gamma_{43}(O_2 - \varepsilon_2) + U_3^*,$$

or

$$O_1 = \beta_{13}^* I_1 + \beta_{23}^* S_1 + \gamma_{43} O_2 + V_1,$$

where $V_1 = U_3^* + \varepsilon_1 - \gamma_{43}\varepsilon_2$. O_2 is obviously an erroneous measure of Y_2 and is negatively correlated with V_1 because of the presence of the measurement error ε_2, in both variables. This negative correlation is in addition to the positive correlation between O_2 and V_1 due to the presence of the simultaneity between O_1 and O_2. Thus the OLS estimation of this equation gives estimates for γ_{43} that are inconsistent and possibly biased toward zero. We can use the education aspiration variable E_2 as an instrumental variable for

O_2 since we have hypothesized by Eq. (10.8) that

$$E_2 = \alpha_2 Y_2 + \varepsilon_4.$$

If ε_4 is independent of ε_1, ε_2, and U_3^*, we have satisfied the requirements for instrumental variables.

To estimate this ambition equation by instrumental variables, we let $Z_1 = (I_1 \ S_1 \ O_2)$ and $Z_2 = (I_1 \ S_1 \ E_2)$. Intelligence and status I_1 and S_1 are treated as exogenous variables and are assumed to be measured without error, so they function as their own instruments. From the covariances given in the appendix to Chapter 9 we derive the estimates

$$
\begin{bmatrix} b_{13}^* \\ b_{23}^* \\ c_{43} \end{bmatrix} = (Z_2'Z_1)^{-1}Z_2'O_1 =
\begin{bmatrix} V_{I_1} & C_{I_1S_1} & C_{I_1O_2} \\ C_{I_1S_1} & V_{S_1} & C_{S_1O_2} \\ C_{I_1E_2} & C_{S_1E_2} & C_{E_2O_2} \end{bmatrix}^{-1}
\begin{bmatrix} C_{I_1O_1} \\ C_{S_1O_1} \\ C_{E_2O_1} \end{bmatrix}
$$

$$
=
\begin{bmatrix} 28.44 & 6.47 & 17.46 \\ 6.47 & 29.84 & 19.19 \\ 16.81 & 18.48 & 97.71 \end{bmatrix}^{-1}
\begin{bmatrix} 27.69 \\ 22.36 \\ 50.08 \end{bmatrix} =
\begin{bmatrix} 0.69 \\ 0.39 \\ 0.32 \end{bmatrix}.
$$

We could derive a parallel equation for O_2 as a function of I_2, S_2, and O_1 and estimate it using E_1 as the instrumental variable. We would then have estimates of both ambition equations.

There are two difficulties with this method however. The first difficulty with using E_2 as the instrument for O_2 is the assumption that the errors in measuring O_2 and E_2, ε_2 and ε_4, are independent. The observations on both variables are measuring the aspirations of a given individual. If there are any omitted influences on aspirations besides ambition or if the two aspiration measures are likely to make similar errors for a given person, the two error terms will not be independent, and E_2 cannot be used as an appropriate instrument. A similar problem holds for using E_1 as an instrument for O_1. If these correlations among error terms are present, we need a different procedure.

Secondly, these IV estimators based on educational aspirations do not use all of the available information in the data. The equations for O_1 and O_2 just derived from the ambition–aspiration model are identical in specification to the equations for occupational aspirations estimated in Chapter 9 before we introduced the unmeasured ambition variables. At that time, we used the exogenous intelligence and status variables to develop the instruments for O_1 and O_2 using two-stage least squares. Since we assumed that these exogenous variables are independent of the various error terms, it is an acceptable estimation method yet gives different estimates as shown in Table 10.4. TSLS is also an appropriate procedure regardless of whether the errors in measuring each person's occupational and educational aspirations are correlated.

TABLE 10.4

Alternative Instrumental Variables Estimates of Ambition Equations [a]

Instruments		Respondent			Peer	
		E_2	I_2 and S_2		E_1	I_1 and S_1
Intelligence	(β_{13}^*)	0.687	0.643	(β_{34}^*)	0.850	0.819
		(0.120)	(0.124)		(0.118)	(0.129)
Status	(β_{23}^*)	0.394	0.348	(β_{44}^*)	0.377	0.348
		(0.119)	(0.124)		(0.113)	(0.122)
Friend's ambition	(γ_{43})	0.321	0.409	(γ_{34})	0.293	0.350
		(0.084)	(0.104)		(0.083)	(0.124)

[a]Asymptotic standard errors in parentheses.

We now have two instrumental variables estimators for the ambition equations; one uses the education aspiration variables as instruments, and the other uses the exogenous intelligence and status variables to create the instrument in a two-stage least squares process.[8] We must now choose between these two estimators. The essence of our problem is that we have an overidentified model. The model "predicts" several different covariations among the observed variables, in this case between occupational and educational aspirations and between occupational aspirations and the exogenous variables. These predicted relationships are functions of the parameters we want to estimate and can be used to compute these estimates. However, due to the sampling variations in the observed covariations we get different estimates of the parameters.[9] Knowing that all the estimates are consistent is

---

[8] The computationally most efficient procedure in either case is a two-stage least squares program. In the case where the number of instruments in Z_2 equals the number of original variables in Z_1, as with using educational aspirations, the estimates obtained from Eq. (10.32) are identical to those obtained from a two-stage least squares program where estimated values for the variables in Z_1 are obtained from the OLS regression on the variables in Z_2 and these estimated values used in the OLS regression with Y. To see this, note that if both Z_1 and Z_2 are $T \times K$, then $\hat{Z}_1 = Z_2(Z_2'Z_2)^{-1}Z_2'Z_1$ and

$$\left(\hat{Z}_1'\hat{Z}_1\right)^{-1} = \left[(Z_1'Z_2)(Z_2'Z_2)^{-1}(Z_2'Z_2)(Z_2'Z_2)^{-1}(Z_2'Z_1)\right]^{-1} = \left[(Z_2'Z_1)^{-1}(Z_2'Z_2)(Z_1'Z_2)^{-1}\right]$$

because both $Z_1'Z_1$ and $Z_2'Z_2$ are $K \times K$ square matrices. The estimated coefficients from the second stage are

$$\hat{b} = \left(\hat{Z}_1'\hat{Z}_1\right)^{-1}\hat{Z}_1'Y = \left[(Z_2'Z_1)^{-1}(Z_2'Z_2)(Z_1'Z_2)^{-1}\right]\left[(Z_1'Z_2)(Z_2'Z_2)^{-1}Z_2'Y\right] = (Z_2'Z_1)^{-1}Z_2'Y.$$

This is just the instrumental variable estimator from Eq. (10.32). Given the general availability of TSLS programs, this is the easiest way to compute the instrumental variables estimates.

[9] We actually could get three estimates of each coefficient by using the intelligence and status variables as separate instruments. The two-stage least squares estimator based on these instruments used in Chapter 9 and shown in Table 10.4 has already combined these two estimates into a single estimate.

of little help in choosing between them. Just as in the previous overidentified situations, we would like to have some way of combining all the separate estimates into one, hopefully better, estimate. This way we could use all of the available sample information to make our estimates.

All the estimation procedures discussed so far are called limited information or single-equation methods because they estimate only one equation of the system at a time. In doing so they do not make use of the information in all the observed variables. There are two important limitations to this approach. In some instances, the information contained in the specification of the entire system and in the complete observed covariance matrix is needed in order to identify some of the parameters in the model. This situation is best illustrated by the discussion at the end of the previous chapter showing how to identify and estimate parts of the occupational and educational aspiration equations from the covariances of the residuals in the reduced form equations. In other situations, the information in the covariances of the endogenous variables can improve some of our estimates simply because they provide additional estimates of some of the structural parameters. This is the overidentified case illustrated with the ambition equations in the previous section. What we want to do is discuss an approach for estimating structural equations that simultaneously uses all the observed variables and that estimates the entire structure at one time. In the case of the occupational and educational aspiration equations of Chapter 9, we would estimate all the coefficients in the occupational and educational aspiration equations concurrently and not successively from the reduced form coefficients and then the residual covariances as in Chapter 9. For the ambition model in the previous section, this approach provides a way to combine the results of the two different instrumental variables methods into a single, theoretically better, estimator.

The procedures to be discussed in this section apply only to overidentified models. Most of the discussion of erroneous or missing variables so far in this chapter has concentrated on exactly identified structures. In the bivariate example of Eq. (10.22), we had six unknowns and six relationships based on the observable variables. In the multivariate case, Z_2 had the same number of variables as Z_1 (and X) so that the vector $(Z_2'Y)$ has just K terms, exactly the number of coefficients we were trying to estimate. (This also means that $Z_2'Z_1$ is a square matrix.) What happens if Z_2 has more variables than X, (and Z_1), meaning that we have found several alternative measures for at least one of the true variables? This in fact is what happened with the ambition model where we could use the intelligence, status, or educational aspiration variables to create the instrument for occupational aspirations.

If Z_2 contains more variables than does X (and Z_1), the instrumental variables procedure in Eq. (10.32) is inappropriate. If there are K^* variables in Z_2, and only K variables in X, with $K < K^*$, then $Z_2'Y$ is $K^* \times 1$ and $Z_2'Z_1$ is a $K^* \times K$ nonsquare matrix that cannot be inverted. By turning Eq. (10.32) around to be $(Z_2'Z_1)b = Z_2'Y$, we can see that if $K^* > K$, there are more equations (K^*) than unknowns (K), which leads to the overidentified case. One approach of course would be to select only K of the variables in Z_2 and then use Eq. (10.32) to estimate β. But which set of K variables do we select? All instrumental variables estimators are consistent, but in small samples different instruments will give different estimated coefficients, as in the ambition example. Thus we get different estimates of β for each different set of K instruments. If all the variables in Z_2 are legitimate instruments for X_1 (and Z_1), we should be able to improve the estimates of β by combining all the instruments in some fashion. One way to do this is to follow the two-stage least squares procedure to get estimated values for each Z_1 as functions of all the variables in Z_2 and then estimate β from these predicted Z_1 variables.[10]

Situations where β is overidentified constitute only one possible problem we face in trying to estimate the entire structure summarized in Eq. (10.22). There will be many cases where we need or want to estimate Σ_X, $\mathbf{A}$, and the error term covariance matrices Σ_{ε_1} and Σ_{ε_2}. The error term covariance matrices are particularly important because they assess how reliable the different measures of X are. The smaller the variances of ε_1 and ε_2, the better the measures and the more confidence we can have in any estimates derived from the observed data. However, the expressions for estimating these additional terms are not easy to solve, and very likely are also overidentified, depending upon the specification of the relationships between Z_2 and X, summarized in $\mathbf{A}$, and the nature of the variances and covariances between each error term in ε_1 and ε_2. The procedure for estimating these additional parameters, as well as the structural coefficients in the overidentified case, are based on the analysis of the covariance matrix of all measured variables. The procedure is based on deriving the expected covariance matrix from the hypothesized structure, ascertaining if it is identified and, if it is, using the observed covariances to estimate the structure. We have seen the elements of this process in the development of the exactly identified instrumental variables approach. This is only one approach however.

10.4   Factor Analysis

The most prominent of the statistical models dealing with unobserved variables is the technique referred to as factor analysis. This method has a long history in psychology, and to a lesser extent in sociology and political science. We shall not spend time discussing this history or the extensive uses of factor analysis, which are quite adequately covered in a number of other textbooks (Harmon, 1967). We want to show how factor analysis can be considered as a proper procedure for estimating certain structural models.

---

[10] This procedure produces estimates that are weighted averages of each alternative estimate obtained by selecting only K of the variables in Z_2 to use as instruments.

The simplest form of the factor analysis model can be considered a measurement model postulating the existence of K_1 unobserved variables measured by K_2 observed variables, with $K_2 \geqslant K_1$. The observed variables are presumably related to the unobserved variables in a systematic fashion which can be represented by a set of structural coefficients, corresponding to the matrix **A** in the previous unobserved variables example [Eq. (10.28)]. The factor analysis model is

$$Z = X\mathbf{L} + \varepsilon, \tag{10.33}$$

where Z is the $T \times K_2$ matrix of observations on the measured variables, X is the $T \times K_1$ matrix of implicit unobserved variables, **L** is the $K_1 \times K_2$ matrix of coefficients relating the observed and unobserved variables [corresponding to **A** in Eq. (10.28)], and ε is the $T \times K_2$ matrix of measurement errors;

$$Z = \begin{bmatrix} Z_{11} & Z_{12} & \cdots & Z_{1K_2} \\ Z_{21} & Z_{22} & \cdots & Z_{2K_2} \\ \vdots & \vdots & & \vdots \\ Z_{T1} & Z_{T2} & \cdots & Z_{TK_2} \end{bmatrix} = \begin{bmatrix} X_{11} & X_{12} & \cdots & X_{1K_1} \\ X_{21} & X_{22} & \cdots & X_{2K_1} \\ \vdots & \vdots & & \vdots \\ X_{T1} & X_{T2} & \cdots & X_{TK_1} \end{bmatrix}$$

$$\times \begin{bmatrix} L_{11} & L_{12} & \cdots & L_{1K_2} \\ L_{21} & L_{22} & \cdots & L_{2K_2} \\ \vdots & \vdots & & \vdots \\ L_{K_1 1} & L_{K_1 2} & \cdots & L_{K_1 K_2} \end{bmatrix} + \begin{bmatrix} \varepsilon_{11} & \varepsilon_{12} & \cdots & \varepsilon_{1K_2} \\ \varepsilon_{21} & \varepsilon_{22} & \cdots & \varepsilon_{2K_2} \\ \vdots & \vdots & & \vdots \\ \varepsilon_{T1} & \varepsilon_{T2} & \cdots & \varepsilon_{TK_2} \end{bmatrix}.$$

Thus each column of Z gives the successive values of one of the observed variables and each column of X represents the implicit values for one of the unobserved, or true, variables. Any row of these matrices corresponds to the values of all the variables for one observation. Each column in **L** indicates how the corresponding observed variables in Z are derived from the unobserved variables. Thus for elements in the second column of Z and **L**, we have

$$Z_{t2} = \sum_{k=1}^{K_1} X_{tk} L_{k2} + \varepsilon_{t2}.$$

In most applications **L** is referred to as the matrix of factor loadings. The objective is to use the observable information, namely the variance–covariance matrix $(1/T)(Z'Z)$, to estimate **L**, $\Sigma_x = (1/T)(X'X)$, and $\Sigma_\varepsilon = (1/T)(\varepsilon'\varepsilon)$.

This is where the analysis of covariance matrices rather than individual observations becomes important. We obviously cannot simply regress Z on to estimate L because X is not observed. We must use our observed covariances $(1/T)(Z'Z)$ to estimate the structure. The expression for this covariance matrix is

$$Z'Z = (L'X' + \varepsilon')(XL + \varepsilon) = L'X'XL + L'X'\varepsilon + \varepsilon'XL + \varepsilon'\varepsilon.$$

If we continue the assumptions about the distribution of the error terms, namely $E(\varepsilon) = 0$, $E(\varepsilon'X) = 0$, and $E(\varepsilon_k \varepsilon_k') = \sigma_{\varepsilon_k}^2 I$ [see Eq. (10.11) and footnote 5], then

$$\Sigma_Z = \frac{1}{T} E(Z'Z) = \frac{1}{T} L'E(X'X)L + \frac{1}{T} E(\varepsilon'\varepsilon) = L'\Sigma_X L + \Sigma_\varepsilon. \quad (10.34)$$

Most factor analytic models further assume that the error terms associated with each measured variable Z_k are independent of the errors associated with the other variables Z_m so that Σ_ε is diagonal ($\sigma_{\varepsilon_{mk}} = 0$ for $m \neq k$). This assumption means that all covariation among the observed variables is accounted for by their relationships with the underlying variables. For example, the expected covariance between Z_i and Z_j for $i \neq j$ is

$$\frac{1}{T} E\left(\sum Z_{ti} Z_{tj}\right) = L_i' \Sigma_X L_j = (L_{1i} L_{2i} \cdots L_{K_1 i}) \Sigma_X \begin{bmatrix} L_{1j} \\ L_{2j} \\ \vdots \\ L_{K_1 j} \end{bmatrix} = \sum_{m=1}^{K_1} \sum_{n=1}^{K_1} L_{mj} L_{ni} \sigma_{x_{mn}},$$

where $\sigma_{x_{mn}} = (1/T)E(\sum X_{tm} X_{tn})$ and is the m, n element of Σ_X, or simply the covariance between the mth and nth unobserved variable. The expected variance of Z_i is

$$\frac{1}{T} E\left(\sum Z_{ti}^2\right) = L_i' \Sigma_X L_i + \sigma_{\varepsilon_i}^2 = \sum_{m=1}^{K_1} L_{mi}^2 \sigma_{x_{mm}} + 2 \sum_{m=1}^{K_1 - 1} \sum_{n=m+1}^{K_1} L_{mi} L_{ni} \sigma_{x_{mn}} + \sigma_{\varepsilon_i}^2.$$

The estimation problem is to use the information contained in the observed covariance matrix $S_Z = (1/T)Z'Z$, which is interpreted as an estimate of Σ_Z, to estimate the unknown elements in L, Σ_X, and Σ_ε. This task poses certain difficulties immediately. In the first place, without further assumptions in the form of restrictions on L, Σ_X, or Σ_ε, the structure is unidentified because there are alternative structures L^* and X^* that give the same expected moments matrix Σ_Z. To see this, note that we can define the arbitrary matrices L^* and Σ_{X*} by appropriate multiplication of L and X by the square matrix F, where F is any $K_1 \times K_1$ nonsingular matrix;

$$L^* = F^{-1}L, \qquad X^* = XF, \qquad \text{and} \qquad \Sigma_{X*} = F'\Sigma_X F. \quad (10.35)$$

In this form $L_{ij}^* = \sum_{n=1}^{K_1} f^{in} L_{nj}$, where L_{ij}^* is the loading in the alternative structure of the jth observed variable on the ith true variable, f^{in} is the (i, n)th element in F^{-1} and L_{nj} is the loading of the jth observed variable on the nth true variable in the original structure. Similar interpretations hold for X^* and Σ_{X^*}. Thus the alternative structure is simply a linear transformation of the original one. Substitution of these new terms into Eq. (10.34), along with Σ_ε, gives the same predicted values for Σ_Z:

$$L^{*'}\Sigma_{X^*}L^* + \Sigma_\varepsilon = (L'F'^{-1})(F'\Sigma_X F)(F^{-1}L) + \Sigma_\varepsilon = L'\Sigma_X L + \Sigma_\varepsilon = \Sigma_Z.$$

This is the classic definition of an unidentified model—an infinite number of possible structures L^* and X^* are consistent with the same observational information.

The only solution to this problem is to place sufficient restrictions on either or both L and Σ_X so that only one structure is consistent with the observed information and the restrictions. Any restrictions we place on L and Σ_X must be maintained in L^* and Σ_{X^*}. For example, if we state that the covariance between the first two unmeasured variables is zero, this means that both $\sigma_{x_{12}}$ and $\sigma_{x_{12}}^*$ must equal zero. Or, if the second observed variable Z_2 is simply the second underlying variable plus an error term $(Z_2 = X_2 + \varepsilon_2)$, $L_{22} = L_{22}^* = 1$, and all other entries in the second column of both L and L^* must equal zero. If we have enough such restrictions, there will be only one set of values for L and Σ_X that will give Σ_Z, and we can try to estimate these from S_Z.

These restrictions are clearly a priori specifications about the model's structure, such as saying that two unobserved variables do not covary or that Z_2 is a function only of X_2. These specifications must be derived from theory or prior knowledge about the relationships among the X's and how the observed variables are measured, just as the identification problem for the simultaneous equation model could be solved only by prior specification of which variables did and did not legitimately belong in each structural equation.

The only way to ascertain whether these prior specifications are sufficient to identify the model is to examine the effects of these restrictions on the matrix F. Restrictions on L and Σ_X, which must be maintained in L^* and Σ_{X^*}, impose certain restrictions on the elements of F. (If L is 2×4 so that F^{-1} is 2×2, show that setting $L_{22} = L_{22}^* = 1$ and $L_{12} = L_{12}^* = 0$ implies that $f^{12} = 0$ and $f^{22} = 1$.) When Σ_ε is diagonal, the model is identified if the restrictions can be satisfied in both structures only by an F that is an identity matrix, implying that $L^* = L$ and $\Sigma_{X^*} = \Sigma_X$. If the specified restrictions can be satisfied by an F other than the identity matrix, the model is not identified.

In most applications of factor analysis, it is assumed that the unobserved variables are orthogonal to each other, so that Σ_X is diagonal, and have unit variances, $\sigma_{X_k} = 1$. These two assumptions mean that $\Sigma_X = I$, and imply that the true variables implicit in the sample have been transformed to have zero means, unit variances, and to be uncorrelated with each other. In other words, the correlation matrix for the true variables is the identity matrix, $R_X = I$. The restriction that $\Sigma_X = I$ reduces the model in Eq. (10.34) to

$$\Sigma_Z = L'L + \Sigma_\varepsilon. \tag{10.36}$$

This restriction does not completely identify the factor model however. With $\Sigma_X = I$, the model is identified up to an orthogonal transformation of the original variables and the factor loadings because it is possible to find a matrix $F \neq I$ such that $L^* = F'L$ and $X^* = XF$ and $\Sigma_{X^*} = F'\Sigma_X F = F'F = I$. (The definition of an orthogonal matrix is that $F'F = I$ or $F' = F^{-1}$.) We shall return to a discussion of such orthogonal transformations since they are a common practice in most factor analysis.

To see how the restriction $\Sigma_X = I$ constrains F but does not identify the structure, consider the case where there are two unobserved variables hypothesized, so that Σ_X and F are 2×2 matrices. The restriction that $F'F = I$ gives

$$F'F = \begin{bmatrix} f_{11} & f_{21} \\ f_{12} & f_{22} \end{bmatrix} \begin{bmatrix} f_{11} & f_{12} \\ f_{21} & f_{22} \end{bmatrix} = \begin{bmatrix} 1 & 0 \\ 0 & 1 \end{bmatrix}$$

or

$$f_{11}^2 + f_{21}^2 = 1, \qquad f_{11}f_{12} + f_{21}f_{22} = 0,$$
$$f_{11}f_{12} + f_{21}f_{22} = 0, \qquad f_{12}^2 + f_{22}^2 = 1.$$

These conditions are satisfied if

$$F = \begin{bmatrix} f_{11} & \mp\sqrt{1 - f_{11}^2} \\ \pm\sqrt{1 - f_{11}^2} & f_{11} \end{bmatrix}$$

for any f_{11} such that $-1 \leqslant f_{11} \leqslant +1$. Thus $X^* = XF$ implies that the observations are linked as

$$X_{11}^* = f_{11}X_{11} \pm \sqrt{1 - f_{11}^2}\, X_{12}, \quad X_{12}^* = \mp\sqrt{1 - f_{11}^2}\, X_{11} + f_{11}X_{12},$$
$$\vdots \qquad\qquad\qquad \vdots$$
$$X_{T1}^* = f_{11}X_{T1} \pm \sqrt{1 - f_{11}^2}\, X_{T2}, \quad X_{T2}^* = \mp\sqrt{1 - f_{11}^2}\, X_{T1} + f_{11}X_{T2},$$

and $L^* = F'L$ implies

$$L_{11}^* = f_{11}L_{11} \pm \sqrt{1 - f_{11}^2}\, L_{21}, \qquad \cdots \qquad L_{1K_2}^* = f_{11}L_{1K_2} \pm \sqrt{1 - f_{11}^2}\, L_{2K_2},$$
$$L_{21}^* = \mp\sqrt{1 - f_{11}}\, L_{11} + f_{11}L_{21} \quad \cdots \quad L_{2K_2}^* = \mp\sqrt{1 - f_{11}^2}\, L_{1K_2} + f_{11}L_{2K_2}.$$

Once we have assumed that $\Sigma_X = I$ as in Eq. (10.36), it is convenient to transform our observational covariance matrix into the correlation matrix for the Z's, R_Z. Now the interpretation of the factor loadings L_{ij} is that they measure the correlation between the observed variable Z_j and the true variable X_i. Because it is assumed that the X's are orthogonal, the sum of the squared values of the loadings in the jth column of L, $\sum_{i=1}^{K_1} L_{ij}^2$, estimates the proportion of the variance in Z_j explained by the true variables X. One minus this value is the proportion accounted for by the error term ε and is the estimate for the corresponding element in Σ_ε.

Once the factor analysis has been reduced to the model shown in Eq. (10.36), there are a number of computational schemes proposed to estimate the values for L and Σ_ε. The most common computational schemes begin with an estimate of Σ_ε. This estimated error matrix is then subtracted from R_Z. Since Σ_ε is diagonal, this amounts to making the diagonals of R_Z less than one and equal to the proportion of the variance in each observed variable which should be accounted for by the unobserved underlying variables. (The diagonal terms in the adjusted correlation matrix are referred to as the commonalities and simply equal one minus the unexplained variance of each observed variable, the appropriate element in Σ_ε.) Once this adjustment to R_Z is made, the true variables X and their relationships to the observed variables, defined by L, are estimated so that the first unobserved variable explains as much of the observed variance in R_Z as possible. The second factor is then chosen so that it is orthogonal to the first factor and explains as much of the remaining variance in R_Z as possible. This process continues for as many factors, or unobserved variables, as one has specified in the model.[11]

Conceptually, this is equivalent to picking the values for L which minimize the expression

$$\text{tr}\left[\left(R_Z - \hat{R}_Z \right)\left(R_Z - \hat{R}_Z \right) \right], \tag{10.37}$$

where tr denotes the trace of a matrix (see Appendix II.4), R_Z the observed correlational matrix, and $\hat{R}_Z$ the predicted matrix; $\hat{R}_Z = \hat{L}'\hat{L} + \hat{\Sigma}_\varepsilon$. Many times this is called the minimum residual, or minres, method because it minimizes the sum of squared deviations of $R_Z - \hat{R}_Z$, taken term by term. In programs where the analyst picks commonalities, R_Z and $\hat{R}_Z$ are the appropriate adjusted correlation matrices.

Most methodological discussions of factor analysis concentrate on two specific issues. The first is the estimation of the proportion of the variance of each observed variable "explained" by the hypothesized underlying variables, the commonality. While textbooks on factor analysis present several ways of obtaining these estimates, all we need to point out for this development is that what one is doing in choosing a communality is a priori estimating the values of Σ_ε, the unique measurement variance of each observed variable, and subtracting that from R_Z so that the model being estimated is $R_Z - \hat{\Sigma}_\varepsilon = L'L$.

The second and more major issue concerns the interpretation of the estimated factor loading $\hat{L}$ and the fact that they may be subject to multiplication by any orthogonal matrix F'. Conventional factor programs, which compute the factors on the basis of maximizing the amount of remaining variance explained by each factor, in effect arbitrarily choose a solution.

---

[11] For description of this and related methods see Harmon (1967) or Lawley and Maxwell (1971).

Most factor programs include options for multiplying L by various orthogonal matrices, or rotating the loadings. The objective here is to raise certain loadings and reduce others so that each observed variable is as uniquely correlated with one underlying variable as possible. There are a variety of definitions for what is meant by "uniquely as possible," and we shall not go into them here; all discussions of factor analysis go into these in detail.[12] We want simply to point out that all one is doing with these routines is mathematically transforming the results of the basic factor analysis model we have been discussing; it is not altering the fundamental model in any way.

The important question to ask at this point is, What social science models are appropriately estimated by this statistical procedure? As presented so far, the technique is strictly exploratory. One has a set of observed variables and is interested in how they correlate with a smaller set of implicit unobserved variables, which are assumed to be orthogonal. We have not been required to state any explicit causal relationships, we do not estimate or test any structure derived from a set of theoretical propositions, and we can evaluate the results only by how much of the variance of each observed variable is accounted for by the estimated underlying unobserved variables. This model, as such, is most appropriate to the case where one has an idea of what the underlying variables are and is interested only in evaluating how well the various observed variables measure each unobserved variable for use in future research. In this situation, one wants to ascertain which measures (observed variables) have the least unique variance and the strongest correlations with the hypothesized unobserved variables. One presumably then bases further research on the information in the "good" measures.

Factor Analysis: Overidentified Errors in Variable Model

It is possible to alter the formulation of the factor analytic model to include a behavioral representation. At the same time, we can consider alternative restrictions to the assumption that $\Sigma_X = I$. Both considerations will make the factor model more appropriate for social science use where one is primarily interested in estimating models with explicitly specified structures. Through such alterations, factor analysis becomes an alternative estimation procedure for the errors in variable model summarized in Eqs. (10.29) and (10.30).

We start with the behavioral expression described in Eq. (10.9), which shows the hypothesized causal relationships between the endogenous variable Y and the set of exogenous variables X,

$$Y = X\beta + U. \tag{10.38}$$

However, X is not observed directly; instead, we observe the erroneous variables in Z_1 and Z_2 whose relationships to X are

$$Z_1 = X + \varepsilon_1 \quad \text{and} \quad Z_2 = X\mathbf{A} + \varepsilon_2.$$

[12] See Harmon (1967) or Lawley and Maxwell (1971).

Equations (10.29) and (10.30) show the assumed relationship between the covariances of the observed variables and those predicted by the above model.

Equation (10.30) fits the standard factor analysis format:

$$\Sigma = \begin{bmatrix} \beta' \\ I \\ A' \end{bmatrix} \Sigma_X \begin{bmatrix} \beta & I & A \end{bmatrix} + \begin{bmatrix} \sigma_U^2 & 0 & 0 \\ 0 & \Sigma_{\varepsilon_1} & 0 \\ 0 & 0 & \Sigma_{\varepsilon_2} \end{bmatrix} = L'\Sigma_X L + Q, \quad (10.39)$$

as long as all the separate error terms in ε_1 and ε_2 are independent (i.e., that Σ_{ε_1} and Σ_{ε_2} are diagonal). The matrix of factor loadings L is simply the partitioned matrix of coefficients $L = (B \quad I \quad A)$. Factor analysis is a useful procedure if Z_2 contains more variables than X (and Z_1) or if we want to estimate A, Σ_X, σ_U^2, Σ_{ε_1}, and Σ_{ε_2} as well as β.

To see that this structure with $Z_1 = X + \varepsilon_1$ is identified, we begin by considering admissible transformations of the structure given by (β^* I $A^*) = L^* = F^{-1}L = F^{-1}[\beta \quad I \quad A]$ and $\Sigma_{X^*} = F'\Sigma_X F$. We are hypothesizing, K_1 unobserved variables, so that Σ_x and F are $K_1 \times K_1$ square matrices, Z_1 is $T \times K_1$ and Z_2 is $T \times K_2$, with $K_2 \geqslant K_1$. β is $K_1 \times 1$, A is $K_1 \times K_2$, and the identity matrix in L is $K_1 \times K_1$. The identification conditions are seen by Eq. (10.35):

$$L^* = F^{-1}[\beta \quad I \quad A] = [F^{-1}\beta \quad F^{-1} \quad F^{-1}A] = [\beta^* \quad I \quad A^*].$$

The constraint that $Z_1 = X + \varepsilon_1$ can be satisfied only if $F^{-1} = I = F$, and the model is identified.

The factor analysis technique used to estimate Eq. (10.39) is not the same as the factor analysis programs in most available standard statistical packages. In the present case, identification is possible because part of the matrix of factor loadings is specified to be an identity matrix. This specification implies that each variable in Z_1 is systematically related to only one of the unobserved exogenous variables and that each unobserved variable is scaled on the same basis as the observed variable in Z_1 to which it is related. This model places no restrictions on X or Σ_X, such as assuming that all X variables are orthogonal with unit variances (i.e., that $\Sigma_X = I$) and is commonly referred to as a restricted or confirmatory factor analysis.

The restriction $\Sigma_X = I$ is unrealistic for most social science modeling. Remember that even though a variable is not observed, it implicitly has values for each observational unit in the sample just as if it were observed. The assumptions required by different statistical procedures must be evaluated in the context of what they are implying about the characteristics of these implicit values for the unobserved variables. The restriction that $\Sigma_X = I$ is unrealistic because of the implausability of finding even true variables in the social sciences orthogonal to each other and having unit variances for any given and available sample. Certainly it is possible with any given data set to transform the variables so that they are orthogonal with unit variances. However, this makes the analysis very sample specific, impossible to replicate for other samples, and not comparable between studies. These are the same arguments against sample specific statistics made in Chapters 1 and 4 and in the discussion of standardized regression coefficients.

The prior assumption that each of the observed variables in Z_1 is related to only one of the unobserved variables in X is a sufficient condition to identify the model so long as both the assumptions about the error terms hold and Z_1 and Z_2 have at least K_1 distinct variables. It is possible however to relax the assumption about the loadings matrix containing an identity matrix. More elaborate structures are possible whereby each observed variable relates to several of the unobserved explanatory variables, but it is necessary that the parts of the loadings matrix relating Z_1 and Z_2 to X contain some zeros. (If $L_{ij} = 0$, the jth observed variable is not a function of the ith true variable.) In other words, each observed variable cannot be related to all of the unobserved variables in order for the model to be identified. This simply means there must be a sufficient number of zero elements in L, and they must be placed in particular places. Unfortunately, there is no easy way to describe the identification requirements for this model the way we did for the simultaneous equation model. One must specify a structure and then work out the details for that structure to ensure that it is identified. The best, but still arduous, way to do this is to investigate whether there is an alternative set of loadings $L^* = F^{-1}L$ that gives the same expected covariance matrix *and* maintains the restrictions placed on L, where F can be any arbitrary nonsingular square matrix.

It is also possible to expand the behavioral model to consider several reduced form relations. In this case, the behavioral model $Y = X\beta + U$ constitutes a set of M equations and β is a $K \times M$ matrix of coefficients. Although it is difficult to envision situations where an exogenous variable is excluded from a reduced form equation, if such were the case, this restriction on the loading matrix can aid in identifying the model, just as restrictions on A do. (In fact we can interpret all the observed variables Y, Z_1, and Z_2 as multiple measures of X in the usual factor analysis sense.)

These restrictions placed on A (and possibly β) must reflect the researcher's judgment about the relationships in the behavioral part of the model (which elements of β are zero), about how the observed variables relate to the underlying variables in terms of which ones measure each unobservable variable (which elements of A are zero) or possibly that some pairs of underlying variables are orthogonal, so that the appropriate element of Σ_X is zero. These restrictions must be based on theoretical arguments or prior empirical findings. Once these restrictions have been applied to the structure, the question is to assess whether any values for F, other than the identity matrix, maintain these restrictions in the alternative structures.

Even once the model is identified through the restrictions placed on it, we still face an estimation problem. We do not want to scale the observed variables by their sample variances to get R_Z; if we do that, we do not get estimates of the population coefficients but simply an equivalent of the standardized regression coefficients. We could choose the coefficients to minimize the expression equivalent to Eq. (10.37) for the covariances of Z rather than for the correlations:

$$\text{tr}\left[\left(S_Z - \hat{S}_Z\right)\left(S_Z - \hat{S}_Z\right)\right].\qquad(10.40)$$

This substitutes one problem for another however. The estimates are no longer sample specific but they do depend upon how each of the observed variables is scaled. If we were to multiply one of the observed variables by a constant, all elements in $S_Z - \hat{S}_Z$ involving that variable will also be multiplied by that constant. This in turn will give these elements greater weight in the minimization process and lead to different estimates for all the coefficients, variances, and covariances. This scaling is not a problem in the conventional regression model because only the coefficient on the variable whose scale is changed is altered. For example, changing an explanatory variable from dollars to thousands of dollars multiplies the coefficient on that variable by a thousand, but leaves all other coefficients unchanged. In the factor analysis model, this alteration changes all the coefficients—the estimates are not scale independent. This is clearly an unsatisfactory situation: which set of coefficients do you want to take as estimates of the population values, the one with income in dollars or in thousands of dollars? This problem is not present if we deal with the correlation matrix because the correlations are unchanged by linear transformations of the variables. This is another reason many people prefer to analyze the correlation rather than the covariance matrix. However, this is unsatisfactory for estimating the population coefficients since the estimates are then sample dependent, and there would have to be an assumption that the sample is representative of the population.

There are alternative criteria for selecting values for L and Σ_ϵ. The most common alternative minimizes the function

$$\log|\hat{S}| + \text{tr}(S\hat{S}^{-1}).\qquad(10.41)$$

where $|\hat{S}|$ denotes the determinant of $\hat{S}$. Scaling one of the variables simply adds a constant to this expression, which leaves the minimization process and the values selected by that process unaffected.[13] Thus it solves the scaling

---

[13] This statement relies upon the property that if one column or one row of a matrix B is multiplied by a scalar c to form a new matrix B^*, then $|B^*| = c|B|$. Further, scalar multiplication enters both S and S^{-1} in opposite ways, so that parameters that minimize $\log|\hat{S}| + \text{tr}(S\hat{S}^{-1})$ also minimize $\log|c\hat{S}| + \text{tr}[(cS)(c\hat{S})^{-1}]$.

problem inherent in Eq. (10.40). The principal feature of this expression, and the criteria from which it is derived, is that if the observed variables Z and Y are normally distributed, the estimates provided by Eq. (10.41) are the maximum likelihood estimates. Thus if one can legitimately assume normality, these estimates have many desirable properties.[14] If the observed variables substantially deviate from the multivariate normal distribution, the only claim that can be made is that we have avoided the scaling problem, but the estimates can no longer be assumed to be the maximum likelihood estimates. They still are consistent, however, so that if one is dealing with large sample sizes these estimates still have some justification.

The discussion to this point establishes two thing. First, the general use of factor analysis as an exploratory technique is very difficult to relate to any underlying behavioral structure. The assumptions commonly introduced, such as $\Sigma_X = I$, and the arbitrary choice of structure through factor rotation inhibit our ability to interpret the factor loadings in a behavioral sense. Secondly, the factor analytic structure and the statistical theory that relates to it may be useful for consideration of the errors in variables problems. By beginning with an underlying behavioral model, it is possible to use factor analytic techniques (in a somewhat modified version) to obtain parameter estimates whenever the underlying parameters are identified. Again, as in the simultaneous equations case, the usefulness of the estimation technique is completely dependent upon whether or not the model is identified.

10.5 Linear Structural Models
and the General Analysis of Covariances

This discussion completes the comparison between the errors in variables or unobserved variables model and factor analysis. There are more complicated models, such as the aspiration model with the unobserved ambition variables, which do not conform to the factor analysis specification. Alternative means are available for estimating these models, but some are fairly ad hoc and useful only for certain applications. The more formal procedures and

---

[14] What does it mean to assume that the observed variables are normally distributed? First, it does not mean that a frequency plot of the T values for each observed variable conform to a normal distribution. The assumption refers to how *each* observed value for any variable is distributed about its own expected value. Thus the assumption is that the values of Z_{ti} generated by all replications are normally distributed for each t and each i. (The formal assumption is that the distribution of the values for the tth observation on all variables is multivariate normal.) What is necessary to meet this requirement is the same assumptions we have been making since Chapter 3. If the values for the exogenous variables are fixed, regardless of whether they are observed or not, and if the various individual errors in U, ε_1, and ε_2 are normally distributed, then the individual values of the observed variables will be normally distributed. Thus the assumptions necessary for maximum likelihood estimates are fixed values for the true exogenous variables and normally distributed error terms.

statistical estimation programs are quite complicated, and the details of these formal procedures are beyond the scope of this book.[15] However, we want to outline their general format and show what types of models can be estimated with them. The formal models are generally referred to as analyses of covariance structures for the simple reason that they estimate the parameters of the formal structural model on the basis of the covariances of the observed variables, just as factor analysis does. The structural models however can be much more general.

The structural model for the errors in the explanatory variables case described in the previous section and estimated by a restricted factor analysis needs to be expanded in two ways to make it completely general. First, note that the model in Eq. (10.38) corresponds to a reduced form model. The initial extension replaces the reduced form behavioral model in Eq. (10.38) by a set of structural equations relating the endogenous and true predetermined variables. This set of structural equations can be either hierarchical or simultaneous, depending upon the behavior being modeled. The second extension treats the case where there are unobserved endogenous variables. This can be interpreted as a model where both the exogenous and endogenous variables are observed with error and there are multiple measures of both. [The model without multiple measures of the endogenous variables is simply the errors in explanatory variables model discussed in the previous section since Y in Eq. (10.38) already contains an error term.] Finally, both extensions can be combined into a single linear structural equation model.

The general linear structural equation model is the least constrained model we have discussed. It is also the most complicated. In particular, because of its generality, one must be careful in specifying the model and considering the identification problem. This model is shown in the equations

$$Y\Gamma + X\beta + U = 0, \tag{10.42}$$

$$Z_X = XA_1 + \varepsilon_1, \tag{10.43}$$

$$Z_Y = YA_2 + \varepsilon_2. \tag{10.44}$$

These equations simply extend the previous errors in variables model in the two directions just mentioned. In this model, Y and X are the matrices of implicit values for the unobserved variables, Z_X is the set of observed variables related to the unobserved predetermined variables, and Z_Y is the matrix of observations of the variables related to the true endogenous variables. The addition of the matrix Γ in Eq. (10.42) makes the behavioral model a set of structural equations, and Eq. (10.44) adds the extension that the true endogenous variables are either not observed or are measured with error. The presence of the error term in Eq. (10.42) indicates that even in the true model, the specified relationships may not account for all behavior. The

---

[15] See Jöreskog (1973) and Jöreskog and van Thillo (1972).

intention is to use the observational information contained in Z_X and Z_Y to estimate the coefficient matrices β, Γ, A_1, and A_2 and the unknown variance–covariance matrices Σ_U, Σ_X, Σ_{ε_1}, and Σ_{ε_2}. [If one is interested only in the reduced form expression for the behavioral model, Eq. (10.42) can be respecified as $Y = X\Pi + V$, where $\Pi = -\beta\Gamma^{-1}$ and $V = -U\Gamma^{-1}$.]

The estimation of the model begins in the same manner as the basic factor analytic model. From Eq. (10.42)–(10.44) we can write the expected variance–covariance matrix for the observed variables, the Z's. We make the same assumptions as previously: that U, ε_1, and ε_2 are distributed independently of X and of each other. For the time being we shall assume that the individual measurement error terms are also independent of each other, meaning that Σ_{ε_1} and Σ_{ε_2}, but not Σ_U, are diagonal. The variance–covariance matrix of the observed variables $Z = (Z_X, Z_Y)$ is[16]

$$S_Z = \frac{1}{T} Z'Z = \frac{1}{T} \begin{bmatrix} Z_X'Z_X & Z_X'Z_Y \\ Z_Y'Z_X & Z_Y'Z_Y \end{bmatrix},$$

$$\Sigma_Z = E(S_Z) = \begin{bmatrix} A_1'\Sigma_X A_1 & -A_1'\Sigma_X \beta\Gamma^{-1}A_2 \\ -A_2'\Gamma'^{-1}\beta'\Sigma_X A_1 & A_2'(\Gamma'^{-1}\beta'\Sigma_X \beta\Gamma^{-1} + \Gamma'^{-1}\Sigma_U\Gamma^{-1})A_2 \end{bmatrix}$$

$$+ \begin{bmatrix} \Sigma_{\varepsilon_1} & 0 \\ 0 & \Sigma_{\varepsilon_2} \end{bmatrix}. \tag{10.45}$$

Even if Σ_{ε_1} and Σ_{ε_2} are diagonal, Eq. (10.45) does not fit the basic factor analysis specification.

In its full form, Eq. (10.45) relates all the unknown parameters, variances, and covariances we have been discussing to the observational data contained in the measured variables as summarized by the variances and covariances in S_Z. We intend to use these measured variances and covariances as the best guess of the variances and covariances predicted, or expected, by the hypothesized model that is summarized by Σ_Z and the righthand side of Eq.

---

[16] Straightforward substitution of Eqs. (10.42)–(10.44) yields

$$S = \begin{bmatrix} A_1'X'XA_1 + A_1'X'\varepsilon_1 + \varepsilon_1'XA_1 + \varepsilon_1'\varepsilon_1 & -A_1'X'X\beta\Gamma^{-1}A_2 - A_1'X'U\Gamma^{-1}A_2 + A_1'X'\varepsilon_2 \\ & -\varepsilon_1'U\Gamma^{-1}A_2 - \varepsilon_1'X\beta\Gamma^{-1}A_2 + \varepsilon_1'\varepsilon_2 \\ \hline -A_2'\Gamma'^{-1}\beta'X'XA_1 - A_2'\Gamma'^{-1}\beta'X'\varepsilon_1 & A_2'\Gamma'^{-1}\beta'X'X\beta\Gamma^{-1}A_2 + A_2'\Gamma'^{-1}\beta'X'U\Gamma^{-1}A_2 \\ -A_2'\Gamma'^{-1}U'XA_1 - A_2'\Gamma'^{-1}U'\varepsilon_1 & +A_2'\Gamma'^{-1}U'X\beta\Gamma^{-1}A_2 + A_2'\Gamma'^{-1}U'U\Gamma^{-1}A_2 \\ +\varepsilon_2'XA_1' + \varepsilon_2'\varepsilon_1 & -\varepsilon_2'X\beta\Gamma^{-1}A_2 - \varepsilon_2'U\Gamma^{-1}A_2 - A_2'\Gamma'^{-1}\beta'X'\varepsilon_2 \\ & -A_2'\Gamma'^{-1}U'\varepsilon_2 + \varepsilon_2'\varepsilon_2 \end{bmatrix}.$$

Equation (10.45) is the expected value of this expression after introducing the assumptions about the error terms $U, \varepsilon_1, \varepsilon_2$. The measurement error variances and covariances are then separated to put the estimating equation in the form of a factor analytical model.

(10.45). The estimation procedure proceeds by selecting the values for the unknown coefficients and variances and covariances that minimize the expression in Eq. (10.41). (The values for $\hat{S}$ are based upon the estimated parameters.) The actual procedure for selecting the estimates that minimize this function involve taking the derivatives of that function with respect to each parameter being estimated and solving this set of simultaneous equations for the values that equate these derivatives to zero. The precise formulas for these derivatives and the computational procedures used to determine the solution are beyond the scope of this textbook and require quite advanced mathematical and computational procedures. The important point is the structure of the model, the ways it can be specified to handle different estimation problems, and the considerations involved in applying the procedure.

We can best illustrate these points by using the method to estimate the various versions of the peer influence model discussed so far. The first version did not include the unobserved ambition variable and hypothesized simultaneous peer influences both on occupational and educational aspirations plus a causal influence from occupational to educational aspirations. This model is repeated in the following equations:

$$P_1 = I_1\beta_{11} + S_1\beta_{21} \qquad\qquad\qquad\qquad\qquad\qquad\qquad + U_1, \quad (10.46)$$

$$P_2 = I_2\beta_{32} + S_2\beta_{42} \qquad\qquad\qquad\qquad\qquad\qquad\qquad + U_2, \quad (10.47)$$

$$O_1 = I_1\beta_{13} + S_1\beta_{23} + P_1\gamma_{13} \qquad\qquad\quad + O_2\gamma_{43} \qquad\qquad + U_3, \quad (10.48)$$

$$O_2 = I_2\beta_{34} + S_2\beta_{44} \quad + P_2\gamma_{24} + O_1\gamma_{34} \qquad\qquad\qquad + U_4, \quad (10.49)$$

$$E_1 = I_1\beta_{15} + S_1\beta_{25} + P_1\gamma_{15} \qquad + O_1\gamma_{35} \qquad\quad + E_2\gamma_{65} + U_5, \quad (10.50)$$

$$E_2 = I_2\beta_{36} + S_2\beta_{46} \quad + P_2\gamma_{26} \qquad + O_2\gamma_{46} + E_1\gamma_{56} \qquad + U_6. \quad (10.51)$$

It is straightforward to see how this model specification fits the form of Eq. (10.45). There are no unobserved or erroneously measured variables. Consequently,

$$Z_X = X = \begin{bmatrix} I_1 & S_1 & I_2 & S_2 \end{bmatrix} \quad \text{and} \quad Z_Y = Y = \begin{bmatrix} P_1 & P_2 & O_1 & O_2 & E_1 & E_2 \end{bmatrix},$$

where the elements inside the brackets represent column vectors of all the observations on the respective variables. The specification that $Z_X = X$ and $Z_Y = Y$ implies

$$\mathbf{A}_1 = I, \qquad \mathbf{A}_2 = I, \qquad \Sigma_{\varepsilon_1} = 0, \qquad \Sigma_{\varepsilon_2} = 0, \qquad \text{and}$$

$$\Sigma_X = \frac{1}{T} \begin{bmatrix} I_1' \\ S_1' \\ I_2' \\ S_2' \end{bmatrix} \begin{pmatrix} I_1 & S_1 & I_2 & S_2 \end{pmatrix} = \frac{1}{T} \begin{bmatrix} I_1'I_1 & & & \\ S_1'I_1 & S_1'S_1 & & \\ I_2'I_1 & I_2'S_1 & I_2'I_2 & \\ S_2'I_1 & S_2'S_1 & S_2'I_2 & S_2'S_2 \end{bmatrix}.$$

(Since Σ_X is symmetrical, the upper triangle is not filled in.) The specification of Eqs. (10.46)–(10.51) results in the following coefficient and error term covariance matrices:

$$\Gamma = \begin{bmatrix} -1 & 0 & \gamma_{13} & 0 & \gamma_{15} & 0 \\ 0 & -1 & 0 & \gamma_{24} & 0 & \gamma_{26} \\ 0 & 0 & -1 & \gamma_{34} & \gamma_{35} & 0 \\ 0 & 0 & \gamma_{43} & -1 & 0 & \gamma_{46} \\ 0 & 0 & 0 & 0 & -1 & \gamma_{56} \\ 0 & 0 & 0 & 0 & \gamma_{65} & -1 \end{bmatrix},$$

$$\beta = \begin{bmatrix} \beta_{11} & 0 & \beta_{13} & 0 & \beta_{15} & 0 \\ \beta_{21} & 0 & \beta_{23} & 0 & \beta_{25} & 0 \\ 0 & \beta_{32} & 0 & \beta_{34} & 0 & \beta_{36} \\ 0 & \beta_{42} & 0 & \beta_{44} & 0 & \beta_{46} \end{bmatrix},$$

$$\Sigma_U = \begin{bmatrix} \sigma_{11} & & & & & \\ \sigma_{12} & \sigma_{22} & & & & \\ \sigma_{13} & 0 & \sigma_{33} & & & \\ 0 & \sigma_{24} & \sigma_{34} & \sigma_{44} & & \\ \sigma_{15} & 0 & \sigma_{35} & 0 & \sigma_{55} & \\ 0 & \sigma_{26} & 0 & \sigma_{46} & \sigma_{56} & \sigma_{66} \end{bmatrix}.$$

This is the same model discussed and estimated in Chapter 9. That discussion showed that the model is identified by combining the information contained in the reduced form coefficients and the reduced form residuals. The estimation at that point proceeded on an equation by equation and residual covariance by residual covariance basis. The present method permits us to combine all of this information into an estimation procedure that simultaneously estimates all the parameters on the basis of all the observational information, thus the name full information.

The coefficients estimated by the full information procedure are shown in Table 10.5, along with the values estimated in Chapter 9 by the single equation method.

We can see that the two methods give quite different estimates for some of the coefficients. The differences are most notable for the influence of parental aspirations on occupational and educational aspirations. However, it is also the case that neither method gives realistic estimates. We shall discuss the assumptions required to interpret the estimated standard errors subsequently. For the time being, however, it is appropriate to point out that the estimated

TABLE 10.5

Full Information and Single Equation Estimates of Simultaneous
Occupation and Education Aspiration Model[a]

		Respondent		Friend	
		Full information	Single equation	Full information	Single equation
Parental aspirations					
Intelligence	(β_{11})	0.11	0.11	(β_{32}) 0.14	0.14
		(0.03)	(0.03)	(0.03)	(0.03)
Status	(β_{21})	0.00	−0.01	(β_{42}) −0.06	−0.08
		(0.032)	(0.03)	(0.03)	(0.03)
Occupational aspirations					
Intelligence	(β_{13})	0.76	0.68	(β_{34}) 0.86	0.88
		(0.27)		(0.30)	
Status	(β_{23})	0.41	0.35	(β_{44}) 0.40	0.31
		(0.13)		(0.18)	
Parental aspiration	(γ_{13})	−1.49	−0.34	(γ_{24}) −0.54	−0.44
		(2.32)		(2.17)	
Peer occupational aspiration	(γ_{43})	0.41	0.41	(γ_{34}) 0.32	0.35
		(0.10)	(0.10)	(0.12)	(0.12)
Educational aspirations					
Intelligence	(β_{15})	0.35	0.56	(β_{36}) 0.31	0.51
		(0.27)		(0.27)	
Status	(β_{25})	0.41	0.53	(β_{46}) 0.28	0.19
		(0.16)		(0.24)	
Parental aspiration	(γ_{15})	−0.75	0.62	(γ_{26}) 0.16	−1.09
		(1.78)		(1.73)	
Occupational aspiration	(γ_{35})	0.58	0.20	(γ_{46}) 0.67	0.51
		(0.25)	(0.60)	(0.27)	(0.78)
Peer educational aspiration	(γ_{65})	0.04	0.20	(γ_{56}) 0.05	0.20
		(0.09)	(0.26)	(0.07)	(0.27)

	Σ_U (full information)					
	U_1	U_2	U_3	U_4	U_5	U_6
U_1	9.86					
	(0.77)					
U_2	0.89	8.92				
	(0.52)	(0.70)				
U_3	19.66	0.00	149.70			
	(22.75)		(92.47)			
U_4	0.00	8.98	−50.92	101.60		
	(19.29)	(20.48)	(40.80)			
U_5	11.99	0.00	10.08	0.00	97.08	
	(17.37)		(52.07)		(43.82)	
U_6	0.00	3.16	0.00	−16.77	7.61	76.54
		(14.81)		(24.29)	(10.54)	(20.54)

[a]Note: standard errors (shown in parentheses) for some coefficients and most error term covariances would not be estimated with the equation by equation method.

standard errors for the parental aspiration coefficients are considerably larger than the estimated coefficients themselves. Thus there is very little we can say about these estimates with any confidence. The major substantive difference in the two methods is that the full information method indicates virtually no simultaneous peer effects on educational aspirations and larger effects of occupational aspirations on educational aspirations than does the single-equation method. The lower estimated standard errors for the full information coefficients are undoubtedly the result of being able to use more sample information than just the reduced form coefficients, namely that contained in $(1/T)(Z'_Y Z_Y)$, to estimate the coefficients. The more information we can use in estimating the coefficient the more confident we can be in that estimate.

The second example is the peer influence model containing the two hypothesized, but unmeasured, ambition variables. This is the model presented at the beginning of this chapter. Due to the very unreliable estimates of the influence of parental ambitions in the model just estimated and for ease of pedagogical presentation and comparison of alternative estimation techniques, the parental ambition variables will be omitted from the model. This is done by substituting the expressions explaining parental ambitions [Eqs. (10.1) and (10.2)] for the actual variable in subsequent equations. This substitution was used in the previous chapter to estimate the peer influence parts of the model. This respecification gives the following model:

$$Y_1 = I_1 \beta_{13}^* + S_1 \beta_{23}^* \qquad\qquad\qquad + Y_2 \gamma_{43} + U_3^*, \qquad (10.52)$$
$$Y_2 = \qquad\quad I_2 \beta_{34}^* + S_2 \beta_{44}^* + Y_1 \gamma_{34} \qquad\quad + U_4^*, \qquad (10.53)$$
$$O_1 = \qquad\qquad\qquad\quad Y_1 \qquad\qquad + \varepsilon_1, \qquad (10.54)$$
$$O_2 = \qquad\qquad\qquad\qquad Y_2 \quad + \varepsilon_2, \qquad (10.55)$$
$$E_1 = \qquad\qquad\qquad Y_1 \alpha_1 \qquad\quad + \varepsilon_3, \qquad (10.56)$$
$$E_2 = \qquad\qquad\qquad\qquad Y_2 \alpha_2 \; + \varepsilon_4. \qquad (10.57)$$

This specification includes the two unobserved ambition variables Y_1 and Y_2 which are simultaneously related. The observed occupational and educational aspiration variables are treated as separate and erroneous measures of ambition. The errors are represented by the terms ε_1 to ε_4. Note that these errors may represent factors other than purely random measurement error. If variables other than ambition influence occupational or educational aspirations, their effects are included in the ε terms.

This specification maintains the previous assumption that the exogenous intelligence and status variables are measured without error, so that $Z_X = X = (I_1 \quad S_1 \quad I_2 \quad S_2)$; $A_1 = I$; and $\Sigma_{\varepsilon_1} = 0$. With $Z_Y = (O_1 \quad O_2 \quad E_1 \quad E_2)$, the remaining terms in Eq. (10.45) are

$$\Gamma = \begin{bmatrix} -1 & \gamma_{34} \\ \gamma_{43} & -1 \end{bmatrix}, \quad \beta = \begin{bmatrix} \beta_{13}^* & 0 \\ \beta_{23}^* & 0 \\ 0 & \beta_{34}^* \\ 0 & \beta_{44}^* \end{bmatrix}, \quad A_2 = \begin{bmatrix} 1 & 0 & \alpha_1 & 0 \\ 0 & 1 & 0 & \alpha_2 \end{bmatrix},$$

$$\Sigma_U = \begin{bmatrix} \sigma_{11} & \sigma_{12} \\ \sigma_{12} & \sigma_{22} \end{bmatrix},$$

$$\Sigma_{\varepsilon_2} = E \begin{bmatrix} \varepsilon_1'\varepsilon_1 & & & \\ \varepsilon_1'\varepsilon_2 & \varepsilon_2'\varepsilon_2 & & \\ \varepsilon_1'\varepsilon_3 & \varepsilon_2'\varepsilon_3 & \varepsilon_3'\varepsilon_3 & \\ \varepsilon_1'\varepsilon_4 & \varepsilon_2'\varepsilon_4 & \varepsilon_3'\varepsilon_4 & \varepsilon_4'\varepsilon_4 \end{bmatrix} = \begin{bmatrix} \sigma_{\varepsilon_{11}} & & & \\ \sigma_{\varepsilon_{12}} & \sigma_{\varepsilon_{22}} & & \\ \sigma_{\varepsilon_{13}} & \sigma_{\varepsilon_{23}} & \sigma_{\varepsilon_{33}} & \\ \sigma_{\varepsilon_{14}} & \sigma_{\varepsilon_{24}} & \sigma_{\varepsilon_{34}} & \sigma_{\varepsilon_{44}} \end{bmatrix}.$$

Σ_{ε_2} is written out in full form because the assumptions one makes about which error covariances are zero are critical to identification. We know the coefficients in Γ, β, and A are identified from the earlier discussions. (They can be determined simply from the reduced form coefficients given by $\Pi = \beta \Gamma^{-1} A_2$.) However, without further knowledge of the error structure, Σ_U and Σ_{ε_2} are unidentified. For example, if we can assume that $\sigma_{\varepsilon_{13}}$, $\sigma_{\varepsilon_{14}}$, $\sigma_{\varepsilon_{23}}$, and $\sigma_{\varepsilon_{24}}$ all equal zero, Σ_U and Σ_{ε_2} are identified. To see the identification problem, note that Σ_U and Σ_{ε_2} together contain thirteen unknowns while $(1/T)(Z_Y'Z_Y)$ contains only ten separate terms that are functions of these unknowns. The model then violates the necessary condition for identifiability.

It is probably not too difficult to justify the assumption that $\sigma_{\varepsilon_{14}}$ and $\sigma_{\varepsilon_{23}}$ equal zero since they refer to the different aspirations of different individuals. The assumptions that $\sigma_{\varepsilon_{13}}$ and $\sigma_{\varepsilon_{24}}$ equal zero are more difficult to accept since they relate to the two aspirations of the same individual. The calculations using education aspirations as instruments for occupational aspirations (and ambitions) made this assumption and we shall continue it here simply to be able to compare the coefficient estimates from the different estimation methods. It should be recognized that this is not good practice however. The assumptions should be made only if they can be justified on the basis of prior knowledge. The three alternative estimates are shown in Table 10.6.

Again, the estimates produced by the alternative procedures show some significant differences. The full information method gives smaller estimates of the effects of peer influences than when the exogenous intelligence and status variables are used to create the instruments. The full information method also gives higher estimates for the influence of family status and lower estimates of the effect of intelligence on ambitions than do the other methods.

TABLE 10.6

Alternative Estimates of Ambition Model

	Respondent				Friend			
	Full information	Instrumental Variables E_2	Instrumental Variables I_2 and S_2		Full information	Instrumental Variables E_1	Instrumental Variables I_1 and S_1	
Ambition								
Intelligence	0.69 (0.10)	0.69 (0.12)	0.64 (0.12)	(β^*_{13}) / (β^*_{34})	0.79 (0.11)	0.85 (0.12)	0.82 (0.13)	
Status	0.52 (0.10)	0.39 (0.12)	0.35 (0.12)	(β^*_{23}) / (β^*_{44})	0.46 (0.07)	0.38 (0.11)	0.35 (0.12)	
Peer Ambition	0.33 (0.09)	0.32 (0.08)	0.41 (0.10)	(γ_{43}) / (γ_{34})	0.31 (0.10)	0.29 (0.08)	0.35 (0.12)	
Education aspiration								
Ambition	1.05 (0.10)			(α_1) / (α_2)	1.03 (0.08)			

Error term covariances[a]

Σ_U

49.91 (7.88)	
−17.67 (8.00)	42.66 (6.73)

Σ_{e_2}

65.35 (8.78)			
15.42 (5.22)	62.92 (7.77)		
0.00	0.00	52.56 (8.88)	
0.00	0.00	5.70 (4.88)	47.04 (7.38)

[a] Estimated for full information model only.

The choice of procedure ultimately comes down to questions of availability, computational ease, and a belief in one method or another in particular circumstances. In general, the single-equation methods are more readily available—all one needs is a two-stage least squares regression program or an OLS package that stores fitted values. These techniques also are considerably cheaper computationally. The instrumental variables method is a linear estimator in that all the equations to be solved to compute the estimates are linear and thus easily and cheaply solved by digital computers. The full information method discussed here requires the solution of a set of nonlinear simultaneous equations, which is not so easily and cheaply done by current computers and programs. In estimating more complex models than we have discussed here, the additional complexity and computational cost may be considerable. These considerations do not relate to the accuracy of one estimator over another however—only to their convenience and cost. Most discussions of full information estimators, which are summarized briefly at the end of the previous chapter, point out that full information methods are better in the abstract and in ideal situations. The method makes maximal use of the available information—both that contained in the data and the prior knowledge used to specify the entire structural system. Single-equation methods consider only the information in the covariances of the exogenous and endogenous variables, as summarized in the reduced form coefficients, and the specification of the equation actually being estimated. However, if some equations in the model are badly misspecified or if some of the observational data are particularly poor, it may be an advantage to exclude these errors from the estimation of some equations. The choice then as to which estimates to believe and which method to use rests with feelings about how good the data are and how well specified the whole system is compared to the individual equations.

The Normality Assumption
and Evaluations of the Estimated Structure

Satisfying the normality assumption provides two additional benefits besides having estimates that are maximum likelihood (the advantages of which are discussed in the statistical review). In the first place, it becomes relatively straightforward to compute the standard errors of the estimated coefficients. These standard errors are computed as the second derivatives of the function in Eq. (10.41). Although the formulas themselves are quite complex, estimates of their values are computed as part of the computational procedure in most maximum likelihood programs. Thus the standard errors, as shown in Tables 10.4 and 10.5, are a standard part of the output. These estimated standard errors are important if we are to develop an idea of how reliable and accurate the estimated coefficients are. For example, the estimated standard errors for the coefficients estimating the influence of parental aspirations are so high as to make interpretation of the coefficients very hazardous and unreliable.

The second advantage gained by meeting the normality assumption is that we can develop a statistical test for the goodness of fit of the hypothesized structure. With some appropriate constants added, Eq. (10.41) is referred to as the log of the likelihood function, and minus twice this value is distributed as a chi-squared statistic for large samples. The degrees of freedom equal the number of observed variances and covariances minus the number of parameters estimated. For example, if we have K_2 observed variables in Z_X and M_2 variables in Z_Y and are estimating N different coefficients, there are $\frac{1}{2}[(K_2 + M_2)(K_2 + M_2 + 1)] - N$ degrees of freedom.

This chi-squared statistic can be used in two ways. In the first instance, it is a test of the hypothesized structure against the null hypothesis that the observed variances and covariances in S_Z could be any value (so long as the matrix satisfies the statistical properties of a covariance matrix). The larger the value of this chi-square, the more the observed covariances deviate from those predicted by the hypothesized and estimated structure. As the chi-squared statistic increases, the more likely we are to reject the null hypothesis that the specified structure generated the data. Conversely, if the deviations between actual and predicted covariances are small, the lower the value of the chi-squared statistic, and the less likely we are to reject the null hypothesis that the observed data can be described by the hypothesized and estimated model. Thus we have a general test of the specified model.

The second use of the chi-squared statistic is to test alternative structures. Consider two alternative models, with N_1 and N_2 estimated parameters respectively, $N_2 > N_1$ (where the N_1 parameters are included in the set of N_2 parameters). The difference in the chi-squared values calculated for each estimated structure is also distributed as a chi-squared with the degree of freedom equal to the difference in the number of estimated parameters, $N_2 - N_1$. If this difference exceeds some critical value for the number of degrees of freedom, we would then reject model one in favor of model two. If the difference in the chi-squared values is less than the critical value, then we cannot reject the first model in favor of the second.

We can illustrate the first application with the two versions of the peer influence model estimated in this chapter. Again we shall restrict our discussion to the version of the model that omits the parental aspiration variable. The observational information used to estimate both models is the same (S_Z), so we are considering which alternative model best describes this observational data. This is by no means the proper way to choose between two arbitrarily selected models. However, in cases where there are prior arguments, theory, or evidence to support each alternative, this method provides a way to test the alternatives statistically. The chi-squared values and the degrees of freedom for each of the peer influence models is shown in Table 10.7. The models' consistencies with the data are about equal. In each case, if

TABLE 10.7
Chi-squared Values for Alternative Peer Influence Models

Model	χ^2	Degrees of freedom	Probability
Ambition	9.554	9	0.40
Simultaneous aspirations	5.854	6	0.44

the estimated model is the true structure, the probability of obtaining the observed or larger deviations from that structure by chance is between 0.40 and 0.45. Thus we could say that in each case the data are quite consistent with the hypothesized structure.

10.6   Conclusion

It is important to review the models developed here, the methods used to estimate these models, and the assumptions required for these estimations. The basic structural model has been expanded considerably to deal with cases of unobserved or true variables. This model has many advantages for social science work where the ability to define and measure important conceptual variables is weak and improving only slowly, at best. To estimate the models needed to further theoretical work and to deal with the substantive problems in these areas, we must postulate models that include such error-laden variables and create means for estimating them. This we have begun in this chapter.

This methodological gain has been purchased at a cost. Models with unobservable variables are substantially more complex than previous models. These more complex models require considerably more information to estimate. At the same time, we are dealing with situations where observable data provide less of this information because important variables are missing or measured erroneously. The researcher must compensate for this lack of observational information with assumptions about the behavioral structure of the model and the relationships among unobserved and observed and true and erroneous variables.

The need for additional information and the evaluation of what assumptions are necessary and appropriate in different models is the identification problem. This additional information must be sufficient to rule out alternative values for the model's parameters which lead to the same expected observational data. The more complex the model and the more unknown terms, the more difficult this becomes. Our modeling efforts are aimed at providing restrictions that exclude the alternative structures. These restrictions, made prior to the statistical analysis, take the form of specific values for some of the terms or stated equalities among them. Because they must be made prior to the analysis, they must come from outside the data at hand, either from the theory behind the model, knowledge about the way the data are collected, or from other empirical estimations. All of these procedures impose large burdens on the researcher to make and justify the prior restrictions that are required. We are in effect substituting researcher's knowledge, in the form of these specific restrictions, for missing observational information.

It is always the case that the more prior information the analyst can provide, the better the analysis and the more that can be accomplished, provided the information is correct and used appropriately. Prior information can be used either of two ways. One way is simply to improve our estimates of the unknown terms in the model. A surplus of information and restrictions, referred to as overidentification, means that we get more than one estimate for some coefficients. If the model specification is correct, the differences in these estimates of the same coefficient arise simply because our observations are samples of behavior that only approximate the true behavior. By appropriately combining these several estimates for a given term, we hope to improve our final estimate. This we did with the full information estimates of the ambition version of the peer influence model. Alternatively, we can use the overidentified parts of the full model to estimate terms that previously had been unidentified. For example, restrictions on some of the error term covariances may make it possible to estimate coefficients that could not be estimated on the basis of the reduced form coefficients. This is the process used to estimate the effect of parental aspirations on occupational and educational aspirations.

We have discussed estimation procedures applicable to some of these models, taking advantage of any prior restrictions. In certain cases the model can be structured to fit the nonrecursive structure described in Chapter 8 and estimated with the techniques described there, instrumental variables and two-stage least squares. In other cases, we must resort to an analysis of the matrix of moments and covariances among the observed variables to estimate the information needed, the subject of this chapter.

I | Appendix: Statistical Review

This appendix summarizes the statistical foundations underlying the discussions in the text. It is not a textbook, or even an outline, for a course in statistical methods. Instead, it quickly reviews and summarizes the statistical concepts central to the developments in this book.

The basic task of statistics, as practiced by most social scientists, is to use observed data to make inferences about the population from which a sample of observations is obtained and about the behavioral processes that generate that population. In general, this amounts to guessing about the value of certain parameters of the population. There are two related aspects to this: *estimation* of the parameters and *hypothesis testing*, or ascertaining the likelihood that a population parameter is some given value or in a given range.

We continually refer to the set of all possible outcomes as the *population* and to the processes underlying the outcomes as the *population model*. For example, if we are interested in the outcomes of rolling a die, the population is made up of the integer values one through six. In such examples of simple discrete events it is possible to write down all possible outcomes. In more complex situations, this is not possible. If we are interested in people's incomes and the relationship between incomes and education and occupation, the possible outcome (income) for any individual in any year may take on an infinite number of values, i.e., any positive number, with some outcomes being more likely than others. The person's income in a given year is simply one outcome from that set of population values. If we were to record the incomes of all individuals in a region, country, or even the universe in a given year we would not have the entire population of outcomes (even though we have the population of individuals) because each person's income in that year is simply one outcome, or value, from the entire set of possible incomes *for that person.* We have sampled from the distributions of income, regardless of whether we have sampled from the universe of people. Our presumption in relating incomes to education and occupation is that the distribution of possible incomes varies *for each individual,* and that these variations in distributions are related to the educational and occupational characteristics of the individual. The purpose of statistical analysis is to use the set of observed outcomes (incomes) for each individual to estimate *how* these variations are related to education and occupation. The first step in learning how to obtain good estimates is to characterize the population of outcomes.

I.1   Probability

The first thing we do in describing a population is construct a variable that assigns a numerical value to each possible outcome of an event. This variable is referred to as a random variable. It may have only two possible values, such as in flipping coins where a head is one and a tail zero. A random variable may also have an infinite number of possible values, such as personal income or journey-to-work distance where any positive real number is a possible outcome. Each outcome of the event in question is now uniquely characterized by the values of the appropriate random variable.

We shall denote by U the set of all possible values of the random variable, and by u_i the value of any particular event. Random variables are of two types: discrete and continuous. Discrete variables can take on only specific prespecified values, such as the outcomes of rolling a die. With continuous variables any number, possibly within a given range, is possible. The previous examples of personal income and journey-to-work distance illustrate continuous random variables.

The first concept associated with the outcome of an event, and the values of the accompanying random variable, is the notion of probability. If we conceive of an implicit frequency distribution of all outcomes within the population, all possible u_i, the probability of a given outcome for the discrete case or the probability of an outcome within a particular range for a continuous variable is simply the proportion of outcomes in the population with that value or within that range. Thus our notion of probability is completely analogous to a frequency distribution for the population.[1]

We write these probabilities as $\text{Prob}(U = u_i) = f(u_i)$, where $\text{Prob}(U = u_i)$ denotes the probability an outcome equals u_i and $f(\cdot)$ is referred to as the probability density function. Discrete density functions have several important properties:

(a) $0 \leqslant f(u_i) \leqslant 1$ for all i,

(b) $\sum_{i=1}^{I} f(u_i) = 1$,

(c) $\text{Prob}(a_1 \leqslant u_i \leqslant b_1) = \sum_{u_i = a_1}^{b_1} f(u_i)$.

The first property simply indicates that probabilities cannot be less than zero nor greater than one, which is in keeping with the notion of probability as a relative frequency. Property (b) provides that if we sum the individual probabilities for all outcomes, this sum must equal one, again staying with the analogy to relative frequencies. The last property provides that the probability of u_i falling between a_1 and b_1 equals the sum of the probabilities that u_i equals each value between a_1 and b_1. For example, in rolling a die, the probability that the outcome falls between 3 and 5 is

$$\text{Prob}(u_i = 3) + \text{Prob}(u_i = 4) + \text{Prob}(u_i = 5) = \text{Prob}(3 \leqslant u_i \leqslant 5).$$

[1] There are other ways to define probability. However, this relative frequency concept, often referred to as objective probability, forms the basis for most formal statistical work and is central to the developments in this book.

Related to the probability density function is the cumulative distribution function which is simply the probability that u_i is less than some specific value. This function is denoted as $F(a_1)$, and

$$F(a_1) = \text{Prob}(u_i \leqslant a_1) = \sum_{u_i = -\infty}^{a_1} f(u_i).$$

A similar set of properties holds for continuous random variables. However, we need to resort to calculus to describe them. Instead of talking about the probability of a given outcome, we define the probability of U being in some small interval du,

$$\text{Prob}(u \leqslant u_i \leqslant u + du) = f(u)\, du.$$

In this case $f(u)$ is a continuous function with the properties

(a) $0 \leqslant f(u)$,

(b) $\int_{-\infty}^{\infty} f(u)\, du = 1$,

(c) $\text{Prob}(a \leqslant u \leqslant b) = \int_{a}^{b} f(u)\, du$.

These properties are completely analogous to those for discrete variables. The value of the probability density function again must be greater than or equal to zero. The total area under the function equals one, meaning that the sum of probabilities is one, and the probability of an outcome falling between points a and b is the area under the probability density function between those two points. There is also the cumulative distribution function, indicating the probability that U is less than or equal to some value a:

$$F(a) = \text{Prob}(u \leqslant a) = \int_{-\infty}^{a} f(u)\, du.$$

Rules for combining the probabilities of the different events depend upon the nature of the events. In particular, the following definitions of different types of events are required.

(1)   Two events are *mutually exclusive events* if the occurrence of one implies the nonoccurrence of the other. For example, in rolling a die, the occurrence of a three precludes the occurrence of any other face of the die. Or, in flipping a coin, the occurrence of a head implies that a tail cannot occur. An example of events that are not mutually exclusive would be the draw of a card from a deck; occurrence of a king does not preclude the occurrence of a spade.

(2)   Two events are *independent* if the occurrence of one outcome does not affect the probability of the occurrence of a second outcome. For example, whether a coin comes up heads on the first flip does not affect the probability of coming up heads on the second flip. Or rolling a three on the first toss of a die does not affect the probability of observing a three on the second toss.

(3)   Two events are *dependent* if the outcome of the second experiment depends upon the outcome of the first. For example, consider drawing cards from a deck and not replacing them. Then the probability of obtaining any value of the second draw from a deck depends upon the card that was drawn the first time. To see this, if we are interested in the probability of an ace on the second draw, $\text{Prob(ace)} = 4/51$ if no ace was drawn the first time and equals $3/51$ if an ace was drawn the first time. (These probabilities are simply the number of aces in the deck on the second draw divided by

the total number of cards in the deck on the second draw, i.e., the relative frequencies of aces.) Note that if the first card was replaced in the deck, the values on the two draws would be independent.

Expected Values

In treating random variables and functions of random variables we are most often concerned with describing particular characteristics of the distribution of possible values. The first is called the mean, or expected value, of the random variable. For discrete variables this is defined as the sum of all possible values of U weighted by the probability of that outcome occurring. Thus if outcome u_1 occurs with probability $f(u_1)$, u_2 with probability $f(u_2), \ldots, u_I$ with probability $f(u_I)$, the expected value of U is

$$E(U) = \sum_{i=1}^{I} u_i f(u_i) = \mu.$$

The expected value of U, written as $E(U)$, is often called the population mean and denoted by the greek letter μ. For continuous variables the population mean is

$$E(U) = \int_{-\infty}^{\infty} u f(u) \, du = \mu.$$

By definition the expected value of a constant is the constant itself, $E(a) = a$, since the outcome a occurs with probability one.

For many purposes we are interested in the expected value of some single-valued function of U, denoted as $E(Y) = E[r(U)]$, where $Y = r(U)$ is the function in question. This expected value is defined as

$$E(Y) = \sum_{i=1}^{I} r(u_i) f(u_i) \qquad \text{(discrete } U\text{)},$$

$$E(Y) = \int_{-\infty}^{\infty} r(u) f(u) \, du \qquad \text{(continuous } U\text{)}.$$

Thus the expected value of the function $Y = r(U)$ is simply the sum of the values of the function for each outcome weighted by the probability of that outcome.

The most commonly used functions are referred to as moments, either about the origin or about the mean. The mth moment about the origin is defined as

$$E(U^m) = \sum_i u_i^m f(u_i) \qquad \text{(discrete)},$$

$$E(U^m) = \int_{-\infty}^{\infty} u^m f(u) \, du \qquad \text{(continuous)}.$$

The mean then is simply the first moment about the origin. The mth moment about the mean is

$$E\{[U - E(U)]^m\} = E[(U - \mu)^m] = \sum (u_i - \mu)^m f(u_i) \qquad \text{(discrete)},$$

$$= \int_{-\infty}^{\infty} (u - \mu)^m f(u) \, du \qquad \text{(continuous)}.$$

The most commonly used moment about the mean is the second moment ($m = 2$), called the variance,

$$\text{var}(U) = E(U - \mu)^2.$$

(The first moment about the mean equals zero, why?) A commonly used transformation of var(U) is its square root. This is called the standard deviation, and denoted as $\sigma_U = \sqrt{\text{var}(U)}$. A common formula for variance is

$$\text{var}(U) = E(U-\mu)^2 = E(U^2 - 2U\mu + \mu^2) = E(U^2) - 2\mu E(U) + \mu^2$$

$$= E(U^2) - 2\mu^2 + \mu^2 = E(U^2) - \mu^2. \qquad (\mu \text{ is a constant})$$

Thus the variance of U is the second moment about the origin minus the mean squared. (This relationship holds for calculating sample statistics as well.)

A second common function is a linear transformation of U, $Y = a + bU$. Now

$$E(Y) = E(a+bU) = \sum (a+bu_i)f(u_i) = \sum af(u_i) + \sum bu_i f(u_i)$$

$$= a \sum f(u_i) + b \sum u_i f(u_i) \qquad \text{[see p. 40]}$$

$$= a + bE(U) \qquad \text{[by property (b) and the definition of the mean]}$$

$$= a + b\mu$$

or

$$E(Y) = E(a+bU) = \int_{-\infty}^{\infty} (a+bu)f(u)\,du = \int_{-\infty}^{\infty} af(u)\,du + \int_{-\infty}^{\infty} buf(u)\,du$$

$$= a\int_{-\infty}^{\infty} f(u)\,du + b\int_{-\infty}^{\infty} uf(u)\,du = a + bE(U) = a + b\mu;$$

also,

$$\text{var}(Y) = E\left[(Y - E(Y))^2\right] = E[(a+bU) - (a+b\mu)]^2$$

$$= E\left[(a+bU-a-b\mu)^2\right] = E[b(U-\mu)]^2$$

$$= E\left[b^2(U-\mu)^2\right] = b^2 \sum_i (u_i - \mu)^2 f(u_i) = b^2 \text{var}(U) \qquad \text{(discrete)},$$

$$= b^2 \int_{-\infty}^{\infty} (u-\mu)^2 f(u)\,du = b^2 \text{var}(U) \qquad \text{(continuous)}.$$

Multivariate Distributions

The preceding discussion has been concerned solely with a single random variable. In most instances, we are interested in events described by the values for two or more random variables. For example, an outcome would be the educational level and income attainment of an individual, or the party preference and voting decision of a given voter. In these cases we are interested in the joint probability distribution, not just the distribution of each variable taken separately. In developing the properties of multivariate distributions, we shall work with bivariate distributions, i.e., the distribution of only two variables. The extensions to three or more variables is very straightforward.

In most respects the properties of multivariate distributions completely parallel those of univariate distributions. We shall denote by U and V the two random variables used to describe the possible outcomes of a given event. The particular outcomes of any one observation are denoted by u_i and v_j. The density function for any outcome is written as $f(u,v)$, with the following properties:

(a) $0 \leqslant f(u_i, v_j)$;

(b) $\sum_i \sum_j f(u_i, v_j) = 1$ (discrete),

$$\int_{-\infty}^{\infty} \int_{-\infty}^{\infty} f(u,v) \, dv \, du = 1 \quad \text{(continuous)};$$

(c) $\mathrm{Prob}(a_1 \leqslant u_i \leqslant b_1, a_2 \leqslant v_j \leqslant b_2) = \sum_{u_i = a_1}^{b_1} \sum_{v_j = a_2}^{b_2} f(u_i, v_j)$ (discrete),

$$= \int_{a_1}^{b_1} \int_{a_2}^{b_2} f(u,v) \, dv \, du \quad \text{(continuous)}.$$

Before discussing the expectations involving multivariate distributions, it is necessary to consider the conditional and marginal distributions of individual random variables. The *marginal distribution* of a random variable indicates how one variable is distributed regardless of the values for the other variable. This is obtained by summing (integrating) over all the values of the second variable in the discrete (continuous) case. Thus, the marginal distribution of U, labeled $g(u)$, is

$$g(u_i) = \sum_j f(u_i, v_j) \quad \text{(discrete)},$$

$$g(u) = \int_{-\infty}^{\infty} f(u,v) \, dv \quad \text{(continuous)}.$$

We are in effect creating the univariate density function for U discussed in the previous section. As can be checked, $g(u)$ satisfies the conditions for a probability density function. It is similarly possible to find a marginal distribution for v, say $h(v)$, from knowledge of $f(u,v)$. However, the reverse is generally not true; from knowledge of $g(u)$ and $h(v)$, it is generally not possible to find $f(u,v)$.

The *conditional distribution* of a random variable indicates the distribution of one variable *given* a value of the other random variable. The conditional density function is found by looking at the joint density function and the marginal density function. Denoting the conditional density of U given V by $g(u|v)$, we have

$$g(u_i|v_j) = \frac{f(u_i, v_j)}{h(v_j)} \quad \text{(discrete)},$$

$$g(u|v) = \frac{f(u,v)}{h(v)} \quad \text{(continuous)}.$$

where $h(v)$ is the marginal distribution of V.

At this point we can introduce the concept of *independence*. Two random variables U and V are independent if the conditional distribution is identical to the marginal distribution for each variable. In other words, U and V are independent if $g(u|v) = g(u)$, or if $h(v|u) = h(v)$. Intuitively, this says that the distribution of U does not depend upon the particular value of V; or, alternatively, that knowledge about the value of V provides no information that would help in predicting the value of U.

An alternative (but equivalent) definition of independence is that U and V are independent if the joint density function can be factored into two functions, one of which contains only U and one only V. In other words, $g(u)$ and $h(v)$ exist such that

$$f(u,v) = g(u) \cdot h(v).$$

When this is the case, the two functions $g(u)$ and $h(v)$ are unique and are simply the marginal distributions of the two variables.

With jointly dependent variables we are also concerned with the relationship between the two variables and the expected value of functions of the two variables. We can extend the operation of expected value developed for the univariate case to multivariate distributions. Let us define any single-valued function of U and V as $r(U,V)$. Then the expected value of $r(U,V)$ is given by

$$E[r(U,V)] = \sum_i \sum_j r(u_i, v_j) f(u_i, v_j) \qquad \text{(discrete)},$$

$$= \int_{-\infty}^{\infty} \int_{-\infty}^{\infty} r(u,v) f(u,v)\, dv\, du \qquad \text{(continuous)}.$$

The simplest case is that of a linear function of U and V, $E(aU + bV) = aE(U) + bE(V)$ since

$$E(aU + bV) = \sum_i \sum_j (au_i + bv_j) f(u_i, v_j) = a \sum_i \sum_j u_i f(u_i, v_j) + b \sum_i \sum_j v_j f(u_i, v_j)$$

$$+ a \sum_i u_i g(u_i) + b \sum_j v_j h(v)$$

$$= aE(U) + bE(V).$$

Another common function is the expected value of a product $E(UV)$ $= \sum_i \sum_j u_i v_j f(u_i, v_j)$. If U and V are independent, then $f(u_i v_j) = g(u_i) \cdot h(v_j)$ and

$$E(UV) = \sum_i \sum_j u_i v_j f(u_i, v_j) = \sum \sum u_i v_j g(u_i) h(v_j)$$

$$= \sum u_i g(u_i) \sum v_j h(u_j) = E(U) \cdot E(V).$$

Thus if two random variables are independent, the expected value of their product is the product of their expected values. It should be clear from the expression for $E(UV)$ that if U and V are not independent, meaning that $f(u,v) \neq g(u) \cdot h(v)$, then this result will not hold.

A very common way of describing the relationship between two random variables is their covariance, defined as

$$\text{cov}(U,V) = E\{[U - E(U)][V - E(V)]\}.$$

Extension of this formula gives

$$\text{cov}(U,V) = E[UV - UE(V) - VE(U) + E(U)E(V)]$$

$$= E(UV) - E(U)E(V) - E(V)E(U) + E(U)E(V)$$

$$= E(UV) - E(U)E(V).$$

If U and V are independent, so that $E(UV)=E(U)E(V)$, then $\text{cov}(U,V)=0$. The reverse is not true, it is possible for the covariance of two variables to be zero, $\text{cov}(U,V)=0$, without their being independent. This happens because the covariance relates only to a linear relationship between the two random variables. There are many nonlinear relationships possible for which $\text{cov}(U,V)$ is zero but for which U and V are not independent.

Lastly we have the variance of a linear function of two random variables, $\text{var}(Y)=\text{var}(aU+bV)=a^2\text{var}(U)+b^2\text{var}(V)+2ab\,\text{cov}(U,V)$. To see this, we have

$$\text{var}(Y)=E[Y-E(Y)]^2=E[aU+bV-aE(U)-bE(V)]^2$$

$$=E\{a[U-E(U)]+b[V-E(V)]\}^2$$

$$=E\{a^2[U-E(U)]^2+b^2[V-E(V)]^2+2ab[U-E(U)][V-E(V)]\}$$

$$=a^2E[U-E(U)]^2+b^2E[V-E(V)]^2+2abE[U-E(U)][V-E(V)]$$

$$=a^2\text{var}(U)+b^2\text{var}(V)+2ab\,\text{cov}(U,V).$$

This also implies that the variance of a sum of independent random variables equals the sum of the variances of the individual variables.

I.2   Theoretical Distributions

Our discussion so far has not related to any specific probability density function, but simply the general form, $f(u)$. If we are to talk about the distribution of a particular estimator and use this distribution to make inferences about population values, we must deal in more specific terms. In other words, given an estimator (or event) and a numerical value of that estimator (an outcome), we want to know something about the probability of getting this outcome under different assumptions about the underlying population. This means we must deal with known density functions. There are a large number of density functions that have been analyzed and estimated. Only a small number of these are used to describe the estimators developed in this book. We shall concentrate our discussion on these distributions. The most important one is the normal distribution. Related to this distribution are the chi-squared, the Student's t, and the F-distributions.

Normal Distribution

The normal distribution is a symmetric, continuous, bell-shaped, distribution which ranges from $-\infty$ to $+\infty$. Figure I.1 shows the density function for the normal distribution.

The probability function is

$$f(X_i)=\frac{1}{\sigma_X\sqrt{2\pi}}\exp\left[-\frac{1}{2}\frac{(X_i-\mu)^2}{\sigma_X^2}\right],$$

where μ is the population mean of X, σ_X is the standard deviation of X, and $\exp=e$ is the base of natural logarithms. The normal distribution is called a two-parameter

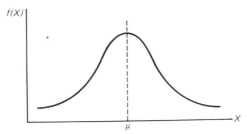

FIGURE I.1   *Normal density function.*

distribution because the probability function can be specified completely by the two parameters μ and σ_X. If we know μ and σ_X, we know everything there is to know about the probability distribution. The normal distribution is symmetric about its mean μ. This also represents the point where the maximum height of the probability function occurs (the mode).

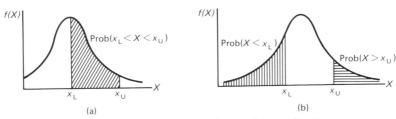

FIGURE I.2   *Probabilities with normal density function.*

As we discussed previously, the probability that $x_L \leqslant X \leqslant x_U$ is the integral of the probability function between x_L and x_U. This is simply the area under the curve between x_L and x_U as illustrated in Fig. I.2a. Figure I.2b illustrates the probability that $X \leqslant x_L$ (vertical striped area) and that $X \geqslant x_U$ (horizontal striped area).

Transformations of the normal distribution are very useful. *Linear transformations of normal variables are also normally distributed.*

This first property states that if X is normally distributed, so is $aX + b$, where a and b are constants and $a \neq 0$. This has important applications since it allows tabulation of one set of normal probabilities to be used to calculate probabilities of any normal distribution. In particular, if X is normally distributed with mean μ and standard deviation σ_X, the transformed variable $Z = (X - \mu)/\sigma_X$ is normally distributed with mean 0 and standard deviation 1. Z is referred to as a *standard* normal variable. (The values for a standard normal variate are shown in Table III.1.) The probabilities for any normal variable can then be ascertained by first transforming the variable into a standard normal variable. Notationally, we usually denote a normal variable with mean μ and variance σ_X^2 as $N(\mu, \sigma_X^2)$; thus, a standard normal variable is distributed $N(0, 1)$.

To illustrate, if X is $N(2, 16)$, what is the probability that a given value for X is greater than 6? This is equivalent to asking what is the probability that $Z \sim N(0, 1)$ is greater than 1, since $Z = (X - 2)/4$ (where $\sim$ implies "is distributed"). From Table

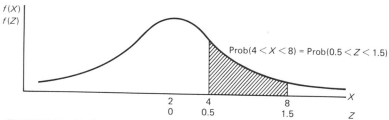

FIGURE I.3   Probabilities with standardized and unstandardized normal variables.

III.1, we find[2] that this probability is 0.159. [This is calculated as $1-(0.341+0.5)$]. As a second example, for $X \sim N(2, 16)$, what is the probability that $4 < X < 8$? This is equivalent to $\text{Prob}(X < 8) - \text{Prob}(X < 4)$; or $\text{Prob}(Z < 1.5) - \text{Prob}(Z < 0.5)$. Thus, $\text{Prob}(4 < X < 8) = 0.933 - 0.692 = 0.241$. This is illustrated in Fig. I.3 where the horizontal scales for both X and Z are shown.

One final example that frequently arises is useful. Consider the probability that X lies outside a given number of standard deviations of its mean. For example, what is the probability that X is more than 2.5 standard deviations from its mean? Since we have not specified the direction away from the mean, we must consider the probability that X is more than 2.5 standard deviations above the mean or below the mean. This is simply the sum of the probabilities in both tails of the distribution, as illustrated by the shaded areas in Fig. I.4. Also, by considering standard deviations about the mean, we are interested in values of Z directly. From Table III.1, $\text{Prob}(Z > 2.5) = 0.5 - 0.494 = 0.006$. By symmetry, the probability of being greater than 2.5 standard deviations in either direction is simply 0.012. Logically then, the probability that a value for X is within 2.5 standard deviations of its mean is $1 - 0.012 = 0.988$.

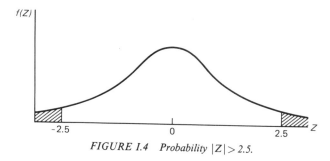

FIGURE I.4   Probability $|Z| > 2.5$.

An alternative question—which often arises in hypothesis testing—is to find the ranges for Z corresponding to specific probabilities. In other words, for what values of Z^* does $\text{Prob}(Z > Z^*) = a$ where a is a given probability value such as 0.05? This can

---

[2] Tables for standard normal variables can be written in several alternative ways. Since the distribution is symmetric, it is necessary only to tabulate one-half of the distribution. Thus, as in Table III.1, the probabilities give the area from the mean (0) to the value of Z. To find the probability of being between $-\infty$ and Z (i.e., less than Z), one adds 0.5 to the probability table. An alternative presentation is to record the values from $-\infty$ to Z. Finally, one can record the probabilities that z is greater than Z; these are simply 1-Prob $(z < Z)$.

be found by entering Table III.1 to find the value of a and then reading the corresponding value for Z. For example, if we are interested in only one tail of the distribution (e.g., that $Z > Z^*$) and the 0.05 level, we find $Z^* = 1.645$. Thus for events distributed according to a standard normal distribution, the probability of getting an outcome greater than 1.645 is 0.05. If we are interested in both tails of the probability distribution and, say the 0.05 level of probability, we find $Z^* = 1.96$. These two situations are illustrated in Fig. I.5a and b.

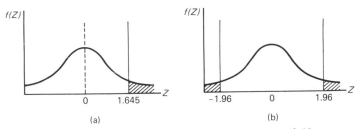

(a)   (b)

FIGURE I.5   Alternative areas equal to probability of .05.

The normal distribution is very important in statistical analysis. Many random variables of interest are normally distributed. Further, the normal distribution offers a reasonable approximation for many random variables that do not strictly follow a normal distribution. In this context, an important theoretical result is the *central limit theorem*, which states that the sample mean computed for a sample of identically distributed independent random variables is normally distributed about the population mean as the sample size becomes large. This holds regardless of the distribution of the sampled variables.[3] The mean of the distribution of the sample mean will be μ and the variance will be σ_X^2 / n, where n is the sample size.

Chi-Squared (χ^2) Distribution

A related distribution is the chi-squared distribution. A chi-squared variable is defined as the sum of squared, independently distributed standard normal variables. If $Z_i \sim N(0, 1)$,

$$\chi^2 = \sum_{i=1}^{n} Z_i^2$$

is distributed as chi-squared. The distribution of χ^2 depends upon n, the number of independent variates going into the sum. This factor n gives the *degrees of freedom*.

The chi-squared distribution is skewed to the right, with values which range from 0 to $+\infty$. Figure I.6 shows the chi-squared distribution. As the degrees of freedom increase, the distribution becomes more symmetrical.

---

[3] The chief restriction required to prove the central limit theorem is that the underlying random variables have a finite mean and variance. There are generalizations of the central limit theorem to allow for random variables drawn from a multivariate distribution and for random variables from different distributions. These generalizations are useful when a random variable is described as being made up of a series of small independent random variables or causes. It would then be natural to expect the sum of these variables to tend toward the normal distribution.

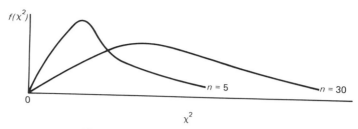

FIGURE I.6   *Chi-squared density function.*

This distribution is important in many statistical applications because one is often interested in statistical tests that involve estimating variances of random variables (which are sums of squared random variables) and in testing hypotheses that involve such estimated variances.

The tabulations of the chi-squared distribution (Table III.2) provide values of χ^2 corresponding to given cumulative probabilities and given degrees of freedom.

Student's t Distribution

The Student's t distribution combines normal and chi-squared variables. If $Z \sim N(0, 1)$ and $S^2 \sim \chi^2(n)$, where n is the degrees of freedom, we define t as

$$t = Z/(S/\sqrt{n}).$$

t follows the Student's t distribution with n degrees of freedom. This distribution arises quite frequently in hypothesis testing.

The shape of the distribution is similar to the normal distribution except that it has "fatter" tails. The t-distribution approaches the normal distribution as the degrees of freedom become large. This is shown in Fig. I.7. (The normal distribution is the limiting distribution of the t-distribution.)

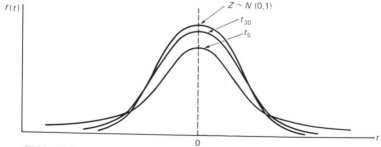

FIGURE I.7   *Relationship of t-distribution to standard normal distribution.*

The general use of the t-distribution is to test hypotheses about normal variables where the variance of the normal variable is unknown and must be estimated. In such a case, the estimated variance (divided by the true variance σ^2) would be distributed as chi-squared divided by the degrees of freedom. Intuitively, the t-statistic allows for the fact that a large value of t might arise from either a large value of Z (the normal variable) or a low estimate of the variance of Z, the chi-squared. As the sample size becomes large, the estimate of the variance becomes better and the probabilities for the normal variable become more closely approximated.

Table III.3 presents tabulated values for the *t*-distribution corresponding to different probability values and different degrees of freedom. For example, Prob($t > 1.753$) with 15 degrees of freedom is 0.05. The bottom row of Table III.3 with ∞ degrees of freedom corresponds to the normal distribution. Note the proximity of the *t*-distribution to the normal distribution at even fewer degrees of freedom such as 60 or 120.

F-Distribution

The final distribution of importance is the *F*-distribution. This is simply defined as the ratio of two independent chi-squared variables, each divided by its degrees of freedom. If we have two independent chi-squared variables S_1^2 and S_2^2 with degrees of freedom n_1 and n_2, we define *F* as

$$F_{n_2}^{n_1} = (S_1^2/n_1)/(S_2^2/n_2).$$

The *F*-distribution is also skewed to the right, as shown in Fig. I.8. *F* is always positive with a range of 0 to $+\infty$. The tabulation of the *F*-distribution in Table III.4 provides values of *F* that correspond to values with 5% and 1% of the distribution remaining in the tail. These values are a function of the degrees of freedom in the numerator (n_1) and the degrees of freedom in the denominator (n_2).

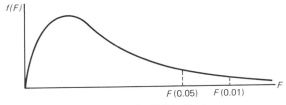

FIGURE I.8   F-Distribution.

The *F*-distribution is used for testing the equality of two estimated variances. This problem frequently occurs when two variances are independently estimated and one wishes to test whether they are equal or not.

I.3   Properties of Estimators

The tasks for which we use statistical methodology relate to estimating parameters of some population distribution and testing hypotheses about these parameters. The general approach is to develop an estimator that allows us to guess about the population values on the basis of sample information. One important facet of this work is understanding the properties of any estimators that we use.

Estimators are based upon a sample of values for random variables. Therefore, the estimators are themselves random variables. Since we often have more than one estimator to choose from, the usefulness of any estimator is based upon its own sampling distribution. This judgment usually relates to expectations for certain parameters of the sampling distribution of the estimator.

We begin with two things: a priori information (or information available before sampling) and sample information. The a priori information is generally introduced in the form of a model, or maintained hypothesis, which specifies certain attributes of the population being considered. These attributes could be about the form of the distribution or reasonable values for certain parameters of the distribution. From a sample of observations from the distribution of a random variable, say X, we shall estimate some parameter of the distribution, say θ. The estimator $\hat{\theta}$ is a function of the observed values of X $(X_1, X_2, \ldots, X_T)$. The first subject to consider is what properties of $\hat{\theta}$ are important.

Small Sample Properties

Consideration of properties of an estimator is divided into two parts: small sample properties and large sample properties. Small sample properties refer to the properties of estimators for any fixed, finite sample size. The next section will consider large sample, or asymptotic, properties. These are the properties observed as the sample size becomes infinitely large, $T \to \infty$.

For any given sample, the estimator will generally be different from the true parameter. This difference $\hat{\theta} - \theta$ is the *sampling error*. One of the first properties of interest is the expected value of this sampling error. This quantity $E(\hat{\theta} - \theta)$ is called the *bias* of an estimator. An estimator is said to be unbiased if $E(\hat{\theta}) = \theta$, or $E(\hat{\theta}) - \theta = 0$. In other words, if the distribution of $\hat{\theta}$ is centered on the true value of the parameter, the estimator is unbiased. This is illustrated in Fig. I.9, which displays the sampling distribution for two estimators of a given population parameter θ. Estimator $\hat{\theta}_1$ is unbiased; its distribution is centered on the true value. Estimator $\hat{\theta}_2$, while having a distribution that includes the true value of θ, is biased because its expected value will not equal the true value.

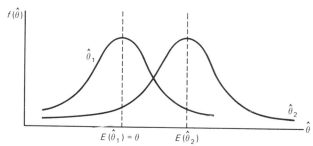

FIGURE I.9   Sample distribution of two alternative estimators of θ.

Unbiasedness is an intuitively appealing property, but one should not make the choice of estimators solely on this criterion. First, there are often several different unbiased estimators to choose from. Second, and more important, unbiasedness provides no information about the dispersion of a given estimator. Without information about the dispersion, we have no way of knowing how close the estimate obtained for a given sample might be to the true parameter.

Thus, the variance of the estimator is also useful in judging an estimator. As before,

$$\text{var}(\hat{\theta}) = \sigma_{\hat{\theta}}^2 = E\left[\hat{\theta} - E(\hat{\theta}) \right]^2 = E(\hat{\theta}^2) - \left[E(\hat{\theta}) \right]^2.$$

The square root of this—a value frequently used in hypothesis testing—is labeled the standard error of the estimator. (This is denoted by $\sigma_{\hat{\theta}}$.)

As in the case of unbiasedness, it is clear that choosing estimators solely on the basis of variance is also unreasonable. If one chooses an estimator of $\hat{\theta} = 17$ regardless of problem or sample, the estimator minimizes the variance ($\mathrm{var}\,\hat{\theta} = 0$) but would not be considered a good estimator.

Both the expected value and the dispersion of an estimator can be combined into a single quantity, mean square error. The mean square error (MSE) is defined as

$$MSE = E(\hat{\theta} - \theta)^2.$$

For the unbiased estimator, $E(\hat{\theta}) = \theta$, so the mean square error simply equals the variance of the estimator. This, however, is not the case for a biased estimator. A useful relationship between MSE, variance, and bias can be found by manipulation of the expression for MSE:

$$MSE = E(\hat{\theta} - \theta)^2 = E\big[(\hat{\theta} - E(\hat{\theta})) + (E(\hat{\theta}) - \theta)\big]^2$$

$$= E\big[\hat{\theta} - E(\hat{\theta})\big]^2 + E\big[E(\hat{\theta}) - \theta\big]^2 + 2E\big[(\hat{\theta} - E(\hat{\theta}))(E(\hat{\theta}) - \theta)\big]$$

$$= \mathrm{var}(\hat{\theta}) + E\big(\bar{\theta} - \theta\big)^2 + 2\big(\bar{\theta} - \theta\big)E\big[\hat{\theta} - E(\hat{\theta})\big],$$

where $E(\hat{\theta}) = \bar{\theta}$. Since the expected value of a constant is the constant and since the expected value of deviations about the mean is zero,[4] the last term is zero, and the equation reduces to

$$MSE = \mathrm{var}\,\hat{\theta} + BIAS^2.$$

As a general statement, it would be desirable to choose the estimator with the smallest mean square error. However, except for unbiased estimators, mean squared error properties are usually unknown and difficult to develop. Therefore, it is customary to restrict attention to a class of estimators such as unbiased estimators. (Since unbiased estimators do not always exist, other classes defined on large sample properties will sometimes be substituted.)

Within the class of unbiased estimators, we define the estimator with the smallest MSE (variance) as *efficient*. This estimator is often called *best*, by which we mean minimum variance. If the minimum variance estimator is unknown, any two unbiased estimators are often compared on the basis of their variance. A comparison of the variances of two unbiased estimators yields the *relative efficiency* of an estimator.

Large Sample Properties

The properties of an estimator as the sample becomes infinitely large are also of interest to us. The small sample properties hold for finite samples and give some indication of the expected performance of an estimator in normal estimation situations. However, in many situations the small sample properties of an estimator are simply unknown. Yet, while the small sample properties might be unknown, it may well be possible to describe the sampling distribution of the estimator as the sample size becomes very large—its asymptotic distributional properties. These asymptotic properties provide alternative choice criteria in situations where the small sample properties are unknown.

The large sample properties of interest are analogous to the small sample properties discussed previously. In large samples, the focus is on asymptotic unbiasedness,

---

[4] θ, $E(\hat{\theta})$, and thus $E(\hat{\theta}) - \theta$ are all constants.

asymptotic variance, and asymptotic efficiency. An additional property—consistency —is in a sense the large sample equivalent to a mean square error criterion.

The fundamental consideration behind asymptotic theory is the shape of the distribution of an estimator as the sample size becomes very large. A distinction must be made early, however. The asymptotic distribution of a variable is not the distribution with an infinite sample size. In many cases, the distribution with an infinite sample size completely collapses to a single point—i.e., the variance of the estimator approaches zero—in which case speaking of a distribution would not make much sense. Instead it is the form the distribution takes as it approaches its limiting case which is of interest. In other words, we want to express $\hat{\theta}$ as a function of T and see what the mean and variance of θ are as $T \to \infty$.

To illustrate this point, consider the sample mean for a sample of size T drawn from any population distribution with a mean μ and variance σ^2. From elementary statistics we know that the sample mean, denoted as $\bar{U} = (1/T)\sum_{t=1}^{T} x_t$, has an expected value of μ and variance σ^2/T, i.e., $E(\bar{U}) = \mu$ and $E(U - \mu)^2 = \sigma^2/T$. The central limit theorem says that as $T \to \infty$, $\bar{U}$ is normally distributed about μ, i.e., $\bar{U} \sim N(\mu, \sigma^2/T)$ as $T \to \infty$, regardless of how the population values are distributed. It is also true that the variance of $\bar{U}$ approaches zero as T approaches infinity. What we can say however is that $\bar{U}$ is asymptotically normal with mean μ and variance σ^2/T.

Certain asymptotic properties are important. An estimator is asymptotically unbiased if $\lim_{T\to\infty} E(\hat{\theta}) = \theta$. In other words, if the asymptotic expectation equals the parameter value, the estimator is asymptotically unbiased. Calculation of the asymptotic variance of an estimator is not, however, accomplished in the same manner as the asymptotic mean. Since an estimator often collapses to a single point as the sample size increases (as with the sample mean $\bar{U}$), it is not useful to define asymptotic variance as the limit of the variance of the estimator as the sample size increases. Instead, it is defined as the variance of the asymptotic distribution of the estimator. Mathematically,

$$\text{asymptotic variance } (\hat{\theta}) = (1/T) \lim_{T\to\infty} E\left[\sqrt{T}\, (\hat{\theta} - \lim E(\hat{\theta})) \right]^2.$$

The key element in this expression is the term $E[\sqrt{T}\,(\hat{\theta} - \lim E(\hat{\theta}))]^2$, the expected squared deviation of $\hat{\theta}$ about its asymptotic mean. Thus the asymptotic variance is the mean squared deviation about the asymptotic mean.

In a manner analogous to the small sample properties, we can define asymptotic efficiency in terms of the asymptotic variances of two estimators. If two estimators are asymptotically unbiased, the one with the smaller asymptotic variance is said to be asymptotically more efficient. The estimator with the smallest asymptotic variance among all asymptotically unbiased estimators is said to be *asymptotically efficient*.

Finally, in the case that the distribution of an estimator does collapse on a given value, we are interested in whether or not that value is the parameter value. In particular, an estimator is *consistent* if

$$\lim_{T\to\infty} \text{Prob}(|\hat{\theta} - \theta| < \varepsilon) = 1, \qquad \text{or} \qquad \text{plim}\,\hat{\theta} = \theta.$$

This says that, as the sample size becomes infinitely large, the probability distribution converges on the true parameter values; i.e., the probability that the estimator is closer to the true value than any arbitrarily small ε is one. This is illustrated in Fig. I.10.

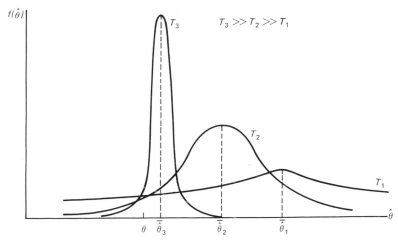

FIGURE I.10   Illustration of consistent estimator.

Consistency is a desirable statistical property and will be considered for many estimators.[5]

I.4   Hypothesis Testing

Statisticians face two related tasks. The first is estimation of certain parameters of a population distribution. The second is testing certain hypotheses about the underlying distribution from which a sample is drawn. *Hypotheses* are simply statements about an underlying population, statements which may be true or false. Sample information is used to test hypotheses, and these tests will usually be in the form of probability statements about the likelihood of a hypothesis being false.

In conducting statistical analyses, it is common to propose several hypotheses. Some of these usually go untested in any given analysis; these are called *maintained hypotheses*. Examples of maintained hypotheses might be that a certain variable is normally distributed or that there is a linear relationship between two variables. The other set of hypotheses are *testable hypotheses*. These are, as the name suggests, the hypotheses to be tested with the sample data.

Logically, testable hypotheses must come in pairs. The pair of hypotheses reveal alternative possibilities for the population. These may be very specific (e.g., that a parameter θ equals a given constant) or they may be quite general (e.g., $\theta > 0$). The first example is a *simple hypothesis*—one where the parameter is hypothesized to equal a single value. The second example is a *composite hypothesis*—one where the parameter is hypothesized to fall somewhere within a range of possible values. As one might suspect, it is possible to design better tests for simple hypotheses than for composite hypotheses. In fact, it is generally required that one hypothesis, the null hypothesis, be

---

[5] While it would seem intuitively that consistency would imply asymptotic unbiasedness, this is not the case. There are some situations where an estimator can be both consistent and asymptotically biased. See Dhrymes (1970). For the most part, such possibilities are unimportant

simple even if the alternative is composite. The *null hypothesis* (denoted H_0) is the hypothesis to be tested, while the *alternative hypothesis* (denoted H_1) states what the population would look like if the null hypothesis were untrue.

Statistical methodology does havė something to say about the choice of which hypothesis is the null hypothesis and which is the alternative. One aspect—that the null hypothesis be simple—was already mentioned. As we shall soon see, statistical theory is geared to rejecting hypotheses, i.e., to stating that a null hypothesis is very unlikely based upon the information in the sample. Implicit in this is that there is a different burden of proof depending upon which hypothesis is the null hypothesis.

The basic scheme followed in hypothesis testing is first to specify a null hypothesis and an alternative hypothesis. Often the set of hypotheses is that one variable does or does not influence another variable; in such a case, the null hypothesis is usually that a parameter relating the two variables is zero against the alternative (composite) hypothesis that the parameter does not equal zero. Then, using sample data a test statistic is constructed summarizing the sample information about the likelihood of the null hypothesis being true. A decision rule is set to decide how to interpret the test statistic. Finally, a decision is made on the basis of the test statistic to reject or not reject the null hypothesis.

The test statistic is simply a random variable, derived from the sample information, that can be used to evaluate the null hypothesis. The decision process used is essentially choosing a range of values of the test statistic for which we shall reject the null hypothesis. This is called the *rejection region*, or *critical region*. The range of values for the test statistic in which we do not reject the null hypothesis is called the *acceptance region*. In formulating the hypothesis test, we must decide upon the boundary between the acceptance and rejection regions. This decision involves both knowledge about the distribution of the test statistic and an evaluation of the costs of making an error.

The use of sample information to evaluate assumed (hypothesized) values of population parameters implies that there is some uncertainty involved. This must be so because the test statistic is itself a random variable. Conceptually, when we are testing a given null hypothesis H_0, we have the following possibilities:

Decision	H_0 true	H_1 true
Reject H_0	Type I error	Correct decision
Accept H_0	Correct decision	Type II error

These two errors are not unrelated. Clearly, we can ensure that we never make a type I error by setting a decision rule to always accept H_0. However, by making the probability of a type I error very small, we have obviously increased the possibility of a type II error, i.e., accepting a false null hypothesis. The decision on the magnitude of the two errors that we are willing to accept depends upon the costs of making either of the wrong decisions.

In the usual social science application, the null hypothesis is believed to be true unless there is strong evidence to the contrary. In other words, we generally wish to ensure that the probability of a type I error (rejecting a true null hypothesis) is low. A customary choice is to set the probability of a type I error at 0.05 (a significance level of 95%). In other words, 5 times out of a hundred when the null hypothesis is true, we will reject it. Other levels of significance are of course possible and their use depends upon how one views the costs of type I and type II errors.

In general, the test statistics that we shall use follow the theoretical frequency distributions discussed previously (normal, Student's t, chi-squared, or F). (More accurately, a maintained hypothesis often is that the test statistics follow one of those distributions.) In such a case, the probability distribution of the test statistic is tabulated, and the task is choosing the critical region of the distribution.

The choice of the critical region depends upon the hypotheses to be tested. Consider testing a null hypothesis that a parameter θ equals a given value, say θ^*. (In many applications, θ^* will be zero.) For an alternative hypothesis of H_1: $\theta \neq \theta^*$, we have not specified a particular range of alternatives; θ could be any value other than θ^* according to H_1. Therefore, we shall consider the possibility that θ is either larger or smaller than θ^*. This is called a two-tailed test. Often, however, the alternative hypothesis is more specific, such as H_1: $\theta > \theta^*$. In this case we wish to consider only the alternative of large values of θ. (When $\theta^* = 0$, this implies considering the alternative of only positive values of θ.) This implies that the critical region lies on only one side of θ^*. This is a *one-tailed* test.

The formulation of the critical region and the critical value of a test statistic can be shown diagrammatically. Assume that we have the hypotheses

$$H_0: \quad \theta = \theta^*, \qquad H_1: \quad \theta \neq \theta^*.$$

For a given test statistic $\hat{\theta}$, with a known distribution, we can plot the probability function under the assumption that θ does in fact equal θ^*. In other words, assume the null hypothesis is true. In such a case, we first choose an amount of type I error that we will tolerate, say 0.05. The question in hypothesis testing is then to determine what values of $\hat{\theta}$ correspond to the cutoffs of the critical region that have 5% of the probability when the null hypothesis is true. In this case, a two-tailed test is appropriate and the critical values and critical region are shown in Fig. I.11.

The decision rule is then to reject H_0 if $\hat{\theta} < \theta_1$ or if $\hat{\theta} > \theta_2$. In situations where H_0 is true, we will make an incorrect decision 5% of the time.

We often have a more specific alternative hypothesis for which a one-tailed test is appropriate. Consider H_1: $\theta > \theta^*$. In this case, the critical region and critical value is represented by Fig. I.12. Here, if $\hat{\theta} > \theta_1$, we would reject H_0.

A specific example will clarify this. Assume that we are testing the hypothesis that the mean of a random variable equals zero. We also know that the random variable X is $N(0, \sigma_X^2)$. The test statistic we shall use is $Z = \bar{X}/(\sigma_X/\sqrt{T})$ which is distributed $N(0, 1)$. If we use a two-tailed test, we would reject the null hypothesis if $|\bar{X}/(\sigma_X/\sqrt{T})| > 1.96$ (‖‖‖ area). With a one-tailed test for H_1: $\mu > 0$, we would reject the null hypothesis if $\bar{X}/(\sigma_X/\sqrt{T}) > 1.65$ ($\equiv$ area). These critical values are shown in Fig. 1.13, which is simply the standard normal distribution.

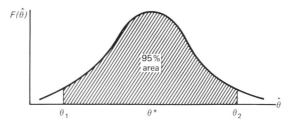

FIGURE I.11   Two-tailed hypothesis test H_0: $\theta = \theta^*$; H_1: $\theta \neq \theta^*$.

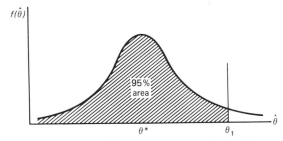

FIGURE I.12   One-tailed hypothesis test H_0: $\theta = \theta^*$; H_1: $\theta > \theta^*$.

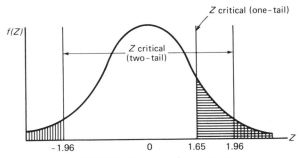

FIGURE I.13   *Normal test statistic.*

This last example shows how the theoretical distributions presented earlier come into play in hypothesis testing. If we know the distribution of the test statistic, we can develop decision rules that tell us whether or not to reject a null hypothesis on the basis of the sample information.

At the same time we should be clear about what the hypothesis test tells us. In rejecting the null hypothesis, we, by implication, accept the alternative hypothesis. However, the alternative hypothesis is often quite vague (e.g., $\mu \neq 0$). Thus, we do not accept a specific value of the parameter. For example, in the previous example, if $\bar{X} = 17$ with $\sigma_X / \sqrt{T} = 5$ and we reject H_0: $\mu = 0$ at the 95% level, we should not conclude that μ in fact equals 17. If we wish a best guess of μ, we would choose this point estimate of 17. But we cannot say with any certainty what the exact value of μ might be. An infinite number of values of μ could yield a sample with $\bar{X} = 17$.

I.5   Maximum Likelihood (ML) Estimation

There are many possible ways to develop estimators for population parameters. For example, we can begin with an estimator, derive its statistical properties, and compare them with the properties of other estimators. This in fact is what we do throughout most of the book. We then pick estimators that meet the following criteria:

(a)   they are unbiased; and
(b)   they have minimum variance among unbiased estimators.

An alternative technique is commonly used in statistics. This is called maximum likelihood estimation. The essence of the method is to choose the underlying parameters of a distribution maximizing the likelihood of yielding the observed sample.

To illustrate this method, consider the following problem. There is an urn with a large number of red and black balls. In ten draws from the urn (done with replacement after each draw) we have three red and seven black balls. What is the proportion of red balls in the urn? An obvious guess is 0.3, but let us show that in fact this guess has a sound statistical basis.

The proportion of red balls in the urn is actually the probability of drawing a red ball on any trial, denoted as π. The probability of getting a black ball, i.e., not getting a red one, is $1 - \pi$. By sampling with replacement we can consider the draws to be independent events. Consequently, the probability of the observed sequence of draws is simply the product of the probabilities of the separate draws. With three red balls and seven black balls, this product is $\pi^3(1 - \pi)^7$. We can consider the ten draws and the specific sequence of red and black balls as one outcome with the above probability.[6] The point of the maximum likelihood method is to choose the value for π that makes the observed outcome of three red balls the most probable outcome. In other words, we want the value for π that maximizes the function $\pi^3(1 - \pi)^7$, called the likelihood function and denoted as $\mathcal{L} = \pi^3(1 - \pi)^7$. Once this value is computed, we treat that as the best guess of the probability of drawing a red ball.

There are several means by which we can determine the value of π that maximizes this likelihood function. One way is to compute $\mathcal{L}$ for different values of π and search for the maximum. This is inefficient and cumbersome, however, particularly in more complicated problems where there are several parameters to estimate. Consequently, we usually resort to the use of calculus to solve for the maximizing values of the parameters. Since taking derivatives of the products in the likelihood expression is usually difficult, the likelihood function is usually transformed into its log:

$$\mathcal{L}^* = \log_e(\mathcal{L}) = 3\log(\pi) + 7\log(1 - \pi).$$

Because the logarithmic function is a monotonic transformation, the value for π that maximizes $\mathcal{L}$ also maximizes $\mathcal{L}^*$. To see this, in Fig. I.14 we have plotted values for $\mathcal{L}$ and $\mathcal{L}^*$ for the red ball problem. Both maximums occur at the value 0.3. We can demonstrate that this is the maximum likelihood point by differentiating $\mathcal{L}^*$ with respect to π and picking the value for π that sets this derivative to zero, which by definition is the maximum for this example (see Appendix 2.1)[7]:

$$\frac{d\mathcal{L}^*}{d\pi} = \frac{3}{\pi} - \frac{7}{1 - \pi}.$$

Solving for $\hat{\pi}$ gives $3(1 - \hat{\pi}) - 7\hat{\pi} = 3 - 10\hat{\pi} = 0$ or $\hat{\pi} = 0.3$.

This procedure can be generalized to more interesting and more complicated situations. The basic approach remains the same however. The likelihood of obtaining the observed sample is derived as a function of the parameters that are to be estimated, then the values for these estimates are selected so as to make the likelihood of getting those observations a maximum.

The general case is developed in the following manner. Let the probability density function for the outcome Y_t be given by $f(Y_t; \Theta)$, where Θ represents the parameters in the probability density function. In the simple linear model $Y_t = \beta_1 + X_t\beta_2 + U_t$, the

---

[6] A different sequence of seven black and three red balls would constitute a different outcome, but with the same probability of occurrence, $\pi^3(1 - \pi)^7$.

[7] The derivative of $\log(X)$ is $1/X$ and the derivative of $\log(1 - X)$ is $-1/(1 - X)$.

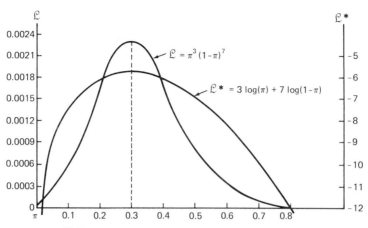

FIGURE I.14   *Likelihood and log likelihood functions.*

probability density function for Y_t is a function of $X_t\beta$ and the variance of U_t, $f(Y_t; \Theta) = f(Y_t; \beta_1 + X_t\beta_2, \sigma_U^2)$. If we have a sample of T observations for Y which constitute a single outcome of the process generating samples of Y_t, and all the observations are drawn independently, the density function for that outcome is

$$f(Y_1, Y_2, \ldots, Y_T) = f(Y_1) \cdot f(Y_2) \cdots f(Y_T) = \prod_{t=1}^{T} f(Y_t) = \prod_{t=1}^{T} f(Y_t; \Theta),$$

where $\prod$ denotes the product of a series of terms (just as Σ denotes the sum of the terms). This expression is also the likelihood function. For the simple estimation model, this is

$$\mathcal{L} = \prod_{t=1}^{T} f(Y_t; \beta_1 + X_t\beta_2, \sigma_U^2).$$

We can easily write the log of this likelihood function as

$$\mathcal{L}^* = \log_e(\mathcal{L}) = \sum_{t=1}^{T} \log f\left[(Y_t; \beta_1 + X_t\beta_2, \sigma_U^2) \right].$$

In order to complete the maximum likelihood calculations, we must pick a specific density function $f(Y_t; \Theta)$ in order to calculate the values for the parameters maximizing the likelihood (or the log likelihood) function. To demonstrate this, we shall proceed with the example of the linear model $Y_t = \beta_1 + X_t\beta_2 + U_t$. From the earlier discussion, we know that if U_t is normally distributed, then Y_t is normally distributed (linear functions of normally distributed variables are also normally distributed). Assuming U_t is $N(0, \sigma_U^2)$ and fixed X_t, Y_t is $N(\beta_1 + X_t\beta_2, \sigma_U^2)$.[8] By the definition of the normal density function

$$f(Y_t; \beta_1 + X_t\beta_2, \sigma_U^2) = \frac{1}{\sqrt{2\pi}} \frac{1}{\sqrt{\sigma_U^2}} \exp \frac{-1}{2\sigma_U^2} (Y_t - \beta_1 - X_t\beta_2)^2,$$

$$\log f(Y_t; \beta_1 + X_t\beta_2, \sigma_U^2) = -\frac{1}{2} \log(2\pi) - \frac{1}{2} \log(\sigma_U^2) - \frac{1}{2\sigma_U^2} (Y_t - \beta_1 - X_t\beta_2)^2.$$

---

[8] $E(Y_t) = E(\beta_1 + X_t\beta_2 + U_t) = E(\beta_1 + X_t\beta_2) + E(U_t) = \beta_1 + X_t\beta_2$ since X_t, B_1, and B_2 are constant for this observation.

Given a sample of T observations, the log of the likelihood function for that sample is

$$\mathcal{L}^* = \sum_{t=1}^{T} \left[-\frac{1}{2} \log(2\pi) - \frac{1}{2} \log(\sigma_U^2) - \frac{1}{2\sigma_U^2}(Y_t - \beta_1 - X_t\beta_2)^2 \right].$$

To find the values for σ_U^2, β_1, and β_2 that maximize this function we need to equate the partial derivatives of $\mathcal{L}^*$ with respect to each parameter to zero:

$$\frac{\partial \mathcal{L}^*}{\partial \beta_1} = \sum_{t=1}^{T} \frac{1}{\sigma_U^2}(Y_t - \beta_1 - X_t\beta_2) = 0,$$

$$\frac{\partial \mathcal{L}^*}{\partial \beta_2} = \sum_{t=1}^{T} \frac{1}{\sigma_U^2}(Y_t - \beta_1 - X_t\beta_2)X_t = 0,$$

$$\frac{\partial \mathcal{L}^*}{\partial \sigma_U^2} = \sum_{t=1}^{T} -\frac{1}{2\sigma_U^2} + \frac{1}{2}\frac{(Y_t - \beta_1 - X_t\beta_2)^2}{(\sigma_U^2)^2} = -\frac{T}{2\sigma_U^2} + \frac{1}{2(\sigma_U^2)^2}\sum(Y_t - \beta_1 - X_t\beta_2)^2 = 0.$$

By multiplying the first two expressions by σ_U^2 we see that they are simply the expressions for the OLS estimators developed in Chapters 2–5. Thus if we assume that each U_t is normally and independently distributed, U_t is $N(0,\sigma_U^2)$ for all t, then the OLS estimator is also the maximum likelihood estimator. To calculate the maximum likelihood estimator for σ_U^2, we substitute the estimates b_1 and b_2 for β_1 and β_2 in the third equation so that $(Y_t - b_1 - X_t b_2)^2 = e_t^2$. Further manipulation gives

$$-\frac{T}{2\hat{\sigma}_U^2} + \frac{1}{2(\hat{\sigma}_U^2)^2}\sum e_t^2 = 0 \quad \text{or} \quad -T\hat{\sigma}_U^2 + \sum e_t^2 = 0 \quad \text{or} \quad \hat{\sigma}_U^2 = \frac{\sum e_t^2}{T}.$$

Thus the maximum likelihood estimator for σ_U^2 differs from the OLS estimator for σ_U^2 by the factor $T/(T-2)$,

$$\hat{\sigma}_U^2 \ (OLS) = \frac{T}{T-2}\frac{\sum e_t^2}{T} = \frac{T}{T-2}\hat{\sigma}_U^2 \ (ML).$$

We show in Chapter 3 that $\hat{\sigma}_U^2$ (OLS) is an unbiased estimator of σ_U^2. Consequently $\hat{\sigma}_U^2$ (ML) is biased. However, as $T \to \infty$, $E[\hat{\sigma}_U^2 \ (ML)] \to \sigma_U^2$. Here is a case of an estimator that is biased in small samples but asymptotically unbiased.

This is one illustration of the use of maximum likelihood estimators. Chapters 7 and 10 provide several more complicated applications. The appeal of the maximum likelihood method results from two properties. One important advantage of the maximum likelihood method resides in the asymptotic properties of the estimators. Under broad conditions, the maximum likelihood estimators are

(a)   consistent,
(b)   asymptotically efficient, and
(c)   asymptotically normal.

The discussion of these properties and their necessary conditions are beyond the scope of this book, but are well covered in other books (Goldberger, 1964, pp. 130–135 and Dhrymes, 1970, pp. 114–129).

In some models the least squares criterion is inappropriate or difficult to apply. For example, in the models discussed in Chapter 7, the least squares criterion cannot be easily applied to variables that are ordinal or categorical in nature. In this instance we resort to the maximum likelihood method. In Chapter 10 we pointed out that the least squares method is difficult to apply to the full information estimation of multiequation systems where we are fitting covariances as well as variances. In that instance, the least squares criterion leads to results that are sensitive to the scaling of the variables. To overcome this difficulty, we again resort to the maximum likelihood procedure to pick the estimated values.

The disadvantage of the maximum likelihood method is that the underlying density functions must both be specified and be an easily manipulated function of the underlying parameters. This is true only for certain functions, such as the normal density function. If the density function cannot be specified, or if it is difficult to take the derivatives with respect to the individual parameters, then the maximum likelihood method does not provide much of an advantage.

II | Appendix: Matrix Algebra

The notation and arithmetic operations used in Chapters 2–4 are known as *scalar* algebra. The rules that are used in scalar algebra are well known and do not have to be repeated. For instance, when rearranging summations we frequently used the rule that $ab + ac = a(b + c)$. However, the notation with the subscripts and summations necessary in the multivariate case becomes extremely cumbersome, particularly when we attempt to achieve any generality to the analysis. In order to overcome the notational problems we resort to a different set of rules included in *matrix* algebra. Since this is usually not included in common curricula, this chapter offers an introduction to matrix algebra. For the purposes of this book, the notation and a few simple rules are all that are needed. In more advanced treatments, a number of specialized theorems become important. For the most part, we shall not consider such detailed elements.

Throughout this appendix, the rules of matrix algebra are demonstrated with simple numerical examples. In addition, to show the relationship between matrix and scalar algebra the development of the estimators b_1, b_2, and b_3 from Chapter 2 will be shown in the examples of matrix manipulations. These latter examples should be followed carefully since they will provide a useful introduction to the discussions in Chapter 5.

II.1 Basic Properties

A matrix is any rectangular array of numbers. Notationally, a matrix is usually denoted by brackets such as matrix A,

$$A = \begin{bmatrix} 10 & 5 & 8 \\ -12 & 1 & 0 \end{bmatrix}.$$

Each of the numbers, or scalars, within a matrix is called an *element*. Elements are referred to by subscripts denoting the row and column position of the element as in a_{rc}. (The row identifier, by convention, comes first.) In matrix A above, $a_{23} = 0$. Matrices are classified by their dimensions. This is called their *order*. A matrix that has m rows and n columns is of order (m, n), or an $m \times n$ matrix. (The latter is read as m by n.) Thus, matrix A is a 2×3 matrix. A square matrix with n rows and n columns is said to be of order n.

A special type of matrix that has many applications is one that has either a single row or a single column. This is called a *vector*. A row vector has only one row such as a_r where

$$a_r = [12 \quad 14 \quad 91].$$

A column vector has a single column such as a_c where

$$a_c = \begin{bmatrix} 1 \\ 0 \\ 3 \end{bmatrix}.$$

Note that a 1×1 matrix or a vector with only one element is a scalar.

Anyone who has tried to organize data for a simple empirical study has already constructed a matrix. If data are tabulated such that the first line represents the first observation on all the variables considered, the second line represents the second observation, and so on, this tabulation constitutes a simple matrix. The rows are the values of all variables for a given observation, and the columns are the set of observations on each variable. In the simple case of Chapter 2, our matrix of observations would be a $T \times 4$ matrix where the first column consists of the T observations on Y and the second column contains the T values for X_1—the constant term, the third the observations on X_2, and the last the observations on X_3.

We also have what are called partitioned matrices, denoted

$$A = \begin{bmatrix} A_{11} & | & A_{12} \\ -- & + & -- \\ A_{21} & | & A_{22} \end{bmatrix},$$

where A_{11}, A_{12}, A_{21}, and A_{22} are themselves matrices. If A is $m \times n$, and A_{11} is $m_1 \times n_1$, and A_{22} is $m_2 \times n_2$ where $m = m_1 + m_2$ and $n = n_1 + n_2$, then A_{12} is $m_1 \times n_2$ and A_{21} is $m_2 \times n_1$, and all we have done is subdivided the matrix A into a set of smaller matrices. (Show that the upper leftmost element in A_{21} is $a_{m_1+1,1}$.) We can divide a matrix into as many partitions as desired—all the way to the point where each element could be considered a 1×1 partition. Thus we could have

$$A = \begin{bmatrix} A_{11} & | & A_{12} & | & \cdots & | & A_{1J} \\ & & & & & & \\ A_{21} & | & A_{22} & | & & | & A_{2J} \\ & & & & & & \\ \cdot & | & \cdot & | & & | & \cdot \\ \cdot & | & \cdot & | & & | & \cdot \\ & & & & & & \\ A_{I1} & | & A_{I2} & | & \cdots & | & A_{IJ} \end{bmatrix}$$

with I times J partitions. Note that submatrices in the same row all have identical number of rows and, similarly, submatrices in the same column all have identical number of columns.

Equality

Two matrices are equal if two properties hold:

(1)   they are of the same order; and
(2)   corresponding elements are equal.

In other words, for two $m \times n$ matrices A and B, $A = B$ if $a_{rc} = b_{rc}$ for $r = 1, \ldots, m$ and $c = 1, \ldots, n$. This implies that our definition of a matrix should be modified to read "an *ordered*" rectangular array of numbers. Let

$$A = \begin{bmatrix} 10 & 5 & 8 \\ -12 & 1 & 0 \end{bmatrix} \quad \text{and} \quad B = \begin{bmatrix} 10 & 8 & 5 \\ -12 & 0 & 1 \end{bmatrix}.$$

By the above definition, $A \neq B$.

II.2   Basic Operations

Addition

Matrix addition is defined for two matrices of the same order. For A and B, both $m \times n$, $A + B = C$ is found by adding corresponding elements. Thus $c_{rc} = a_{rc} + b_{rc}$ for all r and c. Using the previously defined A and B,

$$C = A + B = \begin{bmatrix} 20 & 13 & 13 \\ -24 & 1 & 1 \end{bmatrix}.$$

As in scalar algebra, subtraction is simply negative addition. Thus,

$$D = A - B = \begin{bmatrix} 0 & -3 & 3 \\ 0 & 1 & -1 \end{bmatrix}.$$

It is obvious since matrix addition is defined in terms of scalar addition that

$$(A + B) + C = A + (B + C)$$

when A, B, and C are of the same order.

Scalar Multiplication

A matrix can be multiplied by a scalar by multiplying each element by the scalar. Thus, for any scalar s, the elements of sA are sa_{rc}. For example,

$$10A = \begin{bmatrix} 100 & 50 & 80 \\ -120 & 10 & 0 \end{bmatrix}.$$

In scalar multiplication, $sA = As$.

II.3   Matrix Multiplication

Two matrices may also be multiplied together provided that they are *conformable* for multiplication, or in other words, have the correct dimensions. In order to multiply A times B, the number of columns in A must equal the number of rows in B. The reason becomes apparent when we discuss the rules for multiplication. For $C = AB$ where A is $m \times n$ and B is $n \times p$, each element, c_{rc} is found by multiplying corresponding elements in the rth row of A and the cth column of B and adding them. Since there are the same number of columns in A as rows in B, there are the same number of elements (n) to be multiplied in each. Algebraically, we have

$$c_{rc} = \sum_{j=1}^{n} a_{rj} b_{jc}.$$

This concept is best seen by an example.

$$\begin{bmatrix} 1 & 2 \\ 3 & 4 \end{bmatrix} \begin{bmatrix} 5 & 6 & 7 \\ 8 & 9 & 10 \end{bmatrix} = \begin{bmatrix} 1\cdot5+2\cdot8 & 1\cdot6+2\cdot9 & 1\cdot7+2\cdot10 \\ 3\cdot5+4\cdot8 & 3\cdot6+4\cdot9 & 3\cdot7+4\cdot10 \end{bmatrix} = \begin{bmatrix} 21 & 24 & 27 \\ 47 & 54 & 61 \end{bmatrix}.$$

The order of the resultant matrix from multiplying an $m \times n$ matrix times an $n \times p$ is given by the outside dimensions. In other words, c is $m \times p$.

We can also multiply vectors and matrices as long as they are conformable. For example, if A is an $m \times n$ matrix, we could compute AV if V is an $n \times 1$ column vector. Thus we could write our simple trivariate model as a sum involving the product of a matrix and a vector. Since $Y_t = \beta_1 + \beta_2 X_{t2} + \beta_3 X_{t3} + U_t$ for $t = 1, \ldots, T$, we can use matrix algebra to represent all of the observations at one time. If we define a $T \times 3$ matrix X with the first column all 1's and the second column having our T observations X_{t2}, etc., we can think of our model in terms of matrices and vectors. (The one trick used here is to apply the rule that $\beta_1 = \beta_1 \cdot 1$.) The appropriate vectors are a $T \times 1$ vector of observations on Y, a 3×1 vector of coefficients, and a $T \times 1$ vector of error terms.

$$Y = \begin{bmatrix} Y_1 \\ Y_2 \\ \cdot \\ \cdot \\ \cdot \\ Y_T \end{bmatrix}, \quad X = \begin{bmatrix} 1 & X_{12} & X_{13} \\ 1 & X_{22} & X_{23} \\ \cdot & \cdot & \cdot \\ \cdot & \cdot & \cdot \\ 1 & X_{T2} & X_{T3} \end{bmatrix}, \quad \beta = \begin{bmatrix} \beta_1 \\ \beta_2 \\ \beta_3 \end{bmatrix}, \quad U = \begin{bmatrix} U_1 \\ U_2 \\ \cdot \\ \cdot \\ U_T \end{bmatrix}$$

So that $Y = X\beta + U$ says the same thing as writing out each observation

$$Y_1 = 1 \cdot \beta_1 + X_{12}\beta_2 + X_{13}\beta_3 + U_1,$$

$$Y_2 = 1 \cdot \beta_1 + X_{22}\beta_2 + X_{23}\beta_3 + U_2,$$

$$\vdots$$

$$Y_T = 1 \cdot \beta_1 + X_{T2}\beta_2 + X_{T3}\beta_3 + U_T,$$

only much more conveniently.

Partitioned matrices can be multiplied just as regular matrices, so long as the partitions are conformable in the two matrices. If we have conformable matrices,

$$A = \left[\begin{array}{c|c} A_{11} & A_{12} \\ \hline A_{21} & A_{22} \end{array}\right] \quad \text{and} \quad B = \left[\begin{array}{c|c} B_{11} & B_{12} \\ \hline B_{21} & B_{22} \end{array}\right],$$

then

$$C = AB = \left[\begin{array}{c|c} C_{11} & C_{12} \\ \hline C_{21} & C_{22} \end{array}\right] = \left[\begin{array}{c|c} A_{11}B_{11}+A_{12}B_{21} & A_{11}B_{12}+A_{12}B_{22} \\ \hline A_{21}B_{11}+A_{22}B_{21} & A_{21}B_{12}+A_{22}B_{22} \end{array}\right].$$

Thus if A_{11} is $m_1 \times n_1$, and A_{22} is $m_2 \times n_2$, then B_{11} must be $n_1 \times p_1$ and B_{22} is $n_2 \times p_2$. This process can be extended to cover any number of partitions so long as all matrices and partitions are conformable.

Several algebraic properties are immediately obvious. First, since the rules for conformability require equality of columns in the left hand matrix and rows in the right-hand matrix, AB generally does not equal BA since BA is in general not defined. Even when BA is defined, it equals AB only in very special cases. In other words, the commutative law of scalar algebra does not generally hold, $AB \neq BA$. However, the associative and distributive laws do hold, i.e.,

$$(AB)C = A(BC) \qquad \text{and} \qquad A(B+C) = AB + AC.$$

(You should be able to demonstrate these properties.)

The Identity Matrix

There is a special matrix called the *identity* matrix that is square and has ones on the main diagonal (the diagonal running from upper left to lower right) and zeros elsewhere. It is usually denoted by I or I_n, where n denotes the order of the matrix. For example,

$$I_3 = \begin{bmatrix} 1 & 0 & 0 \\ 0 & 1 & 0 \\ 0 & 0 & 1 \end{bmatrix}.$$

This matrix has some interesting properties in multiplication. For A $(m \times n)$,

$$I_m A = A I_n = A.$$

Multiplication by a conformable identity matrix returns the original matrix unchanged. Our applications of matrix algebra rely heavily on this property of the identity matrix.

II.4   Other Operations

Transpose

In order to increase the generality of the matrix notation, we now introduce the transpose operation. This operation essentially turns a matrix around by making each row into a column. The jth row becomes the jth column in the transpose matrix, implying that the transpose of an $m \times n$ matrix is an $n \times m$ matrix. The normal notation for the transpose of A is A' (read A transpose). Thus, $a_{ij} = a'_{ji}$.

A special matrix, called a symmetric matrix, is often used. This is a matrix where $A = A'$, or each row is identical to the corresponding column.

Also, a useful property is that $(AB)' = B'A'$. To demonstrate this take an earlier example of

$$C = AB = \begin{bmatrix} 1 & 2 \\ 3 & 4 \end{bmatrix} \begin{bmatrix} 5 & 6 & 7 \\ 8 & 9 & 10 \end{bmatrix} = \begin{bmatrix} 21 & 24 & 27 \\ 47 & 54 & 61 \end{bmatrix}.$$

Now

$$C' = \begin{bmatrix} 21 & 47 \\ 24 & 54 \\ 27 & 61 \end{bmatrix} = B'A' = \begin{bmatrix} 5 & 8 \\ 6 & 9 \\ 7 & 10 \end{bmatrix} \begin{bmatrix} 1 & 3 \\ 2 & 4 \end{bmatrix}$$

$$= \begin{bmatrix} 5+16 & 15+32 \\ 6+18 & 18+36 \\ 7+20 & 21+40 \end{bmatrix} = \begin{bmatrix} 21 & 47 \\ 24 & 54 \\ 27 & 61 \end{bmatrix}.$$

Trace

One special operator is the trace operator. This is defined for a square matrix and is simply the sum of the elements on the main diagonal, i.e., for $A(n \times n)$

$$\text{tr} A = \sum_{i=1}^{n} a_{ii}.$$

From this, $\text{tr} A = \text{tr} A'$. It is also true that $\text{tr}(AB) = \text{tr}(BA)$ if A and B are conformable.

II.5   Systems of Linear Equations

One of the most useful aspects of matrix algebra is its ability to represent systems of equations simply. Consider the following set of linear equations, which should look familiar to all students of first year algebra problems:

$$2X_1 + 3X_2 + X_3 = 5,$$

$$X_1 - 2X_2 + 2X_3 = 3, \tag{II.1}$$

$$3X_1 + X_2 - X_3 = 0.$$

This can easily be written as the product of two matrices and set equal to a third,

$$AX = Y, \tag{II.2}$$

where

$$A = \begin{bmatrix} 2 & 3 & 1 \\ 1 & -2 & 2 \\ 3 & 1 & -1 \end{bmatrix} \qquad X = \begin{bmatrix} X_1 \\ X_2 \\ X_3 \end{bmatrix}, \qquad \text{and} \qquad Y = \begin{bmatrix} 5 \\ 3 \\ 0 \end{bmatrix}.$$

This equivalence is easily seen by doing the multiplication and remembering that equality in matrix algebra requires that each corresponding element in two matrices of the same order be equal. The multiplication of A times X yields a 3×1 matrix (a column vector) which must equal the vector Y element by element. The first element of AX is found by multiplying the first row of A (i.e., the individual elements in that row) times the corresponding elements in the vector X and adding. This yields $2X_1 + 3X_2 + X_3$ which, if the equality holds, must equal 5 (the first element of Y). But, this is identical to the first equation in the system of three equations. This is simply the procedure used to write our trivariate model in matrix form.

By the same process, we can represent the equations used to calculate b_2 and b_3 in the trivariate model by series of matrices. Equations (2.12a) and (2.13a), used to find b_2 and b_3 in Chapter 2, are

$$b_2 V_{X_2} + b_3 C_{X_2 X_3} = b_2 V_2 + b_3 C_{23} = C_{YX_2} = C_{Y2},$$

$$b_2 C_{X_2 X_3} + b_3 V_{X_3} = b_2 C_{23} + b_3 V_3 = C_{YX_3} = C_{Y3}.$$

This can be represented by the matrix equation

$$Qb = R, \tag{II.3}$$

where

$$Q = \begin{bmatrix} V_2 & C_{23} \\ C_{23} & V_3 \end{bmatrix}, \qquad b = \begin{bmatrix} b_2 \\ b_3 \end{bmatrix}, \qquad \text{and} \qquad R = \begin{bmatrix} C_{Y2} \\ C_{Y3} \end{bmatrix}.$$

Putting the equations in matrix terms considerably simplifies the notation. However, it does more than that. We are often looking for the solution to the system of equations, i.e., the values for the X's that satisfy the equations in the example, or the values of b_2 and b_3 that minimize the sum of squared errors. This is readily found from a manipulation of these matrices. If we premultiply both sides of Eq. (II.2) by a 3×3 matrix B, we have

$$BAX = BY. \tag{II.4}$$

(Note that it is possible to multiply both sides of an equation by the same matrix and still maintain the equality. However, the matrices must be conformable, and both sides must be multiplied from the same direction, i.e., both premultiplied or both postmultiplied.) If we can find a matrix B such that $BA = I$, Eq. (II.4) becomes

$$BAX = IX = X = BY. \tag{II.5}$$

This implies that BY is the *solution vector* or vector of values solving the equation system of Eq. (II.2).

Similarly, in the trivariate case, if we premultiply both sides of Eq. (II.3) by P, we have

$$PQb = PR,$$

and if we can select P so that $PQ = I$, then we have $Ib = b = PR$, and we have our estimates b_2 and b_3.

The matrix B such that $BA = I$ has its own name, the *inverse* matrix of A. The normal notation of the inverse is A^{-1} (read as A inverse), appealing to the similarity with scalar algebra where $SS^{-1} = S/S = 1$.

Referring to the system of equations in Eq. (II.1), let

$$B = A^{-1} = \begin{bmatrix} 0 & 4/28 & 8/28 \\ 7/28 & -5/28 & -3/28 \\ 7/28 & 7/28 & -7/28 \end{bmatrix}. \tag{II.6}$$

If we multiply A^{-1} defined by (II.6) times A, we get the identity matrix. (You should check this.) This implies that $A^{-1}Y$ will yield the set of X's that satisfy the given equations.

Doing this multiplication yields

$$X = \begin{bmatrix} 12/28 \\ 20/28 \\ 56/28 \end{bmatrix}. \tag{II.7}$$

By substitution it can be seen that this is the solution vector for the equations in Eq. (II.1). Likewise, in the estimation problem, if

$$P = \frac{1}{V_2 V_3 - C_{23}^2} \begin{bmatrix} V_3 & -C_{23} \\ -C_{23} & V_2 \end{bmatrix},$$

we get

$$PQb = b = PR = \frac{1}{V_2 V_3 - C_{23}^2} \begin{bmatrix} V_3 & -C_{23} \\ -C_{23} & V_2 \end{bmatrix} \begin{bmatrix} C_{Y2} \\ C_{Y3} \end{bmatrix}.$$

or

$$b_2 = \frac{V_3 C_{Y2} - C_{23} C_{Y3}}{V_2 V_3 - C_{23}^2}, \qquad b_3 = \frac{V_2 C_{Y3} - C_{23} C_{Y2}}{V_2 V_3 - C_{23}^2},$$

which is the result in Eq. (2.15) and (2.16).

The questions that remain unresolved are under what conditions we can compute an inverse matrix and how we do it. This is the subject of the remaining sections.

II.6   Inverses

Determinant

In order to develop the inverse matrix we must introduce a new concept, the *determinant*. A determinant is a scalar value associated with a square matrix. For small matrices, the determinant is easily computed. The determinant of a 1×1 matrix (a scalar) is the scalar itself. The determinant of a 2×2 matrix A is

$$\det A = \det \begin{bmatrix} a_{11} & a_{12} \\ a_{21} & a_{22} \end{bmatrix} = a_{11}a_{22} - a_{12}a_{21}. \tag{II.8}$$

(The determinant of A is denoted $\det A$, or simply $|A|$.) However, for larger matrices it is easier to develop a general rule than to memorize the formula. The rule derives from the fact that we can calculate a determinant by computing the determinants of a particular set of small matrices. If we can continue reducing the size of the determinant needed, we can eventually reach a scalar or a 2×2 matrix for which we know the rule. In order to develop this reduction process, we first define a cofactor.

Cofactor

For any square matrix of order n, if both the row and column containing any element a_{rc} are eliminated, the remaining matrix is a *minor* of order $n - 1$. In our previous example where

$$A = \begin{bmatrix} 2 & 3 & 1 \\ 1 & -2 & 2 \\ 3 & 1 & -1 \end{bmatrix},$$

$a_{22} = -2$, and the minor of a_{22} is

$$\begin{bmatrix} 2 & 1 \\ 3 & -1 \end{bmatrix};$$

and the minor of $a_{23} = 2$ is

$$\begin{bmatrix} 2 & 3 \\ 3 & 1 \end{bmatrix}.$$

The cofactor of element a_{rc} is defined as $(-1)^{r+c}$ times the determinant of the appropriate minor associated with the element a_{rc}. The first term, $(-1)^{r+c}$, determines the sign; if the sum of the row index and column index is even, the cofactor equals the determinant of the minor, otherwise it equals the negative of this determinant. Thus the cofactor of a_{22} is -5, and the cofactor of a_{23} is $+7$. The entire matrix of cofactors for the matrix A is

$$\operatorname{cof} A = \begin{bmatrix} 0 & 7 & 7 \\ 4 & -5 & 7 \\ 8 & -3 & -7 \end{bmatrix}. \tag{II.9}$$

Knowledge of the cofactors for a matrix provides a general method of computing a determinant. If we multiply each element in any given row times its cofactor and sum them, we have the determinant. In other words,

$$\det A = \sum_{j=1}^{n} a_{ij} \operatorname{cof} a_{ij} \qquad \text{for any } i. \tag{II.10}$$

Equivalently, we can expand along any column as

$$\det A = \sum_{i=1}^{n} a_{ij} \operatorname{cof} a_{ij} \qquad \text{for any } j. \tag{II.11}$$

No matter which row or column is used, the determinant always assumes the same value. For example, expanding the determinant of A along the first row yields

$$\det A = 2 \cdot 0 + 3 \cdot 7 + 1 \cdot 7 = 28.$$

Expanding on the second column,

$$\det A = 3 \cdot 7 + 2 \cdot 5 - 1 \cdot 3 = 28.$$

(Note that the rule for the determinant of the 2×2 matrix can be derived from applying the cofactor expansion rules and using the definition that the determinant of a scalar is the scalar.) We can demonstrate this for

$$Q = \begin{bmatrix} V_2 & C_{23} \\ C_{23} & V_3 \end{bmatrix}$$

where

$$\text{cof } Q = \begin{bmatrix} V_3 & -C_{23} \\ -C_{23} & V_2 \end{bmatrix} \quad \text{and} \quad \det Q = V_2 V_3 - C_{23}^2.$$

Certain properties of determinants are useful. If a single row (or column) of matrix A is multiplied by a scalar c to yield A^*, then

$$\det A^* = c \det A.$$

Also, if A and B are square matrices with $C = AB$, then

$$\det C = \det A \cdot \det B.$$

Adjoint Matrix

The final definition needed in order to develop the inverse matrix is the adjoint matrix, denoted by $\text{adj} A$. The adjoint matrix is the transpose of the matrix of cofactors. Thus, from Eq. (II.9),

$$\text{adj} A = \begin{bmatrix} 0 & 4 & 8 \\ 7 & -5 & -3 \\ 7 & 7 & -7 \end{bmatrix}, \tag{II.12}$$

and $\text{adj} Q$ equals $\text{cof} Q$ since $\text{cof} Q$ is symmetric.

Inverse Matrix

We now have enough to calculate the inverse matrix A^{-1}:

$$A^{-1} = \frac{1}{\det A} (\text{adj} A). \tag{II.13}$$

For the system of equations in (II.1),

$$A^{-1} = \frac{1}{28} \begin{bmatrix} 0 & 4 & 8 \\ 7 & -5 & -3 \\ 7 & 7 & -7 \end{bmatrix} = \begin{bmatrix} 0 & 4/28 & 8/28 \\ 7/28 & -5/28 & -3/28 \\ 7/28 & 7/28 & -7/28 \end{bmatrix}.$$

This is the same matrix introduced before as the inverse. In fact it must be the same since the inverse matrix will be unique. Further, $AA^{-1} = A^{-1}A = I$.

From the definition of the inverse we can get $P = Q^{-1}$, and show how the estimated coefficients b_2 and b_3 were obtained. $Qb = R$ and $PQb = PR$ and $b = PR$ since $PQ = I$

or $P = Q^{-1}$. Now by our method for computing an inverse, i.e., $Q^{-1} = \text{adj}\, Q / \det Q$,

$$Q^{-1} = P = \frac{1}{V_2 V_3 - C_{23}^2} \begin{bmatrix} V_3 & -C_{23} \\ -C_{23} & V_2 \end{bmatrix}.$$

This is precisely the value of P we used in Eq. II.7 to compute b_2 and b_3 and shows that they are the same estimates that were derived in Chapter 2. You should be able to demonstrate that $PQ = I$.

II.7   Existence of an Inverse—Rank

An inverse matrix will not always exist. This is evident from the fact that one cannot always find a unique solution to a set of linear equations. The important question is, What are the conditions where the inverse will not exist?

The first requirement is easily seen when it is noted that we can only take the inverse of a square matrix. Thus if A is $m \times n$ $(m \neq n)$, there can be no A^{-1}. Each row of A corresponds to one of the equations in our simultaneous set and each column of A corresponds to one of the unknowns. Requiring that A be square is equivalent to the old adage that the number of unknowns must equal the number of equations to obtain a unique solution for the set of equations.

This requirement that A be a square matrix is not sufficient to ensure that it can be inverted however. From our formula for inverting the matrix A, note that the adjoint of A is divided by the determinant of A, a scalar. If the value of this determinant is zero, we cannot execute this division, and the inverse of A does not exist, even if it is a square matrix. The determinant is zero if a row or column is a linear combination of any, or all, of the other rows or columns. Again from our example, if one of the equations in our system is a linear function of the other equations we cannot solve for a unique solution. In terms of our matrix A, this condition means that one of the rows of A, say $A_i = (a_{i1}, a_{i2}, \ldots, a_{in})$ is a linear combination of the other rows. This condition is expressed mathematically for l_i not all equal zero as

$$\sum_{i=1}^{m} l_i A_i = \left(\sum_{i=1}^{m} l_i a_{i1}, \sum_{i=1}^{m} l_i a_{i2}, \ldots, \sum_{i=1}^{m} l_i a_{in} \right) = (0, 0, \ldots, 0) = 0, \qquad \text{(II.14)}$$

where $a_{i1}, a_{i2}, \ldots, a_{in}$ are the elements in the respective columns of A. For example, if the third row of A equaled the sum of the first two rows, then the values $l_1 = 1$, $l_2 = 1$, $l_3 = -1$, and $l_4, l_5, \cdots = 0$ would satisfy the condition given in Eq. (II.14).

Before going into more detail on the existence of an inverse, it is useful to define a more general property called *rank*. For any $m \times n$ matrix the rank of the matrix is the order of the largest square submatrix whose determinant is not zero, or the largest set of linearly independent rows [ones for which Eq. (II.14) does not hold]. It is obvious from this that the rank of a matrix cannot be greater than the smaller dimension, i.e., $\min(m, n)$. Further, it can be shown that the rank calculated in terms of linearly independent rows is equivalent to the rank calculated in terms of linearly dependent columns. Thus, if any two columns are multiples of each other, the determinant of a matrix will be zero, and the rank will be less than the number of columns. We can now define when an inverse exists in terms of the rank of the matrix. For A a square matrix of order n, an inverse exists if and only if the rank of A is n. If the rank is less than n, one of the rows is a linear function of the other rows, and the determinant is zero. A matrix for which an inverse exists is called *nonsingular*. A matrix for which an inverse does not exist is called *singular*. (Note that these terms only apply to square matrices.)

Several interesting theorems can be proved concerning rank. If A is nonsingular, then

$$\text{rank of } AB = \text{rank of } B \tag{II.15}$$

for any conformable matrix B. (B does not have to be a square matrix.) Also, if X is $T \times K$ with $K < T$ and rank of $X = K$, then

$$\text{rank of } X'X = K. \tag{II.16}$$

Equation (II.16) implies that $X'X$ has an inverse if the columns of X are linearly independent, i.e., no column is a linear combination of another column in X. It is also true that rank of $X < K$ implies rank of $X'X < K$ and that no inverse exists. This is an important consideration in model estimation.

Finally, two properties of inverses are useful:

$$(AB)^{-1} = B^{-1}A^{-1} \tag{II.17}$$

for A and B square conformable matrices; also, if A is symmetric such that $A = A'$, A^{-1} is also symmetric, implying

$$A^{-1} = (A^{-1})' \tag{II.18}$$

in this special case.

REVIEW QUESTIONS FOR APPENDIX II

1. Compute the matrix product of A and B:

$$A = \begin{bmatrix} 3 & 7 & 7 \\ 9 & 3 & 2 \\ 6 & 7 & 2 \\ 5 & 1 & 0 \end{bmatrix}, \quad B = \begin{bmatrix} 1 & 9 & 3 & 9 & 2 \\ 6 & 5 & 1 & 8 & 4 \\ 5 & 1 & 5 & 4 & 3 \end{bmatrix}.$$

2. Invert the following matrices:

$$A = \begin{bmatrix} 5 & 3 & 1 \\ 9 & 1 & 1 \\ 8 & 9 & 4 \end{bmatrix}, \quad B = \begin{bmatrix} 3 & 4 & 1 & 8 \\ 1 & 1 & 2 & 0 \\ 2 & 7 & 3 & 7 \\ 1 & 0 & 6 & 5 \end{bmatrix}.$$

3. Suppose A_1 is $n_1 \times n_2$, A_2 is $n_2 \times n_3$, and A_3 is $n_3 \times n_4$. Show that

$$(A_1 A_2 A_3)' = A_3' A_2' A_1'$$

using the property that $(AB)' = B'A'$. Check that the matrices on the right-hand side are conformable for multiplication. Extend the result to the matrix products $(A_1 A_2 A_3 A_4)'$.

4. Suppose A and B are conformable for addition. Show that $(A + B)' = A' + B'$.

5. Suppose A_1, A_2, and A_3 are square matrices of the same order. Show that

$$(A_1 A_2 A_3)^{-1} = A_3^{-1} A_2^{-1} A_1^{-1}.$$

6. An important property of determinants is that $|AB| = |A||B|$, for square, conformable matrices A and B. Use this property to show that

$$|A^{-1}| = 1/|A|$$

for nonsingular matrices A. (Hint: compute the determinant of $I = A^{-1}A$.)

III | Appendix: Statistical Tables

TABLE 1

Areas for a Standard Normal Distribution[a]

An entry in the table is the area under the curve, between $z = 0$ and a positive value of z. Areas for negative values of z are obtained by symmetry.

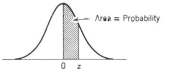

Area = Probability

0   z

Second Decimal Place of z

$\downarrow z$	.00	.01	.02	.03	.04	.05	.06	.07	.08	.09
.0	.0000	.0040	.0080	.0120	.0160	.0199	.0239	.0279	.0319	.0359
.1	.0398	.0438	.0478	.0517	.0557	.0596	.0636	.0675	.0714	.0753
.2	.0793	.0832	.0871	.0910	.0948	.0987	.1026	.1064	.1103	.1141
.3	.1179	.1217	.1255	.1293	.1331	.1368	.1406	.1443	.1480	.1517
.4	.1554	.1591	.1628	.1664	.1700	.1736	.1772	.1808	.1844	.1879
.5	.1915	.1950	.1985	.2019	.2054	.2088	.2123	.2157	.2190	.2224
.6	.2257	.2291	.2324	.2357	.2389	.2422	.2454	.2486	.2517	.2549
.7	.2580	.2611	.2642	.2673	.2703	.2734	.2764	.2794	.2823	.2852
.8	.2881	.2910	.2939	.2967	.2995	.3023	.3051	.3078	.3106	.3133
.9	.3159	.3186	.3212	.3238	.3264	.3289	.3315	.3340	.3365	.3389
1.0	.3413	.3438	.3461	.3485	.3508	.3531	.3554	.3577	.3599	.3621
1.1	.3643	.3665	.3686	.3708	.3729	.3749	.3770	.3790	.3810	.3830
1.2	.3849	.3869	.3888	.3907	.3925	.3944	.3962	.3980	.3997	.4015
1.3	.4032	.4049	.4066	.4082	.4099	.4115	.4131	.4147	.4162	.4177
1.4	.4192	.4207	.4222	.4236	.4251	.4265	.4279	.4292	.4306	.4319
1.5	.4332	.4345	.4357	.4370	.4382	.4394	.4406	.4418	.4429	.4441
1.6	.4452	.4463	.4474	.4484	.4495	.4505	.4515	.4525	.4535	.4545
1.7	.4554	.4564	.4573	.4582	.4591	.4599	.4608	.4616	.4625	.4633
1.8	.4641	.4649	.4656	.4664	.4671	.4678	.4686	.4693	.4699	.4706
1.9	.4713	.4719	.4726	.4732	.4738	.4744	.4750	.4756	.4761	.4767
2.0	.4772	.4778	.4783	.4788	.4793	.4798	.4803	.4808	.4812	.4817
2.1	.4821	.4826	.4830	.4834	.4838	.4842	.4846	.4850	.4854	.4857
2.2	.4861	.4864	.4868	.4871	.4875	.4878	.4881	.4884	.4887	.4890
2.3	.4893	.4896	.4898	.4901	.4904	.4906	.4909	.4911	.4913	.4916
2.4	.4918	.4920	.4922	.4925	.4927	.4929	.4931	.4932	.4934	.4936
2.5	.4938	.4940	.4941	.4943	.4945	.4946	.4948	.4949	.4951	.4952
2.6	.4953	.4955	.4956	.4957	.4959	.4960	.4961	.4962	.4963	.4964
2.7	.4965	.4966	.4967	.4968	.4969	.4970	.4971	.4972	.4973	.4974
2.8	.4974	.4975	.4976	.4977	.4977	.4978	.4979	.4979	.4980	.4981
2.9	.4981	.4982	.4982	.4983	.4984	.4984	.4985	.4985	.4986	.4986
3.0	.4987	.4987	.4987	.4988	.4988	.4989	.4989	.4989	.4990	.4990

[a] Reproduced by permission from R. J. Wonnacott and T. H. Wonnacott, *Econometrics*, Wiley, New York, 1970.

TABLE 2

Upper Percentage Points of the χ^2 Distribution[a]

df	.99	.98	.95	.90	.80	.70	.50	.30	.20	.10	.05	.02	.01	.001
1	.0³157	.0³628	.00393	.0158	.0642	.148	.455	1.074	1.642	2.706	3.841	5.412	6.635	10.827
2	.0201	.0404	.103	.211	.446	.713	1.386	2.408	3.219	4.605	5.991	7.824	9.210	13.815
3	.115	.185	.352	.584	1.005	1.424	2.366	3.665	4.642	6.251	7.815	9.837	11.341	16.268
4	.297	.429	.711	1.064	1.649	2.195	3.357	4.878	5.989	7.779	9.488	11.668	13.277	18.465
5	.554	.752	1.145	1.610	2.343	3.000	4.351	6.064	7.289	9.236	11.070	13.388	15.086	20.517
6	.872	1.134	1.635	2.204	3.070	3.828	5.348	7.231	8.558	10.645	12.592	15.033	16.812	22.457
7	1.239	1.564	2.167	2.833	3.822	4.671	6.346	8.383	9.803	12.017	14.067	16.622	18.475	24.322
8	1.646	2.032	2.733	3.490	4.594	5.527	7.344	9.524	11.030	13.362	15.507	18.168	20.090	26.125
9	2.088	2.532	3.325	4.168	5.380	6.393	8.343	10.656	12.242	14.684	16.919	19.679	21.666	27.877
10	2.558	3.059	3.940	4.865	6.179	7.267	9.342	11.781	13.442	15.987	18.307	21.161	23.209	29.588
11	3.053	3.609	4.575	5.578	6.989	8.148	10.341	12.899	14.631	17.275	19.675	22.618	24.725	31.264
12	3.571	4.178	5.226	6.304	7.807	9.034	11.340	14.011	15.812	18.549	21.026	24.054	26.217	32.909
13	4.107	4.765	5.892	7.042	8.634	9.926	12.340	15.119	16.985	19.812	22.362	25.472	27.688	34.528
14	4.660	5.368	6.571	7.790	9.467	10.821	13.339	16.222	18.151	21.064	23.685	26.873	29.141	36.123
15	5.229	5.985	7.261	8.547	10.307	11.721	14.339	17.322	19.311	22.307	24.996	28.259	30.578	37.697
16	5.812	6.614	7.962	9.312	11.152	12.624	15.338	18.418	20.465	23.542	26.296	29.633	32.000	39.252
17	6.408	7.255	8.672	10.085	12.002	13.531	16.338	19.511	21.615	24.769	27.587	30.995	33.409	40.790
18	7.015	7.906	9.390	10.865	12.857	14.440	17.338	20.601	22.760	25.989	28.869	32.346	34.805	42.312
19	7.633	8.567	10.117	11.651	13.716	15.352	18.338	21.689	23.900	27.204	30.144	33.687	36.191	43.820
20	8.260	9.237	10.851	12.443	14.578	16.266	19.337	22.775	25.038	28.412	31.410	35.020	37.566	45.315
21	8.897	9.915	11.591	13.240	15.445	17.182	20.337	23.858	26.171	29.615	32.671	36.343	38.932	46.797
22	9.542	10.600	12.338	14.041	16.314	18.101	21.337	24.939	27.301	30.813	33.924	37.659	40.289	48.268
23	10.196	11.293	13.091	14.848	17.187	19.021	22.337	26.018	28.429	32.007	35.172	38.968	41.638	49.728
24	10.856	11.992	13.848	15.659	18.062	19.943	23.337	27.096	29.553	33.196	36.415	40.270	42.980	51.179
25	11.524	12.697	14.611	16.473	18.940	20.867	24.337	28.172	30.675	34.382	37.652	41.566	44.314	52.620
26	12.198	13.409	15.379	17.292	19.820	21.792	25.336	29.246	31.795	35.563	38.885	42.856	45.642	54.052
27	12.879	14.125	16.151	18.114	20.703	22.719	26.336	30.319	32.912	36.741	40.113	44.140	46.963	55.476
28	13.565	14.847	16.928	18.939	21.588	23.647	27.336	31.391	34.027	37.916	41.337	45.419	48.278	56.893
29	14.256	15.574	17.708	19.768	22.475	24.577	28.336	32.461	35.139	39.087	42.557	46.693	49.588	58.302
30	14.953	16.306	18.493	20.599	23.364	25.508	29.336	33.530	36.250	40.256	43.773	47.962	50.892	59.703

For larger values of df, the expression $\sqrt{2\chi^2} - \sqrt{2df - 1}$ may be used as a normal deviate with unit variance, remembering that the probability for χ^2 corresponds with that of a single tail of the normal curve.

[a] From Table IV of R. A. Fisher and F. Yates, *Statistical Tables for Biological, Agricultural, and Medical Research*, Longman Group, Ltd., London, 1973 (previously published by Oliver and Boyd, Edinburgh), by permission of the authors and publishers.

TABLE 3

Percentage Points of Student's t Distribution[a]

df	Level of significance for one-tailed test					
	.10	.05	.025	.01	.005	.0005
	Level of significance for two-tailed test					
	.20	.10	.05	.02	.01	.001
1	3.078	6.314	12.706	31.821	63.657	636.619
2	1.886	2.920	4.303	6.965	9.925	31.598
3	1.638	2.353	3.182	4.541	5.841	12.941
4	1.533	2.132	2.776	3.747	4.604	8.610
5	1.476	2.015	2.571	3.365	4.032	6.859
6	1.440	1.943	2.447	3.143	3.707	5.959
7	1.415	1.895	2.365	2.998	3.499	5.405
8	1.397	1.860	2.306	2.896	3.355	5.041
9	1.383	1.833	2.262	2.821	3.250	4.781
10	1.372	1.812	2.228	2.764	3.169	4.587
11	1.363	1.796	2.201	2.718	3.106	4.437
12	1.356	1.782	2.179	2.681	3.055	4.318
13	1.350	1.771	2.160	2.650	3.012	4.221
14	1.345	1.761	2.145	2.624	2.977	4.140
15	1.341	1.753	2.131	2.602	2.947	4.073
16	1.337	1.746	2.120	2.583	2.921	4.015
17	1.333	1.740	2.110	2.567	2.898	3.965
18	1.330	1.734	2.101	2.552	2.878	3.922
19	1.328	1.729	2.093	2.539	2.861	3.883
20	1.325	1.725	2.086	2.528	2.845	3.850
21	1.323	1.721	2.080	2.518	2.831	3.819
22	1.321	1.717	2.074	2.508	2.819	3.792
23	1.319	1.714	2.069	2.500	2.807	3.767
24	1.318	1.711	2.064	2.492	2.797	3.745
25	1.316	1.708	2.060	2.485	2.787	3.725
26	1.315	1.706	2.056	2.479	2.779	3.707
27	1.314	1.703	2.052	2.473	2.771	3.690
28	1.313	1.701	2.048	2.467	2.763	3.674
29	1.311	1.699	2.045	2.462	2.756	3.659
30	1.310	1.697	2.042	2.457	2.750	3.646
40	1.303	1.684	2.021	2.423	2.704	3.551
60	1.296	1.671	2.000	2.390	2.660	3.460
120	1.289	1.658	1.980	2.358	2.617	3.373
∞	1.282	1.645	1.960	2.326	2.576	3.291

[a] Abridged from Table III of R. A. Fisher and F. Yates, *Statistical Tables for Biological, Agricultural, and Medical Research*, Longman Group, Ltd., London, 1973 (previously published by Oliver and Boyd, Edinburgh), by permission of the authors and publishers.

TABLE 4

Upper Percentage Points of the F Distribution [a]

$$p = .05$$

n_2 \ n_1	1	2	3	4	5	6	8	12	24	∞
1	161.4	199.5	215.7	224.6	230.2	234.0	238.9	243.9	249.0	254.3
2	18.51	19.00	19.16	19.25	19.30	19.33	19.37	19.41	19.45	19.50
3	10.13	9.55	9.28	9.12	9.01	8.94	8.84	8.74	8.64	8.53
4	7.71	6.94	6.59	6.39	6.26	6.16	6.04	5.91	5.77	5.63
5	6.61	5.79	5.41	5.19	5.05	4.95	4.82	4.68	4.53	4.36
6	5.99	5.14	4.76	4.53	4.39	4.28	4.15	4.00	3.84	3.67
7	5.59	4.74	4.35	4.12	3.97	3.87	3.73	3.57	3.41	3.23
8	5.32	4.46	4.07	3.84	3.69	3.58	3.44	3.28	3.12	2.93
9	5.12	4.26	3.86	3.63	3.48	3.37	3.23	3.07	2.90	2.71
10	4.96	4.10	3.71	3.48	3.33	3.22	3.07	2.91	2.74	2.54
11	4.84	3.98	3.59	3.36	3.20	3.09	2.95	2.79	2.61	2.40
12	4.75	3.88	3.49	3.26	3.11	3.00	2.85	2.69	2.50	2.30
13	4.67	3.80	3.41	3.18	3.02	2.92	2.77	2.60	2.42	2.21
14	4.60	3.74	3.34	3.11	2.96	2.85	2.70	2.53	2.35	2.13
15	4.54	3.68	3.29	3.06	2.90	2.79	2.64	2.48	2.29	2.07
16	4.49	3.63	3.24	3.01	2.85	2.74	2.59	2.42	2.24	2.01
17	4.45	3.59	3.20	2.96	2.81	2.70	2.55	2.38	2.19	1.96
18	4.41	3.55	3.16	2.93	2.77	2.66	2.51	2.34	2.15	1.92
19	4.38	3.52	3.13	2.90	2.74	2.63	2.48	2.31	2.11	1.88
20	4.35	3.49	3.10	2.87	2.71	2.60	2.45	2.28	2.08	1.84
21	4.32	3.47	3.07	2.84	2.68	2.57	2.42	2.25	2.05	1.81
22	4.30	3.44	3.05	2.82	2.66	2.55	2.40	2.23	2.03	1.78
23	4.28	3.42	3.03	2.80	2.64	2.53	2.38	2.20	2.00	1.76
24	4.26	3.40	3.01	2.78	2.62	2.51	2.36	2.18	1.98	1.73
25	4.24	3.38	2.99	2.76	2.60	2.49	2.34	2.16	1.96	1.71
26	4.22	3.37	2.98	2.74	2.59	2.47	2.32	2.15	1.95	1.69
27	4.21	3.35	2.96	2.73	2.57	2.46	2.30	2.13	1.93	1.67
28	4.20	3.34	2.95	2.71	2.56	2.44	2.29	2.12	1.91	1.65
29	4.18	3.33	2.93	2.70	2.54	2.43	2.28	2.10	1.90	1.64
30	4.17	3.32	2.92	2.69	2.53	2.42	2.27	2.09	1.89	1.62
40	4.08	3.23	2.84	2.61	2.45	2.34	2.18	2.00	1.79	1.51
60	4.00	3.15	2.76	2.52	2.37	2.25	2.10	1.92	1.70	1.39
120	3.92	3.07	2.68	2.45	2.29	2.17	2.02	1.83	1.61	1.25
∞	3.84	2.99	2.60	2.37	2.21	2.09	1.94	1.75	1.52	1.00

Values of n_1 and n_2 represent the degrees of freedom associated with the larger and smaller estimates of variance respectively.

[a] Abridged from Table V of R. A. Fisher and F. Yates, *Statistical Tables for Biological, Agricultural, and Medical Research*, Longman Group, Ltd., London, 1973 (previously published by Oliver and Boyd, Edinburgh), by permission of the authors and publishers.

TABLE 4 (*Continued*)

$p = .01$

n_2 \ n_1	1	2	3	4	5	6	8	12	24	∞
1	4052	4999	5403	5625	5764	5859	5981	6106	6234	6366
2	98.49	99.01	99.17	99.25	99.30	99.33	99.36	99.42	99.46	99.50
3	34.12	30.81	29.46	28.71	28.24	27.91	27.49	27.05	26.60	26.12
4	21.20	18.00	16.69	15.98	15.52	15.21	14.80	14.37	13.93	13.46
5	16.26	13.27	12.06	11.39	10.97	10.67	10.27	9.89	9.47	9.02
6	13.74	10.92	9.78	9.15	8.75	8.47	8.10	7.72	7.31	6.88
7	12.25	9.55	8.45	7.85	7.46	7.19	6.84	6.47	6.07	5.65
8	11.26	8.65	7.59	7.01	6.63	6.37	6.03	5.67	5.28	4.86
9	10.56	8.02	6.99	6.42	6.06	5.80	5.47	5.11	4.73	4.31
10	10.04	7.56	6.55	5.99	5.64	5.39	5.06	4.71	4.33	3.91
11	9.65	7.20	6.22	5.67	5.32	5.07	4.74	4.40	4.02	3.60
12	9.33	6.93	5.95	5.41	5.06	4.82	4.50	4.16	3.78	3.36
13	9.07	6.70	5.74	5.20	4.86	4.62	4.30	3.96	3.59	3.16
14	8.86	6.51	5.56	5.03	4.69	4.46	4.14	3.80	3.43	3.00
15	8.68	6.36	5.42	4.89	4.56	4.32	4.00	3.67	3.29	2.87
16	8.53	6.23	5.29	4.77	4.44	4.20	3.89	3.55	3.18	2.75
17	8.40	6.11	5.18	4.67	4.34	4.10	3.79	3.45	3.08	2.65
18	8.28	6.01	5.09	4.58	4.25	4.01	3.71	3.37	3.00	2.57
19	8.18	5.93	5.01	4.50	4.17	3.94	3.63	3.30	2.92	2.49
20	8.10	5.85	4.94	4.43	4.10	3.87	3.56	3.23	2.86	2.42
21	8.02	5.78	4.87	4.37	4.04	3.81	3.51	3.17	2.80	2.36
22	7.94	5.72	4.82	4.31	3.99	3.76	3.45	3.12	2.75	2.31
23	7.88	5.66	4.76	4.26	3.94	3.71	3.41	3.07	2.70	2.26
24	7.82	5.61	4.72	4.22	3.90	3.67	3.36	3.03	2.66	2.21
25	7.77	5.57	4.68	4.18	3.86	3.63	3.32	2.99	2.62	2.17
26	7.72	5.53	4.64	4.14	3.82	3.59	3.29	2.96	2.58	2.13
27	7.68	5.49	4.60	4.11	3.78	3.56	3.26	2.93	2.55	2.10
28	7.64	5.45	4.57	4.07	3.75	3.53	3.23	2.90	2.52	2.06
29	7.60	5.42	4.54	4.04	3.73	3.50	3.20	2.87	2.49	2.03
30	7.56	5.39	4.51	4.02	3.70	3.47	3.17	2.84	2.47	2.01
40	7.31	5.18	4.31	3.83	3.51	3.29	2.99	2.66	2.29	1.80
60	7.08	4.98	4.13	3.65	3.34	3.12	2.82	2.50	2.12	1.60
120	6.85	4.79	3.95	3.48	3.17	2.96	2.66	2.34	1.95	1.38
∞	6.64	4.60	3.78	3.32	3.02	2.80	2.51	2.18	1.79	1.00

Values of n_1 and n_2 represent the degrees of freedom associated with the larger and smaller estimates of variance respectively.

TABLE 5

Critical Points of the Durbin–Watson Test for Autocorrelation[a]

Sample size = n	Pr = Probability in Lower Tail (Significance Level = α)	k = Number of Regressors (Excluding the Constant)									
		1		2		3		4		5	
		D_L	D_U	D_L	D_U	D_L	D_U	D_L	D_U	D_L	D_U
15	.01	.81	1.07	.70	1.25	.59	1.46	.49	1.70	.39	1.96
	.025	.95	1.23	.83	1.40	.71	1.61	.59	1.84	.48	2.09
	.05	1.08	1.36	.95	1.54	.82	1.75	.69	1.97	.56	2.21
20	.01	.95	1.15	.86	1.27	.77	1.41	.68	1.57	.60	1.74
	.025	1.08	1.28	.99	1.41	.89	1.55	.79	1.70	.70	1.87
	.05	1.20	1.41	1.10	1.54	1.00	1.68	.90	1.83	.79	1.99
25	.01	1.05	1.21	.98	1.30	.90	1.41	.83	1.52	.75	1.65
	.025	1.18	1.34	1.10	1.43	1.02	1.54	.94	1.65	.86	1.77
	.05	1.29	1.45	1.21	1.55	1.12	1.66	1.04	1.77	.95	1.89
30	.01	1.13	1.26	1.07	1.34	1.01	1.42	.94	1.51	.88	1.61
	.025	1.25	1.38	1.18	1.46	1.12	1.54	1.05	1.63	.98	1.73
	.05	1.35	1.49	1.28	1.57	1.21	1.65	1.14	1.74	1.07	1.83
40	.01	1.25	1.34	1.20	1.40	1.15	1.46	1.10	1.52	1.05	1.58
	.025	1.35	1.45	1.30	1.51	1.25	1.57	1.20	1.63	1.15	1.69
	.05	1.44	1.54	1.39	1.60	1.34	1.66	1.29	1.72	1.23	1.79
50	.01	1.32	1.40	1.28	1.45	1.24	1.49	1.20	1.54	1.16	1.59
	.025	1.42	1.50	1.38	1.54	1.34	1.59	1.30	1.64	1.26	1.69
	.05	1.50	1.59	1.46	1.63	1.42	1.67	1.38	1.72	1.34	1.77
60	.01	1.38	1.45	1.35	1.48	1.32	1.52	1.28	1.56	1.25	1.60
	.025	1.47	1.54	1.44	1.57	1.40	1.61	1.37	1.65	1.33	1.69
	.05	1.55	1.62	1.51	1.65	1.48	1.69	1.44	1.73	1.41	1.77
80	.01	1.47	1.52	1.44	1.54	1.42	1.57	1.39	1.60	1.36	1.62
	.025	1.54	1.59	1.52	1.62	1.49	1.65	1.47	1.67	1.44	1.70
	.05	1.61	1.66	1.59	1.69	1.56	1.72	1.53	1.74	1.51	1.77
100	.01	1.52	1.56	1.50	1.58	1.48	1.60	1.46	1.63	1.44	1.65
	.025	1.59	1.63	1.57	1.65	1.55	1.67	1.53	1.70	1.51	1.72
	.05	1.65	1.69	1.63	1.72	1.61	1.74	1.59	1.76	1.57	1.78

[a] Reproduced by permission from R. J. Wonnacott and T. H. Wonnacott, *Econometrics*, Wiley, New York, 1970. Originally abridged from J. Durbin, and G. S. Watson, "Testing for Serial Correlation in Least Squares Regression. II." *Biometrika* **38** (June, 1951), pp. 159–178, with permission of Biometrika Trustees.

References

AMEMIYA, T. AND NOLD, F.   A modified logit model. *Review of Economics and Statistics* **LVII**, No. 2 (May 1975): 255–257.

BERKSON, J.   A statistically precise and relatively simple method of estimating the bioassay with quantal response based upon the logistic function. *Journal of the American Statistical Association* **48** (September 1953): 565–599.

COCHRANE, W. AND ORCUTT, G. H.   Application of least-squares regression to relationships containing auto-correlated error terms. *Journal of the American Statistical Association* **44** (1949): 32–61.

COLEMAN, J. S. *et al.*   *Equality of Educational Opportunity*. Washington D.C.: U.S. Government Printing Office, 1966.

Council of Economic Advisers.   *Economic Report to the President, 1976.* Washington, D.C.: U.S. Government Printing Office, 1976.

DHRYMES, P.   *Econometrics*. New York: Harper and Row, 1970.

DUNCAN, O. D.   *Introduction to Multiequation Models*. New York: Academic Press, 1975.

DUNCAN, O. D., HALLER, A., AND PORTES, A.   Peer influence on aspirations: A reinterpretation. *American Journal of Sociology* **74** (1968): 119–137.

FARRAR, D. E. AND GLAUBER, R. R.   Multicollinearity in regression analysis: The problem revisited. *Review of Economics and Statistics* **49** (February 1967): 92–107.

FISHER, F. M.   *The Identification Problem in Econometrics*. New York: McGraw-Hill, 1966.

FISHER, F. M.   Tests of equality between sets of coefficients in two linear regressions: An expository note. *Econometrica* **38(2)** (March 1970); 361–66.

GOLDBERGER, A. S.   *Econometric Theory*. New York: Wiley, 1964.

GOLDBERGER, A. S.   Efficient estimation in overidentified models: An interpretative analysis, in *Structural Equation Models in the Social Sciences* (Goldberger and Duncan, eds.), pp. 131–152. New York: Seminar Press, 1973.

GOLDBERGER, A. S. AND DUNCAN, O. D. (eds.).   *Structural Equation Models in the Social Sciences*. New York: Seminar Press, 1973.

GOLDFELD, S. M. AND QUANDT, R. E.   Some tests for heteroskedasticity. *Journal of the American Statistical Association* **60** (1965): 539–547.

GOODMAN, L.   The multivariate analysis of qualitative data: Interactions among multiple classifications. *Journal of the American Statistical Association* **60** (1970): 226–256.

GOODMAN, L.   A general model for the analysis of surveys. *American Journal of Sociology* **77** (May 1972): 1035–1085.

GRILICHES, Z.   Errors in variables and other unobservables. *Econometrica* **42** (1974): 971–998.

HANUSHEK, E. *Education and Race*. Lexington, Massachusetts: Lexington Books, 1972.

HARMON, H. H. *Modern Factor Analysis*. Chicago: The University of Chicago Press, 1967.

HIBBS, D. Problems of statistical estimation and causal inference in time series regression models, *Sociological Methodology* (H. L. Costner, ed.). San Francisco: Jossey-Bass, 1974.

HOEL, P. *Mathematical Statistics*, 3rd edition. New York: Wiley, 1962.

JACKSON, J. Issues, party choices, and presidential votes. *American Journal of Political Science* **19** (May 1975): 161–185.

JOHNSTON, J. *Econometric Methods*, 2nd edition. New York: McGraw Hill, 1972.

JÖRESKOG, K. A general method for estimating a linear structural equation system, in *Structural Equation Models in the Social Sciences* (Goldberger and Duncan, eds.), pp. 85–112. New York: Seminar Press, 1973.

JÖRESKOG, K. AND VAN THILLO, M. A general computer program for estimating a linear structural equation system involving multiple indicators of unmeasured variables. Educational Testing Service, Princeton, New Jersey, RB-72-56, December 1972.

KLEIN, L. *An Introduction to Econometrics*. Englewood Cliffs, New Jersey: Prentice-Hall, 1962.

KMENTA, J. *Elements of Econometrics* New York: Macmillan, 1971.

KOYCK, L. *Distributed Lags and Investment Analysis*. Amsterdam: North Holland, 1954.

LAWLEY, D. N. AND MAXWELL, A. E. *Factor Analysis as a Statistical Method*. London: Buttersworth, 1971.

MALINVAUD, E. *Statistical Methods of Econometrics*, 2nd edition. Amsterdam: North Holland, 1970.

McFADDEN, D. Conditional logit analysis of qualitative choice behavior, in *Frontiers in Econometrics* (P. Zarembka, ed.). New York: Academic Press, 1974.

MEYER, J. AND KUH, E. *The Investment Decision: An Empirical Study*. Cambridge, Massachusetts: Harvard University Press, 1957.

MORGENSTERN, O. *On the Accuracy of Economic Observations*, 2nd edition. Princeton, New Jersey: Princeton University Press, 1963.

NERLOVE, M. AND PRESS, S. J. *Univariate and Multivariate Log-linear and Logistics Models*. Santa Monica: The RAND Corporation, 1973.

PECK, J. K. The estimation of a dynamic equation following a preliminary test for autocorrelation. Cowles Foundation Discussion Paper No. 404, Yale University, New Haven, Connecticut, September 1975.

ROBINSON, W. Ecological correlations and the behavior of individuals. *American Sociological Review* **15** (June 1950): 351–357.

RUSSETT, B. *What Price Vigilance*. New Haven: Yale University Press, 1970.

THEIL, H. *Principles of Econometrics*. New York: Wiley, 1971.

THEIL, H. On the estimation of relationships involving qualitative variables. *American Journal of Sociology* **76** (July 1970): 103–154.

TOBIN, J. Estimation of relationships for limited dependent variables. *Econometrica* **26** (January 1958): 24–36.

U.S. Bureau of the Census. *Public Use Sample, 1970*. Washington D.C.: U.S. Bureau of the Census, 1970.

U.S. Department of Commerce. *Statistical Abstract of the United States, 1976*. Washington, D.C.: U.S. Government Printing Office, 1976.

WEISBERG, H. F. Models of statistical relationship. *American Political Science Review* **68** (December 1964): 1638–1655.

WONNACOTT, R. J., AND WONNACOTT, T. H. *Econometrics*. New York: Wiley, 1970.

WORKING, E. J. What do statistical "demand curves" show? *Quarterly Journal of Economics* **41** (February 1927): 212–235.

ZAVOINA, W. AND McKELVEY, R. A statistical model for the analysis of legislative voting behavior. Paper presented to the Annual Meeting of the American Political Science Association, September 1969, New York.

Index

An n following a page number indicates a topic mentioned in a footnote.

QUANTITATIVE STUDIES IN SOCIAL RELATIONS

Consulting Editor: Peter H. Rossi

UNIVERSITY OF MASSACHUSETTS
AMHERST, MASSACHUSETTS